Pharmaceutical Wastewater Treatment Technologies: Concepts and Implementation Strategies

Pharmaceutical Wastewater Treatment Technologies: Concepts and Implementation Strategies

Edited by
Nadeem A Khan, SIrajuddin Ahmed,
Viola Vambol and Sergij Vambol

IWA PUBLISHING

Published by **IWA Publishing**
Unit 104-105, Export Building
1 Clove Crescent
London E14 2BA
Telephone: +44 (0)20 7654 5500
Fax: +44 (0)20 7654 5555
Email: publications@iwap.co.uk
Web: www.iwapublishing.com

First published 2021

British Library Cataloguing in Publication Data
A CIP catalogue record for this book is available from the British Library

ISBN: 9781789061321 (paperback)
ISBN: 9781789061338 (eBook)
ISBN: 9781789061345 (ePUB)

Contents

Chapter 1
Introduction – background and brief history of pharmaceutical wastewater ... **1**
Izhar Ullah, Abdul Rehman, Abid Hussain and Fakhar-ud-Din

Chapter 2
Comprehensive insights into sources of pharmaceutical wastewater in the biotic systems ... **17**
Sergij Vambol, Viola Vambol, Nastaran Mozaffari, Alaa El Din Mahmoud, Nishana Ramsawak, Niloofar Mozaffari, Parisa Ziarati and Nadeem A Khan

Chapter 3
Prioritisation of pharmaceutical in effluent considering its biohazard ... **47**
Aastha Dhingra, Sirajuddin Ahmed, Siddhartha Gautam, Tilemachos K. Koliopoulos, Ahmed Kadhim Hussein and Fateh Mebarek-oudina

Chapter 4
Occurrences: pharmaceutical wastewater in environment ... **69**
Humzah Jamshaid, Fakhar-ud-Din and Izhar Ullah

Chapter 5
***Pharmaceutical presence in effluent detected till date* 95**
Aastha Dhingra, Sirajuddin Ahmed, Siddhartha Gautam, T. Koliopoulos, A. K. Hussein and Fateh Mebarek-oudina

Chapter 6
***Legislative and criminal law aspects of water protection and prevention of pharmaceuticals accumulation in nature* 113**
Viola Vambol, Andrii Shulha, Sergij Vambol, Parveen Sihag, Vitalii Pavlykivskyi, Niloofar Mozaffari, Nastaran Mozaffari, Nadeem A. Khan and Barbara Sawicka

Chapter 7
***Eco-toxicity and health issues associated with the pharmaceuticals in aqueous environments: a global scenario* 145**
Saif Ullah Khan, Hasan Rameez, Farrukh Basheer and Izharul Haq Farooqi

Chapter 8
***Treatment schemes – conventional and dedicated for PhACs treatment* 181**
Pavlo Kozub, Svetlana Kozub, Niloofar Mozaffari, Parveen Sihag, Nastaran Mozaffari, Sergij Vambol, Viola Vambol, Nishana Ramsawak and Nadeem A Khan

Chapter 9
***Effect on treatment schemes pharmaceutical wastewater on conventional treatment plants* 221**
Ahmed Kettab and Fatima Zohra Batana

Chapter 10
***Urban wastewater treatment plants* 239**
Ahmed Kettab and Hanane Bouanani

Chapter 11
Comparison among pharmaceutical wastewater treatment schemes for the removal of particular type of drugs and their residues ***271***
Afzal Husain Khan

Chapter 12
Agricultural water reuse ***305***
Ahmed Kettab and Hanane Bouanani

Chapter 13
Advanced technologies for removing triclosan from wastewater: a state-of-the-art review ***327***
Mariem Chtourou, Albert Galizia, Victoria Salvadó and Hèctor Monclús

Chapter 14
Existing treatment – globally in full scale plants ***353***
Qaisar Mahmood, Muhammad Saqib Khan and Nadia Riaz

Chapter 15
Future scope – expected treatments for Phcs in future ... ***375***
Viola Vambol, Nadeem A Khan, Sergij Vambol and Sirajuddin Ahmed

Index ***385***

Chapter 1

Introduction – background and brief history of pharmaceutical wastewater

Izhar Ullah[1], Abdul Rehman[2], Abid Hussain[1] and Fakhar-ud-Din[3]

[1]*Department of Pharmacy, University of Poonch Rawalakot, AJK Pakistan*
[2]*Department of Microbiology, Kohat University of Science and Technology (KUST), Pakistan*
[3]*Department of Pharmacy, Quaid-e-Azam University Islamabad, Pakistan*

ABSTRACT

Water is an indispensable and vital element for endurance of all living beings, as plenty of fresh and clean water is required for their growth. According to a survey report conducted by the World Health Organization, about 3.4 million people (mostly children less than five years) die per year in developing countries due to unhygienic conditions, poor sanitation and water-borne diseases. Pharmaceutical manufacturing procedures are multistage and batch processes which leads to the production of a huge mass of effluent. In addition, different sources can distribute pharmaceutical active entities and other chemicals to groundwater storage, streams and aquifers. Furthermore, chemicals of daily domestic use, agriculture and industry can penetrate our environment through wastewater. These chemicals/drugs consist of human and veterinary drugs, hormones, disinfectants, fire retardants, detergents, insecticides, plasticizers and antioxidants etc. Active pharmaceutical compounds have a major concern as they have intrinsic biological activity and can interfere with the human normal physiology and can lead to dangerous consequences such as development of multidrug resistant resistant microorganisms in the aquatic environment, potential

Editor(s): Nadeem A Khan, Sirajuddin Ahmed, Viola Vambol and Sergij Vambol
doi: 10.2166/9781789061338_0001

increased toxicity of chemical combinations, as well as metabolites, retardation of nitrite oxidation and methagenosis.

Keywords: Environmental consequences, Pharmaceutical wastewater, Terrestrial hazards of chemical wastewater, Toxic compounds.

Contents

1.1 Overview of Water Crises and Searching for Alternate Resources 2
1.2 Brief Introduction of Wastewater and Their Effects on Environment 3
1.3 Water and Wastewater Scenario in Developing Countries like Pakistan 4
1.4 Pharmaceutical Wastewater 6
1.5 Composition of Pharmaceutical Wastewater 7
1.6 Health Hazard Effects of Pharmaceutical Wastewater 8
1.7 Chronological Outlook of Wastewater Treatment 9
1.8 Biological Wastewater Treatment Technologies 9
1.9 Conclusion 11
References 11

1.1 OVERVIEW OF WATER CRISES AND SEARCHING FOR ALTERNATE RESOURCES

Fresh and clean water is important for the existence of human beings, plants and animals on earth's planet, which offers numerous uses such as water for household consumption, water for recreational activities, water for irrigation purposes and water for neutralizing streams (Aylward *et al.*, 2005). About 97% of the earth's surface is covered by salt water of oceans while the remaining 3% is found in icebergs, glacial masses or is underground and difficult to be available for routine usage. Only a minor fraction of fresh water (about 0.4%) is accessible for human consumption, which is usually sustained by the natural water cycle (Abderrahman, 2000). It has also been reported that in the 21st century, water will be one of the rarest resources on the earth's planet due to the substantial escalation in human population, which in turn will greatly affect aquatic and public health, narrowing socio-economic and agricultural growth/production, especially in developing countries (Bartram & Balance, 1996; Montgomery & Elimelech, 2007).

Water is an indispensable and vital element for the endurance of all living beings, as plenty of fresh and clean water is required for their growth. According to a survey report conducted by the World Health Organization (WHO) in 2015, about 3.4 million people (mostly children less than five years) die per year in developing countries due to unhygienic conditions, poor sanitation and water-borne diseases (Andersson & Bohan, 2001; Craun *et al.*, 2006; Jury & Vaux, 2007; WHO, 2015). Water crises is a global issue nowadays and it has been reported that one in every five people have no access to clean and renewed water resources (Jury & Vaux, 2007). In developing countries, about 50% of the community have no access to standard sanitation (Jayakumar *et al.*, 2009). The Food and Agriculture Organization of the United Nations reported that in many parts of Asia, more than 800 million of the population face severe water shortages, i.e. the possibility of fresh and potable water revenue would be less than 1500 m^3 per person per year, and this may be expected to reach more than 2.5 billion of the population in 2025 (Connor *et al.*, 2017). In order to overcome these challenges of water shortage, there is a need to establish functionable wastewater treatment schemes and projects, especially in developing countries (Yang & Abbaspour, 2007). The use of treated chemical wastewater effluent is divided into three different sections: 1) Plain effluent reuse, where the effluent of one system to be used will be the influent of other system with or without supplementary treatment; b) Programmed usage of treated water, where the untreated wastewater first goes to the treatment facility, such as trickling filter system, constructed wetlands, lagoons and then after treatment, discharge for further reuse; c) Haphazard water reuse, where the wastewater mixes into naturally occurring surface water reservoirs, acting as a source of water for other usages (Asano *et al.*, 2007; USEPA, 2012).

1.2 BRIEF INTRODUCTION OF WASTEWATER AND THEIR EFFECTS ON ENVIRONMENT

Water is the sign of life. It covers about 70% of the total earth crust, out of which less than 1% is normally available for human consumption (Abderrahman, 2000). The quality and quantity of water is continuously affected due to various factors including sedimentation, weathering, precipitation, urbanization and industrialization (Carpenter *et al.*, 1998; Mokaya *et al.*, 2004; Singh *et al.*, 2005). Human activities such as effluent discharge without prior treatment and the use of pesticides and other chemicals in pharmaceutical industries are the major factors for determining the quality of water (Niemi *et al.*, 1990). In Pakistan, water pollution is mainly due to domestic sewage and industrial effluents that produce severe health hazard effects (Aftab *et al.*, 2000; Khan, 1991; Zaman & Ara, 2000). According to the WHO, the unbalanced water supply throughout the world affects almost two out of five individuals and approximately 30% of all diseases and 40% of deaths throughout the world are due to chemically polluted water (Kantawanichkul, 2009). This depletion of water is reflected by the increase in urbanization, industrialization and pollution of fresh water resources (Asano *et al.*, 2007). Globally, about 90–95% of domestic waste and approximately 95% of industrial waste in fresh water comes from urban areas without prior treatment (Hinrichsen *et al.*, 1997).

The waste produced from industrial, municipal and agricultural activities contains large amounts of toxic compounds which disturb the food chain (Fent, 2004). In developing countries like Pakistan, rapid population growth, industrialization and urbanization from the last 20 years have placed stress on water availability (Khan, 1991). The situation is aggravated further by the pollution of fresh water resources due to the discharge of untreated municipal wastewater containing household effluent and pharmaceutical wastes of about 2000 million gallons (7.5708×109 L) of sewage/day directly to a natural drain or open agricultural land (www.wwfpak.org/factsheets_wps.php). The Ministry of Water and Power reported that approximately 12% of the urban wastewater is treated in municipal treatment plants (Government of Pakistan, 2010). The high pollution level of rivers and groundwater due to the discharge of untreated wastewater has led to different environmental consequences such as reduction of biodiversity, increase in water related diseases and decrease in agricultural productivity (Bizier & DeBarry, 2003).

Wastewater is generally defined as a mixture of liquids, polluted with attenuated contents released from household, recreational, trading and industrial sectors (Zhou & Smith, 2002). In addition to this, it may also sometimes contain flooded as well as unclean surface water (Cheremisinoff, 2002). Depending on the sources of production, wastewater has been classified as industrial or municipal wastewater. As the name indicates, industrial wastewater is a type of water generated during manufacturing processes in industrial sectors while the latter is a type of sewage spawned from domestic sites including houses, commercial areas and institutes. The chemistry of both of these are quite different from each other, i.e. domestic wastewater mostly contains carbonaceous compounds such as carbohydrates, proteins, oil and greases etc., while industries produce contaminated wastewater which mostly depends on the activities performed within that particular industry (Cheremisinoff, 2002; Zhou & Smith, 2002). Generally, wastewater contains 99.6% water and 0.4% solids (USEPA, 2007). The solid part of the wastewater is further divided into biodegradable organics (70%) and inorganics (30%). The biodegradable organics in wastewater are usually composed of proteins, carbohydrates (starch, cellulose), fats, oils and grease etc., while the inorganic portion of wastewater is composed of sediments, salts and metals (Tebbutt, 1998).

1.3 WATER AND WASTEWATER SCENARIO IN DEVELOPING COUNTRIES LIKE PAKISTAN

Pakistan's total geographical and territorial area is 796,100 km^2 out of which 27% of the land is used for irrigation purposes. However, the complete data of Pakistan's irrigation is still not clear, i.e. some reports suggested that 70% of the total land area is used for irrigation (Ahmad *et al.*, 2009). Water erosion due to salinity and water shortage are the two key influences that expose the feasibility of irrigated agricultural land and effect yields production in Pakistan. Numerous exertions such as drilling tube-wells, discharge of salts by excess irrigation, use of fertilizers and expensive chemicals and physical methods have been performed to resolve water erosion and salinity issues in Pakistan, but still no progress has been observed in agricultural sectors (Salma *et al.*, 2012). In 2013, while conducting a 30-day trail, the chairman of PASA (Pakistan Agriculture Scientists Association) stated that less than 10% of clean and fresh water reservoirs were available for drinking and other routine activities in Pakistan while the remaining water reservoirs were either exploited or used for irrigation, as shown in Figure 1.1 (https://nation.com.pk/30-Dec-2013/water-storage-capacity-just-for-30-days). According to Hussain *et al.* (2013), more than 33% of the population in Pakistan have no access to safe and clean water reservoirs and as a result more than 700 children lost their life per annum due to water-borne diseases such as cholera, typhoid, diarrhea and polio, which is a serious issue to be resolved.

In many urban areas of Pakistan, domestic wastewater including waste effluent from industries is used to nourish agricultural land due to the unavailability of ground and surface water reservoirs. The use of wastewater for agricultural activity is an approach to utilize chemical wastewater and has considerable benefits, i.e. it is rich in nutrients, which plants require for their growth, and moreover it has the capability to increase soil fertility instead of using fertilizers, which makes irrigation cost effective (Asano *et al.*, 2007; Becerra-Castro *et al.*, 2015; Hameed *et al.*, 2010; Kalavrouziotis & Arslan-Alaton, 2008; Morel, 2006; Viet Anh *et al.*, 2005). Aside from these benefits, industrial wastewater also contains toxic compounds and heavy metals which have extremely adverse effects on the terrestrial and aquatic ecology, environment and public health (Scott *et al.*, 2000; Soller *et al.*, 2003). A set of standard guidelines had been proposed by the WHO in 1989 for the harmless use of wastewater in agricultural sectors in order to secure agrarians and consumers of the said crops irrigated with chemical wastewater. Therefore, it is essential to develop a suitable wastewater treatment technology in order to reduce the concentration of life-threatening

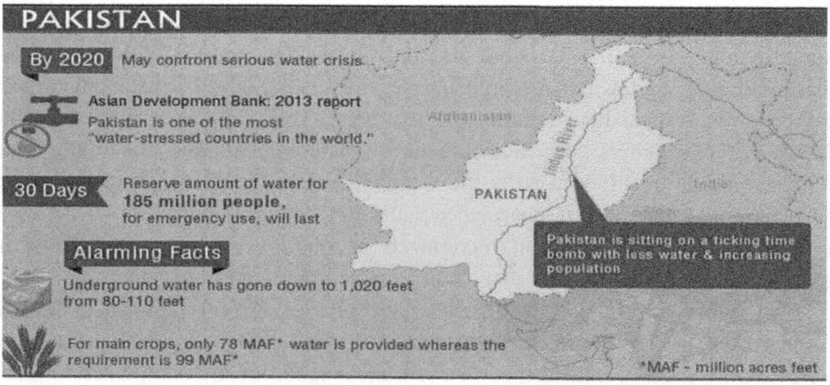

Figure 1.1 Water crises and alarming facts of water reservoirs in Pakistan. Retrieved from www.mapsofworld.com/around-the-world/water-crisis.html.

components of sewage before its usage in agricultural lands. However, in many developing countries, including Pakistan, sewage water is still used as such or after initial treatment for horticulture and agricultural activities, so there is a growing demand to make people aware of the risks of raw sewage to the environment and public health (Feenstra *et al.*, 2000). Different international organizations, in collaboration with national institutes, have now started to make awareness in common people regarding these serious issues of reclaimed wastewater and their possible treatment in the country region (Feenstra *et al.*, 2000).

1.4 PHARMACEUTICAL WASTEWATER

Water is majorly used as raw material for different operation purposes in chemical and pharmaceutical manufacturing such as material processing, production and cooling. The term "process wastewaters" is used in any industry for wastewater coming from the different processes followed in industry. Process wastewater is the wastewater which, during manufacturing or processing, comes in contact with the products, intermediates, byproducts, raw materials or waste products handled in various unit processes or operations (Gadipelly *et al.*, 2014). Pharmaceuticals reach aquatic environments and are prevalent in small concentrations in wastewaters from the drug manufacturing industry, hospital wastewaters, domestic wastewaters and due to the disposal of unused and expired medicines. Many medications are disposed of without being completely metabolized in bodies, leading to biologically active forms mixing with water bodies, which can potentially affect drinking water supplies and human and ecosystem health. Recently, the pharmaceutical manufacturing industry has been rapidly growing in many countries in the world, including Arab countries, resulting in a wide variety of pharmaceutical items for human or animal use. The possibility that pharmaceuticals can enter the environment from a number of different routes and possibly cause untoward effects in biota has been noted in the scientific literature for several decades, but its significance has gone largely unnoticed. Pharmaceutical wastewater research has recently experienced substantial growth and progress, especially in terms of treatment. This can be attributed to the efforts of researchers in environmental, toxicological and pharmaceutical chemistry fields in different parts the world (Saed *et al.*, 2016). In recent years, the occurrence and fate of pharmaceutically active compounds (PhACs) in the aquatic environment has been recognized as one of the emerging issues in environmental chemistry. The disposal of unused medication via the toilet seems to be of minor importance but many of the pharmaceuticals applied in human medical care are not completely eliminated in the human body. Often they are excreted only slightly transformed or even unchanged, mostly conjugated to polar molecules (e.g. as glucoronides). These conjugates can easily be cleaved during sewage treatment and the original PhACs will then be released into the aquatic environment, mostly by effluents from municipal sewage treatment plants. Nowadays, and especially in industrialized countries, strong regulations and advanced manufacturing practices prevent such spills (Heberer, 2002).

1.5 COMPOSITION OF PHARMACEUTICAL WASTEWATER

Pharmaceutical manufacturing has until recently been generally considered to not significantly contribute to environmental pollution on the one hand, and to not pose any environmental threat of consequence on the other. However, recent findings suggest the contrary: certain manufacturing facilities across the globe have been determined to discharge significantly high levels of APIs into the environment, beyond previously established levels, thereby causing pollution (Yakubu, 2017). The pharmaceutical industry wastewater can be roughly classified according to their sources as: (a) municipal wastewater, (b) spent liquors from the fermentation process, (c) chemical waste, (d) condenser waste from evaporation and (e) floor and laboratory washing waste. Manufacturing of pharmaceuticals leads to discharges of chemicals such as buffers, chelators and antibiotics into the aquatic environment. It has been indicated that 15–20% of medicines used in hospitals are potentially bio-accumulative. In order to access the available pharmaceutical waste treatment methods and to develop a

Table 1.1 Most common pharmaceuticals in environment.

S. No.	Class of Drugs	Names of Drugs
1	Antibiotics	Erythromycin, Ofloxacin, Chlortetracycline, Oxytetracycline, Streptomycin, Flumequine, Ciprofloxacin, Trimetoprim, Sulfamethoxazole, Lincomycin, Penicillin, Amoxycillin
2	Steroids and related hormones	17-β-estradiol, Estrone, Diethylstilbestrol, Diethylstilbestrol acetate
3	Anticancer drugs	Cyclophosphamide, Ifosphamide
4	Diuretics	Furosemide
5	Antidepressants	Mianserin
6	Tranquilizers	Diazepam
7	Beta-blockers	Metoprolol, Propranolol, Nadolol, Atenolol, Sotalol and Betaxolol
8	Lipid regulators	Bezafibrate, Gemfibrozil, Clofibric acid and Fenofibrate
9	Anti-inflammatory drugs	Acetylesalicylic acid (Aspirin), Diclofenac, Ibuprofen, Acetaminophen, Metamizol, Codein, Indometacin, Naproxen and Phenazone

comprehensive policy to discharge minimal waste in an environmentally benign manner, pharmaceutical waste may be divided into two categories: 1) waste generated by the medicinal companies and the secondary waste obtained from treatment and recycling plants; and 2) medical waste from hospitals and the domestic sector which substantially contaminates the sewage systems. The second category of waste is the sources that contribute toward active pharmaceutical residues in public sewers and subsequently in the municipal wastewater treatment facilities that are commonly ill-equipped to deal with such specialized waste (Pal, 2018). The most commonly occurring pharmaceuticals in the environment are given in Table 1.1 (Nikolaou *et al.*, 2007).

1.6 HEALTH HAZARD EFFECTS OF PHARMACEUTICAL WASTEWATER

Risks associated with previously unknown, unrecognized, unanticipated, or unsuspected chemical pollutants in the environment have long been a major concern for environmental scientists. A wide variety of sources can deliver pharmaceutical chemicals and drugs to streams, groundwater storage, and aquifers. Chemicals of daily domestic use, agriculture and industry can penetrate our environment through wastewater. These chemicals/drugs consist of human and veterinary drugs, hormones, disinfectants fire retardants, detergents, insecticides, plasticizers and antioxidants etc. (Larsson *et al.*, 2007). Active pharmaceutical compounds are a major concern as they have intrinsic biological activity and can interfere with the human normal physiology and can lead to dangerous consequences (Huerta & Barceló, 2012). It is estimated that approximately half of the pharmaceutical wastewaters produced worldwide are discarded without specific treatment (Enick & Moore, 2007). The presence of pharmaceutical products in the environment has effects such as development of antibiotic resistant microbes in the aquatic environment, retardation of nitrite oxidation and methagenosis, and the potential increased toxicity of chemical combinations and metabolites (Dalrymple *et al.*, 2007; Kümmerer, 2001). Humans and animals can carry drugs to the environment through urine. In addition, pharmaceuticals can be directly disposes to the environment through industry, agriculture and medical treatment centers which can also take part in the entry of drugs/chemicals into fresh water bodies (Klavarioti *et al.*, 2009). On the other hand, veterinary pharmaceuticals contaminate directly soil via manure and surface and ground waters by runoff from fields (Khetan & Collins, 2007). The major pathway for pharmaceutical products mainly distribute to the environment via discharges of their industries wastewater to the wastewater treatment plants (WWTP) and then from municipal effluent, but the level to which these agents are filtered by treatment processes is not clear. Also, the majority of the compounds discharged are non-biodegradable which cannot be efficiently cleared by conventional treatment procedures, leading to high aquatic water contamination (Vieno *et al.*, 2007). So this process wastewater may be a source of high human as well as animal health hazards effects such as the generation of multi-drug resistance in the case of anti-microbial drugs. Wastewater can be a source of various infectious diseases such as Giardiasis, Scariasis and enteropathogenic infections like shigellosis, *Helicobacter pylori,* cholera, *Escherichia coli* infection and typhoid etc. if not treated efficiently. Other toxicities related to the type of process for which wastewater was used and the type of chemical processed in industry includes acute and chronic toxicity, reproductive and carcinogenic, neurotoxic and developmental toxicities.

1.7 CHRONOLOGICAL OUTLOOK OF WASTEWATER TREATMENT

At the start of the 20th century, there was an outbreak of life-threatening water-borne diseases and most of the scientists at that time focused on determining the actual cause and they concluded that it was due to microorganisms present in wastewater that seep into naturally occurring drinking water reservoirs (Henze & Harremoes, 1983). In order to overcome these issues, they started chemical treatment of wastewater in open fields, but soon they realized that this type of sewage treatment would affect the fertility of the soil. Subsequent endeavours were made to withdraw raw wastewater directly into running streams but it was observed that it deteriorated the quality of the existing water. Furthermore, as time goes on, there was a demand to develop suitable methods for wastewater treatment, as population and urbanization increased day by day and so the production of sewage water also increased (Rosen *et al.*, 1998; Spellman, 2000). Therefore, advanced methods were established for the treatment of wastewater in the late 1990s with objectives such as remove suspended and dissolved contents from wastewater that make them toxic and also eradicate disease causing pathogenic microorganisms from wastewater. Later on, further initiatives were made to remove nutrients such as NO_3, NO_2 and PO_4 etc. from wastewater in order to enhance the quality of treated effluent (Barbosa & Sant'Anna, 1989; McGhee, 1991; Metcalf & Eddy, 1991).

1.8 BIOLOGICAL WASTEWATER TREATMENT TECHNOLOGIES

The treatment of wastewater is one of the ultimate processes to eradicate the possible toxicity of compounds present in wastewater. As in this process, multifarious carbonaceous compounds, either by physico-chemical or biological means, were converted into simpler composites that are non-toxic and safe to the aquatic ecosystem (Cheremisinoff, 2002). The pharmaceutical waste comprises of drug residues with high biological oxygen demand (BOD), chemical oxygen demand (COD), pharmaceutically active compounds (PhACs) like hormones, antibiotics, toxic substances, or surfactants, and volatile organic compounds (VOCs) which constitute potential threats for the biosphere (Pal, 2018) and, if untreated waste effluent is discharged as such to the environment, it may lead to the emission of putrid gases, reduction of dissolved oxygen contents from desired levels required by aquatic organisms, and eutrophication of the lakes and streams due to the presence of excessive nutrients which facilitate the growth of unnecessary aquatic plants and algae. Moreover, it contains pathogenic microorganisms belonging to the Enterobacteriaceae family and is responsible for a variety of water-borne diseases in living beings. Because of these reasons, wastewater must be treated to produce an effluent suitable for agricultural or aquaculture reuse, or produce an effluent that can be safely discharged into ground or surface water reservoirs (Spellman, 2000). Many of these pharmaceutical wastes pose a serious threat to human health and the ecosystem because of their continuous accumulation in the ecosystem is even leading to the development of antibiotic-resistant microbial strains. Pharmaceutical industries across the world indiscriminately generate huge quantities of pharmaceutical waste (Pal, 2018).

The existence of well-designed and operated wastewater treatment systems are very rare, especially in developing countries due to lack of funds, lack of well-established pollution control laws and lack of expertise towards these environmental issues. However, a wide-range of consolidated wastewater treatment facilities are available in developing countries which include stabilization ponds, activated sludge, constructed wetlands and trickling filter systems (von Sperling, 1996). Stabilization ponds are aerobic in nature and primarily remove organic compounds present in wastewater by their aerobic oxidation process. Moreover, it has been reported that most of these ponds removed 80–85% of BOD5, 40–45% of ammonia and 70–95% of indicator organisms from wastewater (Shuval *et al.*, 1986). However, the main drawback of stabilization ponds is that they required larger space and extended retention times due to their continuous mode of aeration (von Sperling & de Lemos, 2005). On the other hand, the activated sludge process is also a suspended growth process in which microbial suspension is used for the removal of carbonaceous compounds, nitrogen and phosphorus (Dewil *et al.*, 2006; Mara, 2004), but there are certain shortcomings in this process, i.e. the removal efficiency of pathogenic indicators is not constant and also settling of suspended solids is inappropriate which in turn increases the treatment cost (Doorn *et al.*, 2006).

Constructed wetlands are ecological contrived systems that consume the natural processes comprising wetland foliage, territories, and the accompanying microbial biofilm to treat wastewaters (Vymazal, 2005). It usually provides two types of treatment systems, i.e. free water surface system and sub-surface flow system. The main downsides of constructed wetlands are that they require large land areas, continuous monitoring of operating units and sludge removal. As compared to other conventional treatment methods, fixed-film bioreactors, i.e. trickling filters, provide extended retention times due to high active biomass concentration and thus bear high organic loading rates (Amorim *et al.*, 2005).

Moreover, they require low space, less energy, and less operational and maintenance costs which makes them an attractive option for the treatment of all types of wastewater, especially in developing countries (Cheng & Liu, 2002; Perez & Griebenow, 2001; Vartak *et al.*, 1997).

1.9 CONCLUSION

It was concluded that pharmaceutical wastewater contained compounds and chemicals that are highly toxic and had adverse effects on the aquatic as well as terrestrial ecology. On the basis of literature studies, it was also concluded that most of the compounds used during the manufacturing of pharmaceutical products were highly recalcitrant in nature and were not efficiently removed by conventional (primary, secondary, and tertiary) treatment technologies, leading to an unfavorable accumulation in the aquatic environment. Therefore, it is recommended that a comprehensive policy should be designed by each pharmaceutical industry in order to discharge minimal waste into the environment. Moreover, they may also access the available pharmaceutical waste treatment methods in order to reduce the environmental consequences of pharmaceutical effluent.

REFERENCES

Abderrahman W. A. (2000). Water demand management and Islamic water management principles: A case study. *International Journal of Water Resources Development*, **16** (4), 465–473.

Aftab Z., Ali C. L., Khan A. M., Robinson A. and Irshad I. A. (2000). Industrial Policy and the Environment in Pakistan. United Nations Industrial Development Organization (UNIDO) Report. Industrial Policy and Environment (NC/PAK/97/018).

Ahmad M. U. D., Turral H. and Nazeer A. (2009). Diagnosing irrigation performance and water productivity through satellite remote sensing and secondary data in a large irrigation system of Pakistan. *Agricultural Water Management*, **96**(4), 551–564.

Amorim M. J., Römbke J., Scheffczyk A. and Soares A. M. (2005). Effect of different soil types on the enchytraeids Enchytraeusalbidus and Enchytraeusluxuriosus using the herbicide Phenmedipham. *Chemosphere*, **61**(8), 1102–1114.

Andersson Y. and Bohan P. (2001). Disease Surveillance and Waterborne Outbreaks. IWA Publishing, London, pp. 115–133.

Asano T., Burton H., Leverenz H., Tsuchihashi R. and Tchobanoglous G. (2007). Water Reuse: Issues, Technologies, and Application. McGraw-Hill Professional, New York, p. 1570.

Aylward B., Bandyopadhyay J., Belausteguigotia J. C., Borkey P., Cassar A. Z., Meadors L., Saade L., Siebentritt M., Stein R., Tognetti S. and Tortajada C. (2005). 'Freshwater ecosystem services'. *Ecosystems and Human Well-being: Policy Responses*, **3**, 213–256.

Barbosa R. A. and Sant'Anna G. L. (1989). Treatment of raw domestic sewage in an UASB reactor. *Water Research*, **23**(12), 1483–1490.

Bartram J. and Balance R. (1996). Water Quality Monitoring: A Practical Guide to the Design and Implementation of Fresh Water Quality Studies and Monitoring Programmes. Chapman & Hall, London, p. 383.

Becerra-Castro C., Lopes A. R., Vaz-Moreira I., Silva E. F., Manaia C. M. and Nunes O. C. (2015). Wastewater reuse in irrigation: A microbiological perspective on implications in soil fertility and human and environmental health. *Environment International*, **75**, 117–135.

Bizier P and DeBarry P. (2003). World Water & Environmental Resources Congress June 2003. American Society of Civil Engineers, Philadelphia, USA.

Carpenter S. R., Caraco N. F., Correll D. L., Howarth R. W., Sharpley A. N. and Smith V. H. (1998). Nonpoint pollution of surface waters with phosphorus and nitrogen. *Ecological Applications*, **8**(3), 559–568.

Cheng J. and Liu B. (2002). Swine wastewater treatment in anaerobic digesters with floating medium. *Transactions of the ASAE*, **45**(3), 799.

Cheremisinoff N. P. (2002). Membrane separation technologies. Handbook of Water and Wastewater Treatment Technologies. Butterworth-Heinemann, USA, pp. 335–371.

Connor R., Renata A., Ortigara C., Koncagül E., Uhlenbrook S., Lamizana-Diallo B. M., Zadeh S. M., Qadir M., Kjellén M., Sjödin J. and Hendry S. (2017). The United Nations World Water Development Report 2017. Wastewater: The Untapped Resource. UNESCO, Paris.

Craun M. F., Craun G. F., Calderon R. L. and Beach M. J. (2006). Waterborne outbreaks reported in the United States. *Journal of Water and Health*, **4**(S2), 19–30.

Dalrymple O. K., Yeh D. H. and Trotz M. A. (2007). Removing pharmaceuticals and endocrine disrupting compounds from wastewater by photocatalysis. *Journal of Chemical Technology & Biotechnology: International Research in Process, Environmental & Clean Technology*, **82**(2), 121–134.

Dewil R., Appels L. and Baeyens J. (2006). Energy use of biogas hampered by the presence of siloxanes. *Energy Conversion and Management*, **47**(13), 1711–1722.

Doorn M. R. J., Towprayoon S., Maria S., Vieira M., Irving W., Palmer C., Pipatti, R. and Wang C. (2006). Wastewater treatment and discharge. In 2006 IPCC Guidelines for National Greenhouse Gas Inventories, Volume **5** (Waste). *WMO, UNEP*, pp. 6: 1–6, 28.

Enick O. V. and Moore M. M. (2007). Assessing the assessments: pharmaceuticals in the environment. *Environmental Impact Assessment Review*, **27**(8), 707–729.

Feenstra S., Hussain R. and van der Hoek W. (2000). Health risks of irrigation with untreated urban wastewater in the southern Punjab, Pakistan. Pakistan Program: International Water Management Institute, Lahore, Pakistan, pp. 1–19.

Fent K. (2004). Ecotoxicological effects at contaminated sites. *Toxicology*, **205**(3), 223–240.

Gadipelly C., Pérez-González A., Yadav G. D., Ortiz I., Ibáñez R., Rathod V. K. and Marathe K. V. (2014). Pharmaceutical industry wastewater: review of the technologies for water treatment and reuse. *Industrial & Engineering Chemistry Research*, **53**, 11571–11592.

Government of Pakistan. Ministry of Water and Power. (2010). Pakistan Water Sector Strategy. Executive Summary, Volume **2**. Ministry of Water and Power, Islamabad, Pakistan, pp. 1–94.

Hameed M., Ashraf M., Naz N. and Al-Qurainy F. (2010). Anatomical adaptations of Cynodon dactylon (L.) Pers. from the Salt Range Pakistan to salinity stress. I. Root and stem anatomy. *Pakistan Journal of Botany*, **42**(1), 279–289.

Heberer T. (2002). Occurrence, fate, and removal of pharmaceutical residues in the aquatic environment: a review of recent research data. *Toxicology Letters*, **131**, 5–17.

Henze M. and Harremoes P. (1983). Anaerobic treatment of wastewater in fixed film reactors–a literature review. *Water Science and Technology*, **15**(8–9), 1–101.

Hinrichsen D., Robey B. and Upadhyay U. D. (1997). Solutions for a water-short world, population reports, Series M, No. 14. Johns Hopkins School of Public Health. Population Information Program, Baltimore.

Huerta B., Rodríguez-Mozaz S. and Barceló D. (2012). Pharmaceuticals in biota in the aquatic environment: analytical methods and environmental implications. *Analytical and Bioanalytical Chemistry*, **404**(9), 2611–2624.

Hussain A., Alamzeb S. and Begum S. (2013). Accumulation of heavy metals in edible parts of vegetables irrigated with waste water and their daily intake to adults and children, District Mardan, Pakistan. *Food Chemistry*, **136**(3), 1515–1523.

Jayakumar O. D., Ganguly R., Tyagi A. K., Chandrasekharan D. K. and Krishnan Nair C. K. (2009). Water dispersible Fe3O4 nanoparticles carrying doxorubicin for cancer therapy. *Journal of Nanoscience and Nanotechnology*, **9**(11), 6344–6348.

Jury W. A. and Vaux H. J. (2007). The emerging global water crisis: managing scarcity and conflict between water users. *Advances in Agronomy*, **95**, 1–76.

Kantawanichkul S., Kladprasert S. and Brix H. (2009). Treatment of highstrength wastewater in tropical vertical flow constructed wetlands planted with *Typha angustifolia* and *Cyperus involucratus. Ecological Engineering*, **35**(2), 238–247.

Kalavrouzlotis I. K. and Arslan-Alaton I. (2008). Reuse of urban wastewater and sewage sludge in the Mediterranean countries: Case Studies from Greece and Turkey. *Fresenius Environmental Bulletin*, **17**(6), 625.

Khan F. K. (1991). A geography of Pakistan: environment, people and economy. Oxford University Press, Oxford, UK.

Khetan S. K. and Collins T. J. (2007). Human pharmaceuticals in the aquatic environment: a challenge to green chemistry. *Chemical Reviews*, **107**(6), 2319–2364.

Klavarioti M., Mantzavinos D. and Kassinos D. (2009). Removal of residual pharmaceuticals from aqueous systems by advanced oxidation processes. *Environment International*, **35** (2), 402–417.

Kümmerer K. (2001). Drugs in the environment: emission of drugs, diagnostic aids and disinfectants into wastewater by hospitals in relation to other sources – a review. *Chemosphere*, **45**(6–7), 957–969.

Larsson D. J., de Pedro C. and Paxeus N. (2007). Effluent from drug manufactures contains extremely high levels of pharmaceuticals. *Journal of Hazardous Materials*, **148**(3), 751–755.

Mara D. (2004). To plant or not to plant? Questions on the role of plants in constructed wetlands. In: Joint Session of the Ninth IWA International Conference on Constructed Wetlands and the Sixth IWA International Conference on Waste Stabilization Ponds, Vol. **30**. Avignon, France.

McGhee T. J. (1991). Water Resources and Environmental Engineering. McGraw-Hill, New York.

Metcalf E. and Eddy H. P. (1991). Biological unit processes. Wastewater Engineering: Treatment Disposal Reuse. McGraw-Hill, New York, USA, pp. 411–467.

Mokaya S. K., Mathooko J. M. and Leichtfried M. (2004). Influence of anthropogenic activities on water quality of a tropical stream ecosystem. *African Journal of Ecology*, **42**(4), 281–288.

Montgomery M. A. and Elimelech M. (2007). Water and sanitation in developing countries: including health in the equation. *Environmental Science and Technology*, **24**, 17–24.

Morel A. (2006). Greywater management in low and middle-income countries (No. 628.2 G842 g). Swiss Federal Institute of Aquatic Science and Technology, Dubenforf, CH.

Niemi G. J., DeVore P., Detenbeck N., Taylor D., Lima A., Pastor J., Yount J. D. and Naiman R. J. (1990). Overview of case studies on recovery of aquatic systems from disturbance. *Environmental Management*, **14**(5), 571–587.

Nikolaou A., Meric S. and Fatta D. (2007). Occurrence patterns of pharmaceuticals in water and wastewater environments. *Analytical and Bioanalytical Chemistry*, **387**, 1225–1234.

Pal P. (2018). Treatment and disposal of pharmaceutical wastewater: toward the sustainable strategy. *Separation & Purification Reviews*, **47**(3), 179–198.

Perez C. and Griebenow K. (2001). Improved activity and stability of lysozyme at the water/CH2CI2 interface: enzyme unfolding and aggregation and its prevention by polyols. *Journal of Pharmacy and Pharmacology*, **53**(9), 1217–1226.

Rosen M., Welander T., Löfqvist A. and Holmgren J. (1998). Development of a new process for treatment of a pharmaceutical wastewater. *Water Science and Technology*, **37**(9), 251–258.

Salma S., Shah M. A. and Rehman S. (2012). Rainfall trends in different climate zones of Pakistan. *Pakistan Journal of Meteorology*, **9**(17), 37–47.

Saed H. Z., Zyoud S. H., Al-Jabi S. W., Sweileh W. M. and Awang R. (2016). Contribution of Arab countries to pharmaceutical wastewater literature: a bibliometric and comparative analysis of research output. *Annals of Occupational and Environmental Medicine*, **28**(1), 1–12.

Scott R. L., Shuttleworth W. J., Goodrich D. C. and Maddock III T. (2000). The water use of two dominant vegetation communities in a semiarid riparian ecosystem. *Agricultural and Forest Meteorology*, **105**(1), 241–256.

Shuval H. I., Yekutiel P. and Fattal B. (1986). An epidemiological model of the potential health risk associated with various pathogens in wastewater irrigation. *Water Science and Technology*, **18**(10), 191–198.

Singh K. P., Malik A., Mohan D., Sinha S. and Singh V. K. (2005). Chemometric data analysis of pollutants in wastewater—a case study. *Analytica Chimica Acta*, **532**(1), 15–25.

Soller J. A., Olivieri A. W., Crook J., Cooper R. C., Tchobanoglous G., Parkin R. T., Spear R. C. and Eisenberg J. N. (2003). Risk-based approach to evaluate the public health benefit of additional wastewater treatment. *Environmental Science and Technology*, **37**(9), 1882–1891.

Spellman F. R. (2000). Safe Work Practices for Wastewater Treatment Plants. CRC Press, Boca Raton, Florida, USA.

Tebbutt T. H. Y. (1998). Principles of Water Quality Control, 5th edn. Butterworth-Heinemann, USA.

USEPA. (2007). Bureau of Water Supply and Wastewater Management: Department of Environmental Protection Agency, Wastewater Treatment Plant Operator Training. United State Environmental Protection Agency, Washington, DC.

USEPA. (2012). Guidelines for Water Reuse. United State Environmental Protection Agency, Washington, DC.

Vartak D. R., Engler C. R., McFarland M. J. and Ricke S. C. (1997). Attached-film media performance in psychrophilic anaerobic treatment of dairy cattle wastewater. *Bioresource Technology*, **62**(3), 79–84.

Vieno N., Tuhkanen T. and Kronberg L. (2007). Elimination of pharmaceuticals in sewage treatment plants in Finland. *Water Research*, **41**(5), 1001–1012.

Viet Anh N., Nga P. T., Nhue T. H. and Morel A. (2005). Potential Decentralized Wastewater Management for Sustainable Development from Vietnamese Experience. *Proceedings of the Water Environment Federation*, **2005**(5), 917–946.

von Sperling M. (1996). Comparison among the most frequently used systems for wastewater treatment in developing countries. *Water Science and Technology*, **33**(3), 59–72.

von Sperling M. and de LemosChernicharo C. A. (2005). Biological Wastewater Treatment in Warm Climate Regions, vol. **1** IWA Publishing, London, UK.

Vymazal J. (2005). Horizontal sub-surface flow and hybrid constructed wetlands systems for wastewater treatment. *Ecological Engineering*, **25**(5), 478–490.

World Health Organization (WHO). (2015). Progress on Sanitation and Drinking Water: 2015 Update and MDG Assessment. World Health Organization, Geneva, Switzerland.

Yakubu O. H. (2017). Pharmaceutical wastewater effluent – source of contaminants of emerging concern: phytotoxicity of metronidazole to soybean (Glycine max). *Toxics*, **5**, 1–14.

Yang H. and Abbaspour K. C. (2007). Analysis of wastewater reuse potential in Beijing. *Desalination*, **212**(1), 238–250.

Zaman A. and Ara I. (2000). Rising urbanization in Pakistan, some facts and suggestions. *The Journal NIPA*, **7**, 31–46.

Zhou H. and Smith D. W. (2002). Advanced technologies in water and wastewater treatment. *Journal of Environmental Engineering and Science*, **1**(4), 247–264.

Chapter 2

Comprehensive insights into sources of pharmaceutical wastewater in the biotic systems

Sergij Vambol[1], Viola Vambol[2], Nastaran Mozaffari[3], Alaa El Din Mahmoud[4], Nishana Ramsawak[5], Niloofar Mozaffari[6], Parisa Ziarati[7] and Nadeem A Khan[8]

[1]Life Safety and Law Department, Kharkiv Petro Vasylenko National Technical University of Agriculture, Ukraine
[2]Educational and Scientific Department of Occupational Safety and Health, National Scientific and Research Institute of Industrial Safety and Occupational Safety and Health, Ukraine
[3]Department of Environmental Engineering, Faculty of Natural Resources and Environment, Science and Research Branch, Islamic Azad University, Tehran, Iran
[4]Environmental Sciences Department, Faculty of Science, Alexandria University, 21511, Alexandria, Egypt
[5]Department of Civil Engineering, Queens University, Kingston, Canada
[6]Department of Physics, Faculty of Sciences, Science and Research Branch, Islamic Azad University, Tehran, Iran
[7]Nutrition and Food Sciences Research, Tehran Medical Sciences, Islamic Azad University, Tehran, Iran
[8]Department of Civil Engineering, Jamia Millia Islamia, New Delhi

ABSTRACT

In today's world, the problem of micropollutants penetration is particularly relevant. The protection of the aquatic environment deserves much attention since water is an

Editor(s): Nadeem A Khan, Sirajuddin Ahmed, Viola Vambol and Sergij Vambol
doi: 10.2166/9781789061338_0017

exhaustible resource. This chapter focuses on the mechanisms by which pharmaceutical ingredients enter biotic systems. A detailed analysis of many scientific studies allowed isolating and focusing on some assumptions about the pharmaceutical ingredients' sources in the aquatic environment. Hospitals, households, wastewater treatment plants, aquaculture, horticulture, and animal husbandry were studied as sources that make the main contribution to environmental pollution by pharmaceutical compounds. Each of these suppliers, to a greater or lesser extent, ensures the release of these toxic substances into the environment. The degree of negative impact depends on the level of the country's development, state policy on the natural resources' management, culture and education of citizens, the society's influence on social behaviour, and the like. However, it should be emphasized that there is a closed circulation of pharmaceutical compounds primarily because of the presence of a food chain in nature. At the same time, the prevention of uncontrolled drug uses, of its improper disposal, of the veterinary drug uses to stimulate growth, etc. will significantly reduce the presence of pharmaceutical micro-contaminants in water.

Keywords: Biotic systems, Pharmaceutical ingredients, Polluted water, Sources and penetration mechanisms.

Contents

2.1 Introduction 18
2.2 The Environmental Influences of Pharmaceuticals 27
2.3 Sources of Pharmaceuticals in Water Environments 31
2.3.1 Hospitals 31
2.3.2 Sewage 31
2.3.3 Aquaculture 32
2.3.4 Horticulture and animal husbandry 32
2.3.5 Waste disposal 33
2.3.6 Environmental fate of pharmaceuticals in water environments 34
2.4 Scopus Based Publications for Pharmaceutical Wastewater 35
2.5 Conclusions 35
References 36

2.1 INTRODUCTION

Water is a vital source for all biological objects (Khan *et al.*, 2017). The presence of chemical and microbial pollutants in natural waters is a severe threat to the biosphere and for the health of the public in particular (El Din Mahmoud & Fawzy, 2016; Khan *et al.*, 2017; Mahmoud *et al.*, 2020). Pharmaceuticals (PhAs) are the most common pollutants in the water because 30–90% of them remain in our bodies without decomposition as well as enter the environment after excretion from humans or animals (Mahmoud *et al.*, 2018; Rivera-Utrilla *et al.*, 2013).

The study of PhAs sources is always essential as a first step for pharmaceutical wastewater management because various substances from these sources become components of the chemical composition of the environment. PhAs sources mean the origin of the micropollutant which exists in greater concentrations than the natural ones due to human activities. In this case, their continuous appearance in the environment affects both the health of organisms and the environment (Sirés & Brillas, 2012; Yuan *et al.*, 2009). Both non-prescription and prescription drugs, consisting of assorted compositions, are used in the evasion or cure of diseases (Benotti *et al.*, 2009; Cai *et al.*, 2013; Ratola *et al.*, 2012; Yoon *et al.*, 2007). The prevalently consumptive pharmaceuticals are outlined in Table 2.1.

Why is there a lot of focus on pharmaceutical micropollutants? There is an answer to this question. The physicochemical and biological properties of pharmaceuticals micropollutants make their biological power different when they are dispersed in the surroundings (Fatta-Kassinos *et al.*, 2011). Some of the pharmaceuticals' properties such as polymorphism, their ability to enter the environment from the metabolic process, their complicated chemical structure, and their ability to form multiplex ionization sites all over the molecule after being ionized, distinguish them from other various industrial chemical combinations (Cunningham, 2010; Fatta-Kassinos *et al.*, 2011).

Table 2.1 Prevalently consumptive pharmaceuticals list (Yoon *et al.*, 2007).

Compound	Classification
Amoxicillin	Antibiotic
Ampicillin	
Carbadox	
Ciprofloxacin	
Chloramphenicol	
Erythromycin-H_2O	
Ofloxacin	
Penicillin G	
Sulfachloropyridazine	
Sulfamerazine	
Sulfamethizole	
Sulfamethoxazole	
Tetracycline	
Triclosan	
Trimethoprim	
Diazepam	Anxiolytic
Meprobamate	
Ketoconazole	Antifungal
Ibuprofen	Pain reliever

The distinct chemical structures of drugs, carbohydrates, proteins, organic molecules, and other chemically complex components are large polymers that show their extensive characteristics based on their chemical composition (Fatta-Kassinos *et al.*, 2011). Polymorphism is a property where a molecule can possess more than one crystalline structure which differs in the capability of being dissolved, rate of dissolution, bioavailability, and colours. Their flowability, density, chemical and physical stability, and the point at which they melt, are also dissimilar (Fatta-Kassinos *et al.*, 2011).

Pharmaceuticals can be released into the environment through permeation to soils and sediments, photolysis, a complex combination with metals and organics, chemical oxidation, evaporation, and biodegradation. To distinguish the concentration of a compound in the aquatic or territorial environment, it is important to understand the physicochemical characteristics such as coefficient of octanol/water partition, the pressure exerted by a vapor, dissociation constants or Law constant of Henry (Babic *et al.*, 2007; Fatta-Kassinos *et al.*, 2011). Moreover, the exchange of ions, building a complex combination with metals, surface adsorption of mineral materials, hydrogen bonding, and alliance with organic materials, contribute to the permeation of pharmaceuticals through soil (Fatta-Kassinos *et al.*, 2011). As a result, pH and the ability to be dissolved in water are important properties of pharmaceuticals in polymorphic forms, because soil organic materials have negative charges, while pharmaceutical compounds are visible in cationic form. Solubility also has a relation to pH (Fatta-Kassinos *et al.*, 2011). These issues may affect the transfer of drugs as well as the evaluation of impacts on the environment, which underestimates the significance of biological effects like toxicity (Fatta-Kassinos *et al.*, 2011).

Drugs developed from non-homogeneous groups combined with distinct molecular weight and structure are used for specific biological processes (Kümmerer, 2009). The human body can be infiltrated by drugs that are potent enough to travel through the epithelial cell linings' lipid bilayer. Lipophilicity, charge, the capability of hydrogen-bonding, and size are properties that affect the permeability of drugs to biological membranes (Fatta-Kassinos *et al.*, 2011). Hydrogen bonding and dipole-dipole effects are intermolecular interactions of the lipophilicity property and a physicochemical property that determines the drug permeability in biological systems (Kaliszan, 1992; Mälkiä *et al.*, 2004). Additionally, the drugs excreted by humans and animals, or through trash disposal after expiration, contribute more to water pollution than any industrial pharmaceuticals (Mahmoud, 2020).

Aquatic places including surface waters, sewage from wastewater treatment plants and groundwater have been widely exposed to pharmaceuticals that finally enter the drinking water supplies (Benotti *et al.*, 2009; der Beak *et al.*, 2016; Guan *et al.*, 2017). It was inaccurately presumed that chemical accumulations ended in sewage treatment plants or landfills. However, it can be transferred in drinking-water sources through niches by leaching (Mahmoud,

2020; Ziarati *et al.*, 2020). Studies have shown over 100 different pharmaceutical types found with high concentrations in aquatic places in some countries (Benotti *et al.*, 2009). These compounds stem from the drug family of antibiotics, anticancer, cardiovascular, diagnostic aid-adsorbable organic halogen compounds, and analgesic (Klavarioti *et al.*, 2009; Nikolaou *et al.*, 2007; Rivera-Utrilla *et al.*, 2013); which helps in the identification of reaction mechanisms of pharmaceutical wastes and deep comprehension of the pharmaceutical life cycle, specifically manufacturing, consumption, excretion and half-life (Mahmoud, 2020).

There are reasonable assumptions that households are the significant source of flowing pharmaceuticals to the environment and, as a result, into surface waters. According to the authors' assumptions, Figure 2.1 depicts the route of pharmaceuticals to surface water in an illustrative way.

Homeownership is primarily a source of unused pharmaceuticals that, after the expiration date, turn into waste and need to be disposed of. According to an

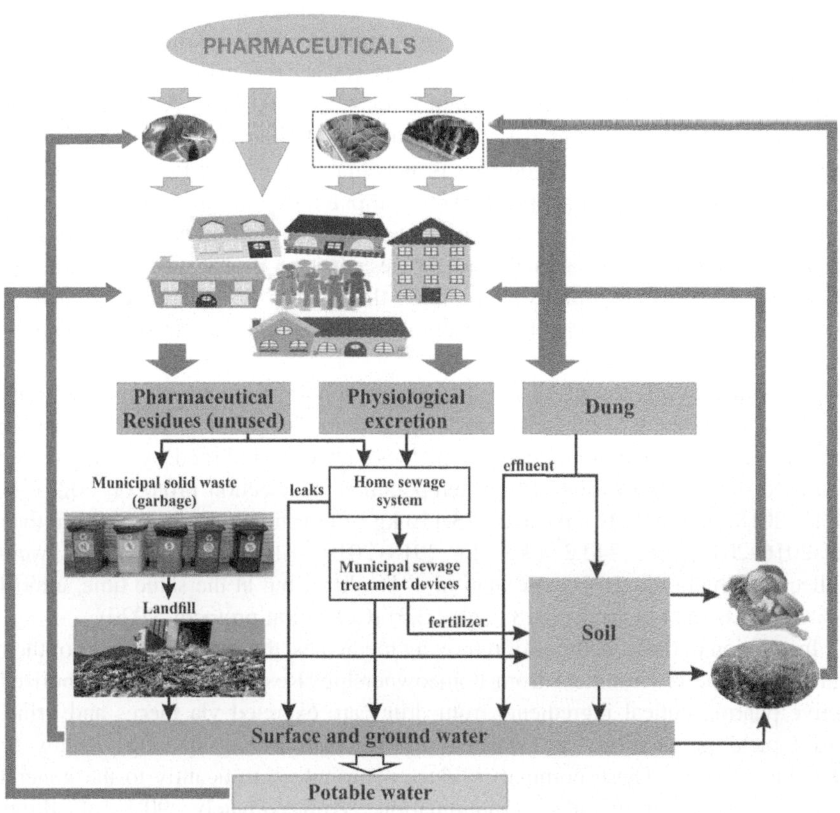

Figure 2.1 Presumptive environmental pharmaceutical turnover.

investigation from 400 households, mainly from south-eastern England, 63.2% of unused pharmaceuticals are thrown into household waste, 11.5% flushed into sewers, and a small number of residents take them to urban landfills, which inevitably ensures the leaching of active pharmaceutical ingredients into the environment (Bound & Voulvoulis, 2006; Vambol *et al.*, 2018; Vambol *et al.*, 2019). Approximately 60% of Belarusians throw generated drug waste into the trash, and about 8% of them throw it in the sewers (Egorova, 2016). Given the report (Oliver & Chapman, 2003), pharmaceuticals in the USA enter sewage in the ratio of 65% from homeowners and 35% from various medical institutions. The situation in Germany and Austria is to some extent better because only 25% of the inhabitants of Germany and 33% of the inhabitants of Austria flush unnecessary drugs into the sewers (Kümmerer, 2008).

In Ukraine, the destruction of unusable drugs issue has been generally regulated for pharmaceutical companies, medical organizations, and pharmacies; hence, if counterfeit and substandard drugs are found, they can be returned to the suppliers, or can get a written-off certificate for expired and unsuitable drugs to send them for disposal. In pharmacy chains, this procedure was centralized after approval of rules for the disposal and destruction of medicines in 2015, by Order of Health Ministry No. 242 in Ukraine. However, a system for the collection of unsuitable drugs does not currently exist in Ukraine. Additionally, the information to illustrate the risk of the ingress of pharmaceutical sewage and household waste into the environment is unavailable (Gubin & Evseeva, 2017).

Interestingly, 52.8% of people consume all purchased drugs; therefore, they do not have a residue to remove. Approximately 30.7% of people store drugs until their expiration date (but after that medicines must be destroyed), and 12.2% of them get rid of them immediately after the course of treatment (Bound & Voulvoulis, 2006). Methods of disposal of drugs by population in some countries are presented in Table 2.2.

Unfortunately, at present Australia has been known as a country that has centrally provided a collection of medicines for its population for years. In compliance with regulatory and EPA requirements, medicines that were returned by residents are disposed of by incineration at high temperatures (Collection protocol, 2020). In 2017–2018, pharmacists collected 823.116 kg of unwanted drugs, 14% more than in 2016–2017, i.e. 742.274 kg. In 2018–2019, 816.242 kg of drugs were collected, which was 10% more than in 2016–2017, but at the same time, almost 1% were less than in 2017–2018 (Table 2.3) (Collection protocol, 2020).

In addition to these methods of removing medicines, there are other ways of their entrance to the environment from homeownership. Residues of non-metabolized active pharmaceutical ingredients from drugs are excreted via faeces and urine, and topical preparations are washed off the skin during bathing (Ruhoy & Daughton, 2008). These nominal residues contribute significantly to the general situation of environmental contamination. Approximately 90% of drugs consumed (not for local use) are changed by the body, and therefore, 7.3 units of

Table 2.2 Population behaviour regarding unused and expired drugs.

Country		Population Behaviour Regarding Unused and Expired Drugs			Reference
		Through Sewer (%)	**Through Household Waste (%)**	**Through a Pharmacy (or Doctor) (%)**	
Eastern and Southeast Europe	Saint-Petersburg, Russia	14	82	Insignificant amount	Trofimova Ye and Get'man (2013)
	Poland	68	—	30	Rogowska *et al.* (2019)
	Lithuania	—	80	—	Kruopienė and Dvarionienė (2010)
	Belarus	8	60	—	Egorova (2016)
	Ukraine	100	—	For the population there is no centralized drug collection program	Gubin and Evseeva (2017)
	Serbia	—	—	4	Kusturica *et al.* (2012)
Northern and Western Europe	Germany	25		—	Kümmerer (2008)
	Austria	33		—	Kümmerer (2008)
	United Kingdom	11.5	63.2	21.8	Bound and Voulvoulis (2006)
	Portugal	—	—	69	Dias-Ferreira *et al.* (2016)
	Sweden	—	3	43	Persson *et al.* (2009)
Asia	Hong Kong	—	53.9	—	Chung Sh-sh and Brooks (2019)
	Saudi Arabia	16.6	62.9	7.9	Shaaban *et al.* (2018)
Africa	Ga South (Ghana)	—	80	—	Udofia *et al.* (2017)
Australia	Australia	—	—	For the population there is centralized drug collection program	Collection protocol, (2020)

Table 2.3 Data of national, not-for-profit program "Return Unwanted Medicines" (Collection protocol, 2020).

Month	ACT	NSW	NT	QLD	SA	TAS	VIC	WA	Monthly Total	Quarter
Date range 2016–2017 (KGS)										
July 2016	217	28,018	0	11,515	0	0	15,868	4352	59,970	—
August 2016	0	9308	0	11,965	18,401	3522	14,394	5710	63,300	—
September 2016	0	18,621	0	12,003	0	0	16,092	5179	51,895	175,164
October 2016	2037	17,171	0	10,515	9453	3389	14,577	4859	62,001	—
November 2016	1133	27,047	2692	11,539	8935	0	15,204	5856	72,406	—
December 2016	0	9247	0	11,345	9085	0	14,654	3847	48,178	182,584
January 2017	0	18,550	0	11,877	0	3621	15,382	6672	56,102	—
February 2017	1569	17,245	0	12,031	8821	0	17,342	4407	61,415	—
March 2017	0	28,260	0	12,887	8954	0	14,198	4979	69,278	186,794
April 2017	0	9544	0	10,443	9012	0	15,236	4654	48,889	—
May 2017	0	29,247	0	14,255	9161	3538	18,316	6350	80,866	—
June 2017	2058	17,853	3372	12,769	9271	0	18,038	4617	67,978	197,733
Totals year	7014	230,111	6064	143140	91093	14070	189301	61482	742274	—
Per month	585	19,176	505	11928	7591	1173	15775	5124	61856	—
Growth%	0.5	8.7	–22.1	5.3	–0.5	–0.5	7.1	2.4	5.3	—
Date range 2017–2018 (KGS)										
July 2018	0	29,439	0	12,962	9189	3617	14,440	4767	74,414	—
August 2018	1184	17,956	0	11,949	9319	0	22,081	6509	68,998	—
September 2018	0	18,966	0	11,830	9055	3050	15,833	4652	63,386	206,797
October 2018	0	28,505	0	10,705	0	0	17,360	6202	62,772	—
November 2018	1559	17,393	0	13,345	9127	3598	15,864	5625	66,511	—
December 2018	0	19,169	0	12,389	9325	0	15,459	3325	59,667	188,949
January 2019	1102	17,701	0	12,924	18,166	0	20,135	6997	77,025	—

February 2019	0	28,645	213	12,393	0	3590	17,857	5390	68,088	—
March 2019	1509	8124	0	13,633	0	0	16,800	5183	45,249	190,362
April 2019	0	28,663	0	10,581	18,319	0	15,752	4724	78,039	—
May 2019	0	28,622	217	13,536	8995	3743	19,508	6972	81,593	—
June 2019	1311	27,367	0	15,630	9288	0	18,251	5530	77,377	237,009
Totals year	6665	270,550	430	151,874	100,783	17,598	209,340	65,876	823,116	—
Per month	555	22,546	36	12,656	8399	1467	17,445	5490	68,593	—
Growth%	–5.0	17.6	–92.9	6.1	10.6	25.1	10.6	7.1	10.9	—
Date range 2018–2019 (KGS)										
July 2018	654	18,534	5391	12,234	9153	0	17,549	6637	70,152	—
August 2018	0	28,312	358	13,393	9019	3610	19,744	5095	79,531	—
September 2018	2303	16,863	261	10,317	0	3815	13,818	4735	52,112	201,794
October 2018	0	28,278	227	9648	11,775	0	18,710	6227	74,865	—
November 2018	0	18,957	0	16,984	9439	3572	16,745	5530	71,227	—
December 2018	0	9689	287	9555	9242	0	13,390	4110	46,273	192,365
January 2019	2268	26,395	0	13,864	9289	0	19,354	7405	78,575	—
February 2019	0	28,443	211	11,279	9364	3004	17,167	5732	75,200	—
March 2019	0	19,281	4432	12,474	8969	0	13,857	6057	65,070	218,845
April 2019	2209	26,136	0	14,171	0	0	22,891	5990	71,397	—
May 2019	0	18,659	0	13,119	9181	3674	15,633	5280	65,546	—
June 2019	0	19,078	0	12,605	9111	0	20,134	5367	66,295	203,238
Totals year	7434	258,625	11,167	149,644	94,542	17,675	208,990	68,165	816,242	—
Per month	620	21,552	931	12,470	7879	1473	17,416	5680	68,020	—
Growth%	11.5	–4.4	2497.0	–1.5	–6.2	0.4	–0.2	3.5	–0.8	—

Note: ACT = Australian Capital Territory; NSW = New South Wales; NT = Northern domain; QLD = Queensland; SA = South Australia; TAS = Tasmania; VIC = Victoria; WA = Western Australia.

the active component enter the wastewater system (Huschek *et al.*, 2004). This contributes to municipal wastewater treatment plants producing solid sludge containing a significant amount of active pharmaceutical ingredients. The European Union generates millions of tons of sewage sludge yearly (Bound & Voulvoulis, 2006). An extensive list of pharmaceuticals at sewage treatment plants in Sweden was identified in the wastewater and is the source of pharmaceutical substances in the marine environment over the past decade (Falås *et al.*, 2012). The properties of the drug, sorption with organic material, surface adsorption with mineral materials, exchange of ions, etc., have great impacts on the concentration of pharmaceuticals in the sewage sludge (Diaz-Cruz *et al.*, 2003; Khan *et al.*, 2019). There are current trends in the disposal of silt sediments (which remain after the treatment of domestic wastewater) through using as a fertilizer for agricultural land, or by disposal in landfills. Consequently, the active pharmaceutical ingredients from the sludge will enter the environment and the degradation of environmental components will continue (that is, the likelihood of contamination of soil and groundwater and adverse effects on plants and animals increases significantly) (Jjemba, 2002).

Poultry farms, livestock farms, and fisheries should be given substantial attention as a pharmaceutical source in the environment. For the cultivation of animals and fishes, both on small private farms and an industrial scale, hormonal preparations, antibiotics, and other drugs can be used to stimulate the growth and development of animals, improve multiple fertility, digestibility of feed, and accelerate puberty. To boost animals' growth, significant amounts of antibiotics are applied as a feed additive (Kaczala & Blum, 2016). Up to 75% of these drugs can be excreted naturally from animals. The use of veterinary pharmaceuticals to stimulate growth is prohibited in the European Union. However, in Europe, as well as in North America and China, it was found that fertilizer and soil can contain from 30 to 100 pharmaceutical materials. It was also shown that 4–10 different substances are present in Turkey, South Korea, Brazil, Australia, and Malaysia (aus der Beek *et al.*, 2016). Based on a report article (Richardson *et al.*, 2005), 8% of the total use of antibiotics in Australia is for veterinary aims and 56% of it is mixed with foods. In Belarus, one of the pharmaceutical sources which releases into the environment is animal husbandry and fish farming; while the bulk of animal waste (manure) is used as fertilizers on the land (Center for Environmental Solutions, 2018; Egorova, 2016). Similarly, the existence of pharmaceuticals in aquaculture and agricultural products in Asian countries shows the use of wastewater in irrigation of field crops (underdeveloped countries generally use wastewater for these purposes) (aus der Beek *et al.*, 2016). There are few publications on this topic in many other regions so few data sets are available. A notable mention is the improper disposal of carcasses of livestock and poultry because the degradation of them ensures the percolation of used pharmaceuticals (among them ammonia, nitrate, chloride, and fecal pathogens) into the soil and groundwater (Kemper *et al.*, 2008; Schneider *et al.*, 2007; Yuan *et al.*, 2013).

Table 2.4 Commonly found medicines in wastewater treatment plant with their side effects (Kostich *et al.*, 2014).

Medicine	Harmful Effects
Metoprolol	Low heart rate, hypotension, irregular rhythm of heartbeat
Ibuprofen	High blood pressure, heart attack, vertigo
Valsartan	Vertigo, low blood pressure, diarrhea, pain in joint
Carbamazepine	Unable to produce blood cells, low blood sodium, paroxysm, congenital malformations
Lisinopril	Serious allergic reaction, vertigo, headache, low blood pressure
Hydrochlorothiazide	Low blood potassium, podagra, high blood sugar
Sertraline	Anxiety, insomnia, suicidal ideation
Propranolol	Insomnia, nightmares, premature birth
Acetaminophen	Hepatic disease, asthma
Ciprofloxacin	Tendonitis, paroxysm, headache
Ofloxacin	Nausea, headache, insomnia, disorder of nervous system
Ranitidine	Vertigo, somnolence, insomnia, irregular heartbeat, hepatic insufficiency
Sulfamethoxazole	Nausea, vomiting, skin rashes, anorexia
Furosemide	Hypotension, light sensitivity, high blood sugar, hearing loss
Atenolol	Type 2 diabetes

It is alarming that the uncontrollable overdose of some medicines such as analgesic, antibiotics, and forbidden drugs such as cocaine and opioids are increasing in domestic waters (Mahmoud, 2020; Wu *et al.*, 2009). Although some of the increased drugs into waters may disappear, they can be harmful to public health (Kostich *et al.*, 2014; Mahmoud, 2020). Table 2.4 represents commonly found medicines in wastewater treatment plants with their side effects for human health.

Interestingly, the first announcement and investigation of the existence of drugs in the marine environment go back to the 1970s (Fatta-Kassinos *et al.*, 2011; Fielding *et al.*, 1981; Garrison *et al.*, 1976; Norpoth *et al.*, 1973; Richardson & Bowron, 1985; Tabak & Brunch, 1970). The environmental challenges and impacts of pharmaceutical compounds in marine environments, sediments, soils, and sludge have been indicated in a significant number of publications (Fatta-Kassinos *et al.*, 2011; Rivera-Utrilla *et al.*, 2013).

2.2 THE ENVIRONMENTAL INFLUENCES OF PHARMACEUTICALS

In many countries worldwide, pharmaceuticals are present in the environment as a contaminant and have been extensively studied (Khan *et al.*, 2017; Mahmoud *et al.*,

2016). According to World Health Organization, PhAs concentrations are typically $<0.1\ \mu g\ L^{-1}$ in groundwater, surface waters, and partially treated water while they are $<0.05\ \mu g\ L^{-1}$ in treated water. From these investigations, numerous water sources receiving effluents of wastewater contain pharmaceuticals at nominal concentrations. A comprehensive assessment by a German-based commission carried out in 2014 found that of the 713 drugs tested, 631 were higher than their detection limits (Khan *et al.*, 2017). All sources of water, including surface water, groundwater, drinking water, and even other areas such as soil, are places in which these pharmaceuticals were found (Beek *et al.*, 2016; Khan *et al.*, 2017).

These levels were found to be below the permissible dosage limits which make its effect on the ecosystem difficult to monitor. Because organisms such as sea lice and algae are so small and not noticeable, unlike many other aquatic organisms, they often go unnoticed and the effect of pharmaceuticals on them is often not considered (Khan *et al.*, 2017). Consequently, these only skyrocket to the forefront when there is a threat to a population, usually a larger animal. This was evident in 1996 and 2007 in India when cattle were treated with high doses of the anti-inflammatory drug, diclofenac, for fever and pain (Khan *et al.*, 2017; Swan *et al.*, 2006). The unhealed cattle were left in the field and vultures devoured the cattle which led to millions of them being killed due to kidney failure as they were susceptible to this high dosage leading to their near extinction (Beek *et al.*, 2016; Khan *et al.*, 2017). This was also observed in Europe and Africa (Khan *et al.*, 2017).

Many marine species, including fishes and invertebrates, have been susceptible to the exposure of pharmaceuticals (Benotti *et al.*, 2009; Boxall *et al.*, 2012; Ikehata *et al.*, 2006). Many wastewater treatment plants (WWTPs) are unable to remove some of the more resistant pharmaceutical compounds due to their lack of design (Rivera-Utrilla *et al.*, 2013). This may be exacerbated during flooding or sewer overflow (Klavarioti *et al.*, 2009; Ort *et al.*, 2010; Pauwels & Verstraete, 2006; Rivera-Utrilla *et al.*, 2013; Verlicchi *et al.*, 2015). Painkillers such as diclofenac and others have shown to be harmful to marine and aquatic life even at low doses and, as such, the impacts of these on the environment at trace levels are being investigated extensively (Akmehmet Balcıoglu & Otker, 2003; Fatta-Kassinos *et al.*, 2011; Klavarioti *et al.*, 2009; Rivera-Utrilla *et al.*, 2013; Tambosi *et al.*, 2010).

Ethinylestradiol, an active ingredient in contraceptives (Khan *et al.*, 2017), the most studied pharmaceutical, has been found to alter the sexual development in female fishes. This has been observed downstream of many WWTPs around the world where female fishes are intersex, leading them to near extinction (Jobling *et al.*, 2005; Khan *et al.*, 2017; Kidd *et al.*, 2007). Ethinylestradiol and other potentially toxic pollutants may have contributed to the significant decrease in fishes in the past 10–15 years (Khan *et al.*, 2017).

The food chain is being disrupted due to fishes such as the European perch being altered by drugs like oxazepam with concentrations as low as $1.8\ \mu g/L^{-1}$. This has caused a change in their eating patterns as well as their behaviour (Brodin *et al.*,

2013; Khan *et al.*, 2017). Similarly, the American fathead minnow fish have shown altered behaviours when exposed to low concentrations of the antidepressant fluoxetine (Di Poi *et al.*, 2014; Khan *et al.*, 2017; Weinberger & Klaper, 2014) and it has also been found to cause physiological changes in water fleas (Campos *et al.*, 2012; Khan *et al.*, 2017). For CBZ (an anti-epilepsy drug), with low concentrations of 7.85 μg/mL, the Atlantic salmon was shown to have altered brain physiology (Hampel *et al.*, 2014; Khan *et al.*, 2017). A drug widely used in infectious disease therapy, antibiotic ciprofloxacin (CIP) (Hartmann *et al.*, 1998; Khan *et al.*, 2017; Martins *et al.*, 2008) biodegrades slowly (Hirsch *et al.*, 1999; Khan *et al.*, 2017) and this contributes to many deleterious implications such as a feeling of nausea in human beings, stomatitis and leukopenia (Cox *et al.*, 2002; Khan *et al.*, 2017; Kümmerer *et al.*, 2000). The chronic presence of CIP develops a resistance of bacteria in aqueous media (Diwan *et al.*, 2010; Khan *et al.*, 2017) and contributes to adverse changes in the water quality (Chang *et al.*, 2010; Khan *et al.*, 2017).

Other than all living organisms, pharmaceuticals have found to be harmful to plants as well, especially on aquatic plants and cyanobacteria (Khan *et al.*, 2017). It damages the chloroplast and affects the metabolic pathway with bacteria. Plants have been found to be more susceptible to fluoroquinolones which affects the production of chloroplast (Brain *et al.*, 2008; Khan *et al.*, 2017). Similar effects occur with tetracycline including p-aminoglycosides, macrolides, linosamides, and pleuromutilins, in that it alters the metabolic pathway. The sterol biosynthesis is affected by other types of pharmaceuticals such as statin-type blood lipid regulators (Brain *et al.*, 2008; Khan *et al.*, 2017). Pharmaceuticals were created to annihilate pathogenic bacteria. However, it was found that the growth, development, and variety of microbial can be influenced by these pharmaceuticals in the entire ecosystem. They can also cause some bacteria to develop a resistance against these antibiotics (Drillia *et al.*, 2005; Khan *et al.*, 2017) causing the effectiveness of these antibiotics to be lowered. This is evident in the use of sulfamethoxazole where it was previously quite effective being used alone, now it is combined with trimethoprim to achieve the same aim for which it was created (Drillia *et al.*, 2005; Khan *et al.*, 2017). These pharmaceuticals must be properly treated before disposal into the environment (Khan *et al.*, 2017) as they can eventually pose a health risk to humans by disrupting endocrine receptors even at low levels (Halling-Sørensen *et al.*, 1998; Khan *et al.*, 2017; Larsson *et al.*, 1999).

When treated wastewater is reused for first and second-generation agriculture, the plants uptake the pharmaceuticals and this can hinder their development and growth. However, researchers are unable to determine if the negative effect is due to direct uptake or because the antimicrobial function of the microorganisms in the soil affects its symbiosis (Chander *et al.*, 2005; Fatta-Kassinos *et al.*, 2011). This occurs because the antibiotics in the soil negatively affect the development of the plants by affecting the soil function, leading to a shortage of food for the soil fauna (protozoa, microarthropods, and nematodes) and slower decomposition of plant residues as well as slower denitrification which leads to decreased nutrient recycling time (Fatta-Kassinos *et al.*, 2011; Migliore *et al.*, 1998). Due to risk assessments carried out on first-generation crops and due to the chronic effects and allergenic potential of these antibiotics, the risks cannot be deemed as negligible (Fatta-Kassinos *et al.*, 2011; Kumar *et al.*, 2005).

With the widespread entity of antibiotics in the environment, it stunts the development of naturally occurring positive microbes which simultaneously contributes to increasing antibiotic-resistant human pathogens (Fatta-Kassinos *et al.*, 2011; Mahmoud, 2020; O'Connor & Aga. 2007; Stolker *et al.*, 2004). Kümmerer (Fatta-Kassinos *et al.*, 2011; Kümmerer, 2009) found that if the crops are irrigated with non-treated water containing bacteria or genes that are resistant to antibiotics, human beings will receive these bacteria through food and water. This occurs because wastewater is a route in which antibiotics, as well as resistant genes, enter the natural bacterial ecosystem with non-pathogenic bacteria serving as a basin of resistant genes (Fatta-Kassinos *et al.*, 2011).

The resistance of bacteria develops due to industrially produced antimicrobial agents present on the bacteria (Baquero *et al.*, 2008; Fatta-Kassinos *et al.*, 2011). Because of a high biological connection and generation of alteration, genetic evolution occurs. Going past mutational occurrences, genetic variations occur due to genetic recombination which usually results from exchanges between organisms within a population (Fatta-Kassinos *et al.*, 2011). Antibiotic resistance develops through four main reactors. According to Baquero *et al.* (2008), the first is therapeutic or preventative antibiotics from human and animal microbiota. Second is farmland, clinics, long-term care facilities, or any place where the most vulnerable individuals are subjected to exchange of bacteria. The third is from wastewater or any biological reactor type from sewage treatment plants, compost toilets or lagoons, as well as from hospitals and farms where different bacteria organisms can mix and genetically react. The fourth reactor is the soil or marine environment in which other reactors mix and counteract with environmental organisms (Baquero *et al.*, 2008). In all reactors, especially the fourth one, water plays a significant role. The chance of lowering the antibiotic resistance's evolution is based on human's capability of controlling the flow of active antimicrobial agents, biological information as well as bacterial clones. This is crucial in wastewater discharge and reuse (Baquero *et al.*, 2008; Fatta-Kassinos *et al.*, 2011).

At low concentrations, antibiotics act as signalling agents in microbial environments (Fatta-Kassinos *et al.*, 2011). Numerous amounts of antibiotics and disinfectants which disrupts chloroplast were found to have common receptors, found in plants for replication (fluoroquinolones), transcription-translation (tetracyclines, macrolides, lincosamides, aminoglycosides, pleuromutilins), folate biosynthesis (sulfonamides and possibly trimethoprim), fatty acid synthesis (triclosan), and sterol biosynthesis (azoles, statins) (Baquero *et al.*, 2008; Fatta-Kassinos *et al.*, 2011; Fajardo *et al.*, 2008; Fajardo & Martínez, 2008). The ecological impacts on the release of fatty acid synthesis (triclosan) in freshwater have been a topic of focus in recent studies (Capdevielle *et al.*, 2008) while ciprofloxacin has been found to affect stream microbial communities and colonized plant aging in leaves (Maul *et al.*, 2006). Water safety and supply are important and understanding sources of contaminants are crucial. As such, sources of contamination have been the focus of a new study (Rivera-Utrilla *et al.*, 2013).

2.3 SOURCES OF PHARMACEUTICALS IN WATER ENVIRONMENTS

2.3.1 Hospitals

One of the principal contributors of pharmaceuticals into the environment is hospitals (Rivera-Utrilla *et al.*, 2013). The wastewaters from hospitals contain many active ingredients from drugs, and microorganisms such as those from residues, radionuclides, solvents, and disinfectants in various concentrations due to excretion, laboratory analysis and research can range from analgesics and antibiotics to blood pressure medication and hormones and often includes many over the counter drugs as well (Rivera-Utrilla *et al.*, 2013). In most instances there is a direct discharge of wastewater without pre-treatment to municipal wastewater treatment plants (WWTPs) (Fatta-Kassinos *et al.*, 2011; Kümmerer, 2009). Verlicchi *et al.* found that the concentration and characteristics of the effluent being discharged are dependent on the size of a hospital (Rivera-Utrilla *et al.*, 2013; Verlicchi *et al.*, 2012) and has become a challenge due to wastewater discharges containing chemicals and other microbial agents into WWTPs (Chitnis *et al.*, 2004; Ikehata *et al.*, 2006; Rivera-Utrilla *et al.*, 2013) being resistant to regular wastewater treatment (Emmanuel *et al.*, 2001; Rivera-Utrilla *et al.*, 2013; Verlicchi *et al.*, 2015) and also due to hospitals being the main source of pathogens and iodinated X-ray contrast media (Emmanuel *et al.*, 2001; Larsson *et al.*, 2007; Rivera-Utrilla *et al.*, 2013; Verlicchi *et al.*, 2015).

2.3.2 Sewage

Sewage effluent is one of the principal pharmaceutical sources that enters the marine environment (Gaw *et al.*, 2014; Jafari *et al.*, 2016; Shokri *et al.*, 2016; Zahirnejad *et al.*, 2017; Ziarati *et al.*, 2018). Technology and physico-chemical characteristics determine the removal rate of pharmaceuticals in WWTPs, which can range from less than ten to maximum efficiency in some cases (Kümmerer, 2009; Larsson *et al.*, 2007). Patient use at homes, discharges from hospitals, as well as wastewater from pharmaceutical manufacturing companies are some sources of human pharmaceuticals (Daughton & Ternes, 1999; Gaw *et al.*, 2014; Jafari *et al.*, 2016; Larsson *et al.*, 2007; Shokri *et al.*, 2016; Zahirnejad *et al.*, 2017; Ziarati *et al.*, 2018). Sewage can discharge into the marine environment through some sources such as ocean and coastal outfalls for WWTPs, rivers and sewer overflows receiving WWTP effluents (Gaw *et al.*, 2014; Jafari *et al.*, 2016; Larsson *et al.*, 2007; Mahmoud & Fawzy, 2015; Mahmoud *et al.*, 2018; Shokri *et al.*, 2016; Zahirnejad *et al.*, 2017; Ziarati *et al.*, 2018). Sewage can also come from ships and boats and can be significant, especially from cruise liners. Effluents from small boats may not necessarily obtain pre-treatment before discharge (Larsson *et al.*, 2007). It is worth mentioning that groundwater contaminated with sewage is a pharmaceutical source in seacoasts as it was found that a coastal aquifer in the Yucatan Peninsula, Mexico, contained pharmaceuticals which discharged through municipal sewage (Larsson *et al.*, 2007; Metcalfe *et al.*, 2011). Enhancing the pharmaceutical pollution in groundwater through irrigation by reused domestic wastewater was also found in Mallorca (Larsson *et al.*, 2007; Rodríguez-Navas *et al.*, 2013).

Septic tanks and small decentralized systems are mainly used in rural areas as well as its outskirts in many coastal holiday destinations throughout the world (Larsson *et al.*, 2007; Withers *et al.*, 2013). This can contribute to leakage into ground and surface water and is a potential source for pharmaceuticals in coastal waters, depending on the effectiveness of the treatment and soil capacity (Dougherty *et al.*, 2010; Larsson *et al.*, 2007; Zhao *et al.*, 2010).

2.3.3 Aquaculture

With the rapidly increasing production of seafood, with over 90% in Asia being produced through aquaculture (Larsson *et al.*, 2007; Sapkota *et al.*, 2008), an array of medicine, some even registered for use in humans, are used to control and prevent diseases in the aquaculture environment with up to 75% of the administered dietary dose being lost to its environment. This loss process includes excretion of unprocessed drugs and drug metabolites as well as the dispersal of non-ingested pills (Grigorakis & Rigos, 2011; Larsson *et al.*, 2007). Wild fishes that feed on leftover food in the same environment can possibly further spread these pharmaceuticals and its biotransformation. Farm ponds containing a high concentration of pharmaceuticals are another source of antibiotics which can ooze, unload, and enter coastal waters (Larsson *et al.*, 2007).

2.3.4 Horticulture and animal husbandry

Rearing of animals and horticulture along the coastline and riverbanks can contribute to pharmaceuticals entering the waterways (Jia *et al.*, 2011; Kümmerer, 2009; Larsson *et al.*, 2007) as antibiotics are added to animal feeds and are also used in freshwater to treat diseases in barns or areas that rear a large number of animals (Kemper, 2008; Larsson *et al.*, 2007). Even though it is banned in Europe, there are some parts of the world that still use low dosages of antibiotics in the feed as growth promoters (Du & Liu, 2012; Larsson *et al.*, 2007). Oxytetracycline and streptomycin are still being used on horticulture crops in some countries as well (Kümmerer, 2009; Larsson *et al.*, 2007). One further source of pharmaceuticals entering agriculture systems is from the use of biosolids to farmlands as a fertilizer (Kümmerer, 2009; Larsson *et al.*, 2007).

2.3.5 Waste disposal

Other pharmaceutical sources flowing into the environment are due to improper disposal of household and clinical medicinal waste leading to leachate from coastal land and sea-fills, as well as waste disposal in coastal areas into the marine environment (Larsson *et al.*, 2007). Along with pharmaceutical ingredients, heavy metals are leached, which also enter groundwater and then surface water. It is known that heavy metal contamination of water samples changes the quality of water samples, and therefore it becomes necessary to use the most advanced methods for analyzing surface water quality (Fathinezhad *et al.*, 2020; Yin *et al.*, 2019). In addition, there are a large number of different nanomaterials in landfills, which, after the end of their service life, lose nanoparticles during decomposition (Vambol *et al.*, 2017; Vambol *et al.*, 2017). These nanoparticles, in turn, having great penetrating power (Dastan *et al.*, 2016; Suchikova *et al.*, 2018), can contribute to the rapid transfer of other toxic substances into natural waters. Manufacturing waste from pharmaceuticals, sewage sludge and fertilizer have been cleared from the sea in some regions (Larsson *et al.*, 2007; Lee & Arnold, 1983; Son *et al.*, 2011). There has also been seepage of pharmaceuticals on Mallorca Island which is up to 27,000 $\mu g\ L^{-1}$ (Larsson *et al.*, 2007; Rodríguez-Navas *et al.*, 2013).

A significant universal problem is the unreasonable disposal of waste, and the undesirable management of waste in developing countries. Moreover, conventional methods of water and soil treatment are generally costly and power-consuming, and the extra expenses for removal of toxic materials from foul water and soils inhibit repudiation from being undertaken, especially in low-income areas (Jafari *et al.*, 2016; Shokri *et al.*, 2016; Zahirnejad *et al.*, 2017; Ziarati *et al.*, 2018). The pH of wastewater produced by pharmaceutical laboratories for education and research varies from acidic to alkaline. For instance, the pH range of 9–10 is for an alkaline waste stream from a synthetic organic pharmaceutical plant, whereas pH of acidic waste streams can be 0.8 (Zahirnejad *et al.*, 2017), and the Nutrition and Food Industries Research's pH range from Tehran Medical Science, Islamic Azad University, Iran is 2–4. Nevertheless, since all types of streams of waste generated from the laboratories of pharmaceutical research are approximately either alkaline or acidic, neutralization is required for them before biological remediation (Ziarati *et al.*, 2018).

2.3.6 Environmental fate of pharmaceuticals in water environments

When pharmaceuticals are disposed of into the marine environment, it can negatively impact the entire ecosystem by degrading and sorbing to suspended particulate matter (SPM) and sediments and can even lead to a build-up of tissues in aquatic organisms (Larsson *et al.*, 2007; Ramirez *et al.*, 2009). Some solid matrices such as soils and sediments have a high variation in the sorption process of pharmaceuticals (Jelic *et al.*, 2011; Williams *et al.*, 2009) and, among them, SPM has a high tendency for sorption of cations due to the existence of ionic interactions. Therefore, new structures will be built-up through the sorption of pharmaceuticals by SPM (Pan *et al.*, 2009; Ramirez *et al.*, 2009; Tolls, 2001).

Secondary evidence on the environmental fate of pharmaceuticals in freshwater is not necessarily the same for the marine environment as their physico-chemical conditions may differ and the parameters such as salinity, pH, temperature, and organic substances influence the pharmaceutical fate (Dhingra *et al.*, 2020; Khan *et al.*, 2020; Larsson *et al.*, 2007; Weigel *et al.*, 2002).

Some pharmaceuticals remain resistant, although microbial transformation can happen in the water under specific conditions (Khan *et al.*, 2021; Lahti & Oikari, 2011; Mazhar *et al.*, 2020). For instance, diclofenac can be detected in the water without having any sign of degradation (Joss *et al.*, 2006; Khan *et al.*, 2019; Khan *et al.*, 2020; Quintana *et al.*, 2005), while it can have biotransformation under special conditions (Carballa *et al.*, 2007; Gröning *et al.*, 2007; Yu *et al.*, 2006). Increased pH of seawater can also impact the dissociation of some pharmaceuticals. Ascorbic acid does not deteriorate in the acidic area, whereas it engages in photo-oxidation in the alkaline pH range (Ahmad *et al.*, 2016; Grissom *et al.*, 1993). The different reactions of pharmaceuticals under the influence of light are either photo-degradation reaction with the creation of free radicals or photosensitization reaction with the transformation of intermolecular energy (Ahmad *et al.*, 2016). Light attenuation can alter the role of photodegradation in the removal mechanism in shallow freshwater and coastal waters due to the composition of the water (Choong *et al.*, 2006; Ge *et al.*, 2010; Larsson *et al.*, 2007), as research has shown that fresh and saline environments can have various environmental fates of pharmaceuticals. Considering the work by Choong *et al.* (2006), prochlorperazine can remain in the depths of the sea for 14 days; however, it deteriorates due to the process of sunlight in shallow water. This can be observed in the degradation of ibuprofen in fresh and seawater (Weigel *et al.*, 2004) and prochlorperazine which was shown to be less stable in freshwater as compared to that of seawater (Choong *et al.*, 2006). Thus, we should be more conscious of drug detachment that is persistent in the water.

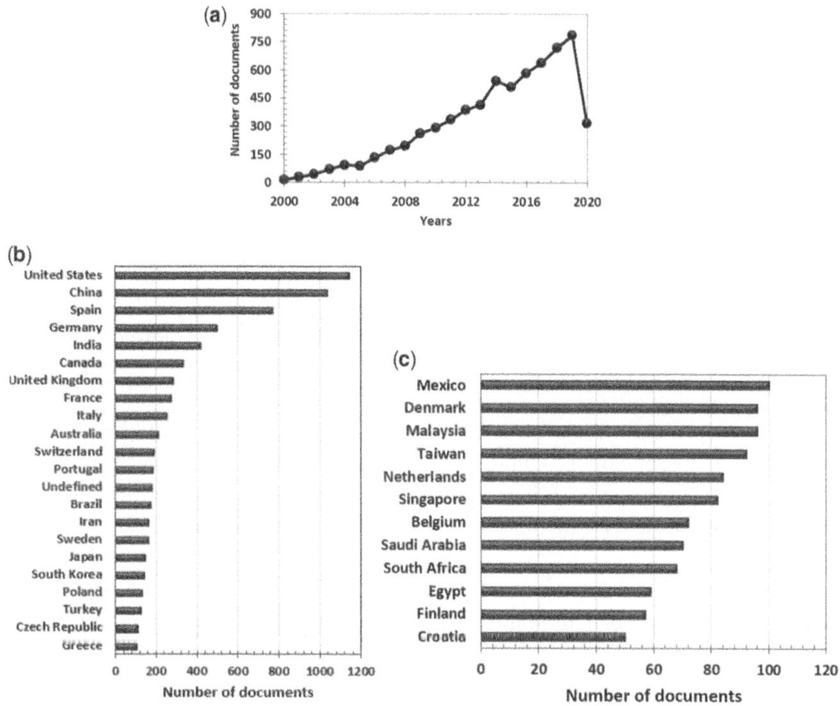

Figure 2.2 (a) The number of published documents for pharmaceutical wastewater, (b) Top-countries having more than 100 documents, and (c) Rest of the countries having less than 100 documents (Scopus 4.2020).

2.4 SCOPUS BASED PUBLICATIONS FOR PHARMACEUTICAL WASTEWATER

We found that there are approximately 6832 documents about pharmaceutical wastewater in the Scopus database. We focus on the published documents from 2000 to 2020 for the graphical representation of these documents. The continuous increment of the published documents in the last 20 years is shown in Figure 2.2(a). Figure 2.2(b) displays the top countries that have a contribution to more than 100 publications in pharmaceutical wastewater up to 2020. Figure 2.2(c) shows the rest of the countries. The top country is the USA with 1140 documents, and after that is China with 1031 documents. Spain and Germany are the leading European countries regarding pharmaceutical issues.

2.5 CONCLUSIONS

In the modern world, the penetration of pharmaceutical micropollutants is especially relevant. The protection of the aquatic environment deserves much attention since water is an exhaustible resource. Studies have shown that in some countries more than 100 different types of pharmaceuticals with a high concentration are found in water. Water bodies, including surface water, wastewater from wastewater treatment plants, and groundwater, were widely exposed to pharmaceuticals, which ultimately end up in drinking water supplies. The disposal of unused drugs into the sewer, the remains of un-metabolized active pharmaceutical ingredients from drugs that are excreted in the faeces and urine, as well as topical preparations that are washed off the skin during bathing, contribute to the fact that municipal treatment facilities produce solid sludge, containing a significant amount of active pharmaceutical ingredients. Subsequently, this sludge is used as a fertilizer, the active pharmaceutical ingredients enter food and are leached through the soil into groundwater and then used as drinking water. Consequently, pharmaceutical ingredients re-enter the human body.

Due to the presence of active pharmaceutical ingredients in natural waters, the food chain is disrupted, as some species change under the influence of such drugs, namely, their eating habits and behaviour change. Also, some drugs can affect the growth, development, and diversity of plants and microbes throughout the ecosystem.

Pharmaceuticals must be properly processed before they are released into the environment, as they can ultimately pose a health hazard, destroying endocrine receptors even at low levels.

REFERENCES

Ahmad I., Ahmed S., Anwar Z., Sherza M. A. and Sikorski M. (2016). Photostability and photostabilization of drugs and drug producys. *International Journal of Photoenergy*, **2016**(7), 1–19.

Akmehmet Balcıoglu L. and Otker M. (2003). Treatment of pharmaceutical wastewater containing antibiotics by O_3 and O_3/H_2O_2 processes. *Chemosphere*, **50** (1), 85–95.

aus der Beek T., Weber F. A., Bergmann A., Hickmann S., Ebert I., Hein A. and Küster A. (2016). Pharmaceuticals in the environment—Global occurrences and perspectives. *Environmental Toxicology and Chemistry*, **35**(4), 823–835.

Babic S., Horvat A. J. M., Pavlovic D. M. and Kastelan-Macan M. (2007). Determination of pKa values of active pharmaceutical ingredients. *Trends in Analytical Chemistry*, **26** (11), 1043–1061.

Baquero F., Martínez J. L. and Cantón R. (2008). Antibiotics and antibiotic resistance in water environments. *Current Opinion in Biotechnology*, **19**, 260–265.

Beek T. A. D., Weber F.-A., Bergmann A., Hickmann S., Ebert I., Hein A. and Küster A. (2016). Pharmaceuticals in the environment—global occurrences and perspectives. *Environmental Toxicology and Chemistry*, **35**(4), 823–835.

Benotti M. J., Trenholm R. A., Vanderford B. J., Holady J. C., Stanford B. D. and Snyder S. A. (2009). Pharmaceuticals and Endocrine Disrupting Compounds in U.S. Drinking Water. *Environmental Science and Technology*, **43**(3), 597–603.

Bound J. and Voulvoulis N. (2006). Household disposal of pharmaceuticals as a pathway for aquatic contamination in the United Kingdom. *Environmental Health Perspectives*, **113** (12), 1705–1711.

Boxall A. B. A., Rudd M. A., Brooks B. W., Caldwell D. J., Choi K., Hickmann S., Innes E., Ostapyk K., Staveley J. P., Verslycke T., Ankley G. T., Beazley K. F., Belanger S. E., Berninger J. P., Carriquiriborde P., Coors A., Deleo P. C., Dyer S. D., Ericson J. F., Gagné F., Giesy J. P., Gouin T., Hallstrom L., Karlsson M. V., Larsson D. G., Lazorchak J. M., Mastrocco F., McLaughlin A., McMaster M. E., Meyerhoff R. D., Moore R., Parrott J. L., Snape J. R., Murray-Smith R., Servos M. R., Sibley P. K., Straub J. O., Szabo N. D., Topp E., Tetreault G. R., Trudeau V. L. and Van Der Kraak G. (2012). Pharmaceuticals and personal care products in the environment: what are the big questions? *Environmental Health Perspectives*, **120**(9), 1221–1229.

Brain R. A., Hanson M. L., Solomon K. R. and Brooks B. W. (2008). Aquatic plants exposed to pharmaceuticals: effects and risks. In: Reviews of Environmental Contamination and Toxicology, D.M. Whitacre (ed.), Springer, vol. **192**, pp. 67–115.

Brodin T., Fick J., Jonsson M. and Klaminder J. (2013). Dilute concentrations of a psychiatric drug alter behavior of fish from natural populations. *Science*, **339**, 814–815.

Cai Z., Dwivedi A. D., Lee W.-N., Zhao X., Liu W., Sillanpää M., Zhao D., Huang C.-H. and Fu J. (2013). Application of nanotechnologies for removing pharmaceutically active compounds in water: development and future trends. *Environmental Science: Nano*, **00**, 1 20.

Campos B., Piña B. and Barata C. C. (2012). Mechanisms of action of selective serotonin reuptake inhibitors in Daphnia magna. *Environmental Science and Technology*, **46**(5), 2943–2950.

Capdevielle M., van Egmond R., Whelan M., Versteeg D., Hofmann-Kamensky M., Inauen J., Cunningham V. and Woltering D. (2008). Consideration of exposure and species sensitivity of triclosan in the freshwater environment. *Integrated Environmental Assessment and Management*, **4**(1), 15–23.

Carballa M., Omil F., Ternes T. and Lema J. M. (2007). Fate of pharmaceutical and personal care products (PPCPs) during anaerobic digestion of sewage sludge. *Water Research*, **41**, 2139–2150.

Center for Environmental Solutions. (2018). Medicines in the environment of the Republic of Belarus. Overview of the situation (from 07.23.2018). Available from: https://ecoidea.by/ru/media/3626.

Chander Y., Kumar K., Goyal S. M. and Gupta S. C. (2005). Antibacterial activity of soil-bound antibiotics. *Journal of Environmental Quality*, **34**, 1952–1957.

Chang X., Meyer M. T., Liu X., Zhao Q., Chen H., Chen J.-A., Qiu Z., Yang L., Cao J. and Shu W. (2010). Determination of antibiotics in sewage from hospitals, nursery and slaughter house, wastewater treatment plant and source water in Chongqing region of Three Gorge Reservoir in China. *Environmental Pollution*, **158**, 1444–1450.

Chitnis V., Chitnis S., Vaidya K., Ravikant S., Patil S. and Chitnis D. S. (2004). Bacterial population changes in hospital effluent treatment plant in central India. *Water Research*, **38**, 441–447.

Choong A. M. F., Teo S. L.-M., Leow J. L., Koh H. L. and Ho P. C. L. (2006). A preliminary ecotoxicity study of pharmaceuticals in the marine environment. *Journal of Toxicology and Environmental Health Part A*, **69**(21), 1959–1970.

Chung Sh-sh and Brooks B. W. (2019). Identifying household pharmaceutical waste characteristics and population behaviors in one of the most densely populated global cities. *Resources, Conservation and Recycling*, **140**, 267–277.

Collection protocol. (2020). The National Return & Disposal of Unwanted Medicines Limited ABN 79 082 871 663. Available from: https://returnmed.com.au/collections/.

Cox C. E., Marbury T. C., Pittman W. G., Brown G. L., Auerbach S. M., Fox B. C. and Yang J. Y. (2002). A randomized, double-blind, multicenter comparison of gatifloxacin versus ciprofloxacin in the treatment of complicated urinary tract infection and pyelonephritis. *Clinical Therapeutics*, **24**, 223–236.

Cunningham V. I. (2010). Pharmaceuticals in the environment. *Annual Review of Environment and Resources*, **35**, 23–33.

Dastan D., Panahi S. L., Yengantiwar A. P. and Banpurkar A. G. (2016). Morphological and electrical studies of titania powder and films grown by aqueous solution method. *Advanced Science Letters*, **22**(4), 950–953.

Daughton C. G. and Ternes T. A. (1999). Pharmaceuticals and personal care products in the environment: agents of subtle change? *Environmental Health Perspectives*, **107**(Suppl. 6), 907–938.

der Beak T. A., Weber F. A., Bergmann A., Hickmann S., Ebert I., Hein A. and Kuster A. (2016). Pharmaceuticals in the environment – Global occurrences and perspectives. *Environmental Toxicology and Chemistry*, **35**(4), 823–835.

Dhingra A., Khan N. A., Ahmed S., Gautam S., Vambol S., Vambol V. and Kovalenko S. (2020). Investigation of medical institutions in India as a source of surface water pollution. World Review of Science, Technology and Sustainable Development (in press).

Dias-Ferreira C., Valente S. and Vaz J. (2016). Practices of pharmaceutical waste generation and discarding in households across Portugal. *Waste Management & Research*, **34**(10), 1006–1013.

Diaz-Cruz M. S., Lopez de Alda M. J. and Barcelo D. (2003). Environmental behavior and analysis of veterinary and human drugs in soils, sediments and sludge. *TrAC Trends in Analytical Chemistry*, **22**, 340–351.

Di Poi C., Bidel F., Dickel L. and Bellanger C. (2014). Cryptic and biochemical responses of young cuttlefish *Sepia officinalis* exposed to environmentally relevant concentrations of fluoxetine. *Aquatic Toxicology*, **151**, 36–45.

Diwan V., Tamhankar A. J., Khandal R. K., Sen S., Aggarwal M., Marothi Y., Iyer R. V., Sundblad-Tonderski K. and Stålsby-Lundborg C. (2010). Antibiotics and antibiotic-resistant bacteria in waters associated with a hospital in Ujjain, India. *BMC Public Health*, **10**, 1–8.

Dougherty J. A., Swarzenski P. W., Dinicola R. S. and Reinhard M. (2010). Occurrence of herbicides and pharmaceutical and personal care products in surface water and groundwater around Liberty Bay, Puget Sound, Washington. *Journal of Environmental Quality*, **39**(4), 1173–1180.

Drillia P., Dokianakis S. N., Fountoulakis M. S., Kornaros M., Stamatelatou K. and Lyberatos G. (2005). On the occasional biodegradation of pharmaceuticals in the activated sludge process: the example of the antibiotic sulfamethoxazole. *Journal of Hazardous Materials*, **122**, 259–265.

Du L. and Liu W. (2012). Occurrence, fate, and ecotoxicity of antibiotics in agro-ecosystems. *A review.* Agronomy for Sustainable Development, **32**(2), 309–327.

Egorova I. (2016). "Medicinal" water: how the remnants of potions are in our glass. Center for Environmental Solutions. Available from: https://ecoidea.by/ru/article/1514.

El Din Mahmoud A. and Fawzy M. (2016). Bio-based methods for wastewater treatment: green sorbents. In: Phytoremediation: Management of Environmental Contaminants, A. A. Ansari, S. S. Gill, R. Gill, G. R. Lanza and L. Newman (eds), Springer International Publishing, Cham, vol. **3**, pp. 209–238.

Emmanuel E., Blanchard J.-M., Keck G. and Perrodin Y. (2001). Chemical, biological and ecotoxicological characterisation of hospital effluents. Déchets Sciences et Techniques, revue francophone d'écoloécologie industrielle, No 222è me trimestre, pp. 31–33.

Fajardo A. and Martínez J. L. (2008). Antibiotics as signals that trigger specific bacterial responses. *Current Opinion in Microbiology*, **11**(2), 161–167.

Fajardo A., Martínez-Martín N., Mercadillo M., Galán J. C., Ghysels B., Matthijs S., Cornelis P., Wiehlmann L., Tümmler B., Baquero F. and Martínez J. L. (2008). The neglected intrinsic resistome of bacterial pathogens. *Plos ONE*, **3**(2), e1619.

Falås P., Andersen H. R., Ledin A. and Jansen J. L. C. (2012). Occurrence and reduction of pharmaceuticals in the water phase at Swedish wastewater treatment plants. *Water Science and Technology*, **66**(4), 783–791.

Fathinezhad M., AbbasiTarighat M. and Dastan D. (2020). Chemometrics heavy metal content clustering of water samples using electrochemical data of modified carbon paste electrode. Environmental Nanotechnology, Monitoring and Management, **14**, 100307–100315. DOI: https://doi.org/10.1016/j.enmm.2020.100307.

Fatta-Kassinos D., Meric S. and Nikolaou A. (2011). Pharmaceutical residues in environmental waters and wastewater: current state of knowledge and future research. *Analytical and Bioanalytical Chemistry*, **399**(1), 251–275.

Fielding M., Gibson T. M., James H. A., McLoughlin K. and Steel C. P. (1981). Organic Micropollutants in Drinking Water. Water Research Centre (Great Britain), London, UK.

Garrison A. W., Pope J. D. and Allen F. R. (1976). Identification and analysis of organic pollutants in water. In: Identification and Analysis of Organic Pollutants in Water, C. H. Keith (ed.), Ann Arbor Science Publisher, the University of Michigan, USA, pp. 517–566.

Gaw S., Thomas K. V. and Hutchinson T. H. (2014). Sources, impacts and trends of pharmaceuticals in the marine and coastal environment. *Philosophical Transactions of the Royal Society B Biological Sciences*, **369**, 20130572.

Ge L., Chen J., Wei X., Zhang S., Qiao X., Cai X. and Xie Q. (2010). Aquatic photochemistry of fluoroquinolone antibiotics: kinetics, pathways, and multivariate effects of main water constituents. *Environmental Science and Technology*, **44**(7), 2400–2405.

Grigorakis K. and Rigos G. (2011). Aquaculture effects on environmental and public welfare – the case of Mediterranean mariculture. *Chemosphere*, **85**(6), 899–919.

Grissom C. B., Chagovetz A. M. and Wang Z. (1993). Use of viscosigens to stabilize vitamin B12 solutions against photolysis. *Journal of Pharmaceutical Sciences*, **82**(6), 641–643.

Gröning J., Held C., Garten C., Claussznitzer U., Kaschabek S. R. and Schlömann M. (2007). Transformation of diclofenac by the indigenous microflora of river sediments and identification of a major intermediate. *Chemosphere*, **69**, 509–516.

Guan Y. D., Wang B., Gao Y. X., Liu W., Zhao X. L., Huang X. F. and Yu J. H. (2017). Occurrence and fate of antibiotics in the aqueous environment and their removal by constructed wetlands in China: a review. *Pedosphere*, **27**(1), 42–51.

Gubin YI and Evseeva LV (2017). Good pharmacy practice and treatment of pharmaceutical species in Ukraine. Quality management in pharmacy: materials of the XI scientific-practical. conf. z international participation, Kharkiv, pp. 45–46. Available from: http://dspace.nuph.edu.ua/bitstream/123456789/15130/1/45-46.pdf.

Halling-Sørensen B., Nielsen S. N., Lanzky P., Ingerslev F., Lützhøft H. H. and Jørgensen S. (1998). Occurrence, fate and effects of pharmaceutical substances in the environment – a review. *Chemosphere*, **36**, 357–393.

Hampel M., Bron J. E., Taggart J. B. and Leaver M. J. (2014). The antidepressant drug carbamazepine induces differential transcriptome expression in the brain of Atlantic salmon, *Salmo salar*. *Aquatic Toxicology*, **151**, 114–123.

Hartmann A., Alder A. C., Koller T. and Widmer R. M. (1998). Identification of fluoroquinolone antibiotics as the main source of umuC genotoxicity in native hospital wastewater. *Environmental Toxicology and Chemistry*, **17**, 377–382.

Hirsch R., Ternes T., Haberer K. and Kratz K.-L. (1999). Occurrence of antibiotics in the aquatic environment. *Science of the Total Environment*, **225**, 109–118.

Huschek G., Hansen P. D., Maurer H. H., Krengel D. and Kayser A. (2004). Environmental risk assessment of medicinal products for human use according to European Commission recommendations. *Environmental Toxicology*, **19**(3), 226–240.

Ikehata K., Naghashkar N. J. and Ei-Din M. G. (2006). Degradation of aqueous pharmaceuticals by ozonation and advanced oxidation processes: a review. *Ozone Science and Engineering*, **28**(6), 353–414.

Jafari A., Amin G. and Ziarati P. (2016). Potential of Echium ameonum Fisch & Mey in removing heavy metals from pharmaceutical effluent. *Bioscience & Biotechnology Research Asia*, **13**(3), 1585–1594.

Jelic A., Gros M., Ginebreda A., Cespedes-Sánchez R., Ventura F., Petrovic M. and Barcelo D. (2011). Occurrence, partition and removal of pharmaceuticals in sewage water and sludge during wastewater treatment. *Water Research*, **45**, 1165–1176.

Jia A., Hu J., Wu X., Peng H., Wu S. and Dong Z. (2011). Occurrence and source apportionment of sulfonamides and their metabolites in Liaodong Bay and the adjacent Liao River Basin, North China. *Environmental Toxicology and Chemistry*, **30**(6), 1252–1260.

Jjemba P. K. (2002). The potential impact of veterinary and human therapeutic agents in manure and biosolids on plants grown on arable land: a review. *Agriculture, Ecosystems & Environment*, **93**, 267–278.

Jobling S., Williams R., Johnson A., Taylor A., Gross-Sorokin M., Nolan M., Tyler C. R., van Aerle R., Santos E. and Brighty G. (2005). Predicted exposures to steroid estrogens in UK rivers correlate with widespread sexual disruption in wild fish populations. *Environmental Health Perspectives*, **114**, 32–39.

Joss A., Zabczynski S., Göbel A., Hoffmann B., Löffler D., McArdell C. S., Ternes T. A., Thomsen A. and Siegrist H. (2006). Biological degradation of pharmaceuticals in municipal wastewater treatment: proposing a classification scheme. *Water Research*, **40**, 1686–1696.

Kaczala F. and Blum S. E. (2016). The occurrence of veterinary pharmaceuticals in the environment: a review. *Current Analytical Chemistry*, **12**(3), 169–182.

Kaliszan R. (1992). Quantitative structure-retention relationships. *Analytical Chemistry*, **64** (11), 619A–631A.

Kemper N. (2008). Veterinary antibiotics in the aquatic and terrestrial environment. *Ecological Indicators*, **8**(1), 1–13.

Kemper N., Färber H., Skutlarek D. and Krieter J. (2008). Analysis of antibiotic residues in liquid manure and leachate of dairy farms in Northern Germany. *Agricultural Water Management*, **95**, 1288–1292.

Khan A., Wang J., Li J., Wang X., Chen Z., Alsaedi A., Hayat T., Chen Y. and Wang X. (2017). The role of graphene oxide and graphene oxide-based nanomaterials in the removal of pharmaceuticals from aqueous media: a review. *Environmental Science and Pollution Research*, **24**, 7938–7958.

Khan N. A., Ahmed S., Vambol S., Vambol V. and Farooqi I. (2019). Field hospital wastewater treatment scenario. *Ecological Questions*, **30**(3), 57–69.

Khan N. A., Khan S. U., Ahmed S., Farooqi I. H., Hussain A., Vambol S. and Vambol V. (2019). Smart ways of hospital wastewater management, regulatory standards and conventional treatment techniques: a short review. *Smart and Sustainable Built Environment*. https://doi.org/10.1108/SASBE-06-2019-0079.

Khan A. H., Khan N. A., Khan S. A., Dhingra A., Khan S. U., Mohammadi A. A., Changani F., Yousefi M., Alam S., Vambol S., Vambol V., Khursheed A. and Ali I. (2020). Application of advanced oxidation processes followed by different treatment technologies for hospital wastewater: a case study of multispecialty hospitals. *Journal of Cleaner Production*, **269**, 122411–122424.

Khan N. A., Ahmed S., Farooqi I. H., Ali I., Vambol V., Changani F., Yousefi M., Vambol S., Khan S. U. and Khan A. H. (2020). Occurrence, Sources and Conventional Treatment Techniques for various antibiotics present in hospital wastewaters: a critical review. *Trends in Analytical Chemistry*, **129**, 115291–115935.

Khan N. A., Ahmed S., Farooqi I. H., Vambol S., Vambol V. and Koloskov V. (2021). Operational parameters optimisation for simultaneous removal of two drugs in hospital wastewater – RSM approach. *International Journal of Environment and Waste Management* (in press).

Kidd K. A., Blanchfiel P. J., Mills K. H., Palace V. P., Evans R. E., Lazorchak J. M. and Flick R. W. (2007). Collapse of a fish population after exposure to a synthetic estrogen. *Proceedings of the National Academy of Sciences*, **104**(21), 8897–8901.

Klavarioti M., Mantzavinos D. and Kassinos D. (2009). Removal of residual pharmaceuticals from aqueous systems by advanced oxidation processes. *Environment International*, **35** (2), 402–417.

Kostich M., Batt A. L. and Lazorchak J. M. (2014). Concentrations of prioritized pharmaceuticals in effluents from 50 large wastewater treatment plants in the US and implications for risk estimation. *Environmental Pollution*, **184**, 354–359.

Kruopienė J. and Dvarionienė J. (2010). Management of environmental risks in the life cycle of human pharmaceuticals in Lithuania. *Environmental Research, Engineering & Management*, **2**(52), 41–47.

Kumar K., Gupta S. C., Baidoo S. K., Chander Y. and Rosen C. J. (2005). Antibiotic uptake by plants from soil fertilized with animal manure. *Journal of Environmental Quality*, **34** (6), 2082–2085.

Kümmerer K. (2008). Pharmaceuticals in the Environment Sources, Fate, Effects and Risks. Springer, Berlin Heidelberg.

Kümmerer K. (2009). Antibiotics in the aquatic environment – A review – Part II. *Chemosphere*, **75**(4), 435–441.

Kümmerer K. (2009). Antibiotics in the aquatic environment: a review. I. *Chemosphere*, **75**, 417–434.

Kümmerer K. (2009). The presence of pharmaceuticals in the environment due to human use – present knowledge and future challenges. *Journal of Environmental Management*, **90**(8), 2354–2366.

Kümmerer K., Al-Ahmad A. and Mersch-Sundermann V. (2000). Biodegradability of some antibiotics, elimination of the genotoxicity and affection of wastewater bacteria in a simple test. *Chemosphere*, **40**, 701–710.

Kusturica P. M., Sabo A., Tomic Z., Horvat O. and Solak Z. (2012). Storage and disposal of unused medications: knowledge, behavior, and attitudes among Serbian people. *International Journal of Clinical Pharmacy*, **34**, 604–610.

Lahti M. and Oikari A. (2011). Pharmaceuticals in settleable particulate material in urban and non-urban waters. *Chemosphere*, **85**, 826–831.

Larsson D., Adolfsson-Erici M., Parkkonen J., Pettersson M., Berg A., Olsson P.-E. and Förlin L. (1999). Ethinyloestradiol – an undesired fish contraceptive? *Aquatic Toxicology*, **45**, 91–97.

Larsson D. G. J., de Pedro C. and Paxeus N. (2007). Effluent from drug manufactures contains extremely high levels of pharmaceuticals. *Journal of Hazardous Materials*, **148**, 751–755.

Lee W. Y. and Arnold C. (1983). Chronic toxicity of oceandumped pharmaceutical wastes to the marine amphipod Amphithoe valida. *Marine Pollution Bulletin*, **14**(4), 150–153.

Mahmoud A. E. D. (2020). Nanomaterials: Green synthesis for water applications. In: Handbook of Nanomaterials and Nanocomposites for Energy and Environmental Applications, O. Kharissova, L. Martínez and B. Kharisov (eds), Springer, Cham.

Mahmoud A. E. D. and Fawzy M. (2015). Statistical methodology for Cadmium (Cd (II)) removal from wastewater by different plant biomasses. *Journal of Bioremediation and Biodegradation*, **6**, 1–7.

Mahmoud A. E. D., Fawzy M. and Radwan A. (2016). Optimization of cadmium (CD2+) removal from aqueous solutions by novel biosorbent. *International Journal of Phytoremediation*, **18**(6), 619–625.

Mahmoud A. E. D., Stolle A. and Stelter M. (2018). Sustainable synthesis of high-surface-area graphite oxide via dry ball milling. *ACS Sustainable Chemistry Engineering*, **6**(5), 6358–6369.

Mahmoud A. E. D., Stolle A., Stelter M. and Braeutigam P. (2018). Adsorption Technique for Organic Pollutants using Different Carbon Materials. Abstracts of Papers of the American Chemical Society, Washington, DC, USA.

Mahmoud A. E. D., Franke M., Stelter M. and Braeutigam P. (2020). Mechanochemical versus chemical routes for graphitic precursors and their performance in micropollutants removal in water. *Powder Technology*, **366**, 629–640.

Mälkiä A., Murtomäki L., Urtti A. and Kontturi K. (2004). Drug permeation in biomembranes: in vitro and in silico prediction and influence of physicochemical properties. *Journal of Pharmaceutical Sciences*, **23**(1), 13–47.

Martins A. F., Vasconcelos T. G., Henriques D. M., Frank C. S., König A. and Kümmerer K. (2008). Concentration of ciprofloxacin in Brazilian hospital effluent and preliminary risk assessment: a case study. *Clean – Soil, Air, Water*, **36**, 264–269.

Maul J. D., Schuler L. J., Beiden J. B., Whiles M. R. and Lydy M. J. (2006). Effects of the antibiotic ciprofloxacin on stream microbial communities and detritivorous macroinvertebrates. *Environmental Toxicology and Chemistry*, **25**(6), 1598–1606.

Mazhar M. A., Khan N. A., Ahmed S., Khan A. H., Hussain A., Changani F. and Vambol V. (2020). Chlorination disinfection by-products in Municipal drinking water – A review. *Journal of Cleaner Production*, **273**, 123159–123171.

Metcalfe C. D., Beddows P. A., Bouchot G. G., Metcalfe T. L., Li H. and Van Lavieren H. (2011). Contaminants in the coastal karst aquifer system along the Caribbean coast of the Yucatan Peninsula, Mexico. *Environmental Pollution*, **159**(4), 991–997.

Migliore L., Civitareale C., Cozzolino C., Casoria P., Brambilla G. and Gaudio L. (1998). Laboratory models to evaluate phytotoxicity of sulphadimethoxine on terrestrial plants. *Chemosphere*, **37**, 2957–296.

Nikolaou A., Meric S. and Fatta D. (2007). Occurrence patterns of pharmaceuticals in water and wastewater environments. *Analytical and Bioanalytical Chemistry*, **387**(4), 1225–1234.

Norpoth K., Nehrkorn A., Kirchner M., Holsen H. and Teipel H. (1973) Investigations on the problem of solubility and stability of steroid ovoluation inhibitors in water, waste water and activated sludge. *Zentralblatt fuer Bakteriologie Parasitenkunde Infektionskrankheiten und Hygiene Erste Abteilung Originale Reihe B Hygiene Krankenhaushygiene Betriebshygiene Praeventive Medizin*, **156**(6), 500–51.

O'Connor S. and Aga D. S. (2007). Analysis of tetracycline antibiotics in soil: advances in extraction, clean-up, and quantification. *TrAC Trends in Analytical Chemistry*, **26**, 456–465.

Oliver D. and Chapman A. I. (2003). Pharmaceutical Waste Survey, Final Report. Local Hazardous Waste Management Program in King County, Seattle, WA.

Ort C., Lawrence M. G., Reungoat J., Eaglesham G., Carter S. and Keller J. (2010). Determining the fraction of pharmaceutical residues in wastewater originating from a hospital. *Water Research*, **44**(2), 605–615.

Pan B., Ning O. and Xing B. (2009). Part V – Sorption of pharmaceuticals and personal care products. *Environmental Science and Pollution Research*, **16**, 106–116.

Pauwels B. and Verstraete W. (2006). The treatment of hospital wastewater: an appraisal. *Journal of Water and Health*, **4**(4), 405–416.

Persson M., Sabelström E. and Gunnarsson B. (2009). Handling of unused prescription drugs – knowledge, behaviour and attitude among Swedish people. *Environment International*, **35**, 771–774.

Quintana J. B., Weiss S. and Reemtsma T. (2005). Pathways and metabolites of microbial degradation of selected acidic pharmaceutical and their occurrence in municipal wastewater treated by a membrane bioreactor. *Water Research*, **39**, 2654–2664.

Ramirez A. J., Brain R. A., Usenko S., Mottaleb M. A., O'Donnell J. G., Stahl L. L., Wathen J. B., Snyder B. D., Pitt J. L., Perez-Hurtado P., Dobbins L. L., Brooks B. W. and

Chambliss C. K. (2009). Occurrence of pharmaceuticals and personal care products in fish: results of a national pilot study in the United States. *Environmental Toxicology and Chemistry*, **28**(12), 2587–2597.

Ratola N., Cincinelli A., Alves A. and Katsoyiannis A. (2012). Occurrence of organic microcontaminants in the wastewater treatment process. A mini review. *Journal of Hazardous Materials*, **239–240**, 1–18.

Richardson M. L. and Bowron J. M. (1985). The fate of pharmaceutical chemicals in the aquatic environment. *Journal of Pharmacy and Pharmacology*, **37**(1), 1–12.

Richardson B. J., Lam P. K. S. and Martin M. (2005). Emerging chemicals of concern: pharmaceuticals and personal care products [PPCPs] in Asia, with particular reference to Southern China. *Marine Pollution Bulletin*, **50**, 913–920.

Rivera-Utrilla J., Sánchez-Polo M., Ferro-García M. Á., Prados-Joya G. and Ocampo-Pérez R. (2013). Pharmaceuticals as emerging contaminants and their removal from water. A review. *Chemosphere*, **93**(7), 1268–1287.

Rodríguez-Navas C., Björklund E., Bak S. A., Hansen M., Krogh K. A., Maya F., Forteza R. and Cerdà V. (2013). Pollution pathways of pharmaceutical residues in the aquatic environment on the island of Mallorca, Spain. *Archives of Environmental Contamination and Toxicology*, **65**(1), 56–66.

Rogowska J., Zimmermann A., Muszyńska A., Ratajczyk W. and Wolska L. (2019). Pharmaceutical household waste practices: preliminary findings from a case study in Poland. *Environmental Management*, **64**, 97–106.

Ruhoy I. S. and Daughton C. G. (2008). Beyond the medicine cabinet: an analysis of where and why medications accumulate. *Environment International*, **234**, 1157–1169.

Sapkota A., Sapkota A. R., Kucharski M., Burke J., McKenzie S., Walker P. and Lawrence R. (2008). Aquaculture practices and potential human health risks: current knowledge and future priorities. *Environment International*, **34**(8), 1215–1226.

Schneider M. K., Stamm C. and Fenner K. (2007). Selecting scenarios to assess exposure of surface waters to veterinary medicines in Europe. *Environmental Science and Technology*, **41**, 4996–4676.

Shaaban H., Alghamdi H., Alhamed N., Alziadi A. and Mostafa A. (2018). Environmental contamination by pharmaceutical waste: assessing patterns of disposing unwanted medications and investigating the factors influencing personal disposal choices. *Journal of Pharmacology and Pharmaceutical Research*, **1**(1), 1–7.

Shokri F., Ziarati P. and Mousavi Z. (2016). Removal of selected heavy metals from pharmaceutical effluent by aloe vera L. *Biomedical and Pharmacology Journal*, **9**(2), 705–713.

Sirés I. and Brillas E. (2012). Remediation of water pollution caused by pharmaceutical residues based on electrochemical separation and degradation technologies: a review. *Environmental International*, **40**, 212–229.

Son S., Wang M. and Shon J.-K. (2011). Satellite observations of optical and biological properties in the Korean dump site of the Yellow Sea. *Remote Sensing of Environment*, **115**(2), 562–572.

Stolker A. A. M., Niesing W., Hogendoorn E. A., Versteegh J. F. M., Fuchs R. and Brinkman U. A. T. (2004). Liquid chromatography with triple-quadrupole or quadrupole-time of flight mass spectrometry for screening and confirmation of residues of pharmaceuticals in water. *Analytical and Bioanalytical Chemistry*, **378**(4), 955–963.

Suchikova Y., Bogdanov I., Vambol S., Vambol V. and Kondratenko O. (2018). Investigation of the porous GaP layers' chemical composition and the quality of the tests carried out. *Journal of Achievements in Materials and Manufacturing Engineering*, **86**(2), 49–60.

Swan G. E., Cuthbert R., Quevedo M., Green R. E., Pain D. J., Bartels P., Cunningham A. A., Duncan N., Meharg A. A., Oaks J. L., Parry-Jones J., Shultz S., Taggart M. A., Verdoorn G. and Wolter K. (2006). Toxicity of diclofenac to Gyps vultures. *Biology Letters*, **2**(2), 279–282.

Tabak H. H. and Brunch R. L. (1970). Steroid hormones as water pollutants I. Metabolism of natural and synthetic ovulation inhibiting hormones by micro-organisms of activated sludge and primary settled sludge. In: Developments in Industrial Microbiology, American Institute of Biological Sciences, Washington, DC, USA, **11**, 367–376.

Tambosi J. L., Yamanaka L. Y., José H. J., Moreira R. D. F. P. M. and Schröder H. F. (2010). Recent research data on the removal of pharmaceuticals from sewage treatment plants (STP). *Química Nova*, **33**(2), 411–420.

Tolls J. (2001). Sorption of veterinary pharmaceuticals in soils: a review. *Environmental Science and Technology*, **35**, 3397–3406.

Trofimova Ye O. and Get'man M. A. (2013). Survey of visitors to St. Petersburg pharmacies on the problem of disposal of unused drugs. Remedy. *Journal of the Russian Market of Medicines and Medical Equipment*, **9**, 65–69.

Udofia E. A., Gulis G. and Fobil J. (2017). Solid medical waste: a cross sectional study of household disposal practices and reported harm in Southern Ghana. *BMC Public Health*, **17**, 464–475.

Vambol S., Bogdanov I., Vambol V., Suchikova Y., Lopatina H. and Tsybuliak N. (2017). Research into effect of electrochemical etching conditions on the morphology of porous gallium arsenide. *Eastern-European Journal of Enterprise Technologies*, **6/5** (90), 22–31.

Vambol S., Vambol V., Bogdanov I., Suchikova Y. and Rashkevich N. (2017). Research of the influence of decomposition of wastes of polymers with nano inclusions on the atmosphere. *Eastern-European Journal of Enterprise Technologies*, **6/10**(90), 57–64.

Vambol S., Vambol V., Sobyna V., Koloskov V. and Poberezhna L. (2018). Investigation of the energy efficiency of waste utilization technology, with considering the use of low-temperature separation of the resulting gas mixtures. *Energetika*, **64**(4), 186–195.

Vambol S., Vambol V., Sundararajan M. and Ansari I. (2019). The nature and detection of unauthorized waste dump sites using remote sensing. *Ecological Questions*, **30**(3), 43–55.

Verlicchi P., Al Aukidy M., Galletti A., Petrovic M. and Barcelo D. (2012). Hospital effluent: investigation of the concentrations and distribution of pharmaceuticals and environmental risk assessment. *Science of the Total Environment*, **430**, 109.

Verlicchi P., Al Aukidy M. and Zambello E. (2015). What have we learned from worldwide experiences on the management and treatment of hospital effluent? An overview and a discussion on perspectives. *Science of the Total Environment*, **514**, 467–491.

Weigel S., Kuhlmann J. and Hühnerfuss H. (2002). Drugs and personal care products as ubiquitous pollutants: occurrence and distribution of clofibric acid, caffeine and DEET in the North Sea. *Science of the Total Environment*, **295**(1–3), 131–141.

Weigel S., Berger U., Jensen E., Kallenborn R., Thoresen H. and Hühnerfuss H. (2004). Determination of selected pharmaceuticals and caffeine in sewage and seawater from Tromsø/Norway with emphasis on ibuprofen and its metabolites. *Chemosphere*, **56**, 583–592.

Weinberger J. and Klaper R. (2014). Environmental concentrations of the selective serotonin reuptake inhibitor fluoxetine impact specific behaviors involved in reproduction, feeding and predator avoidance in the fish *Pimephales promelas* (fathead minnow). *Aquatic Toxicology*, **151**, 77–83.

Williams M., Ong P. L., Williams D. and Kookana R. S. (2009). Estimating the sorption of pharmaceuticals based on their pharmacological distribution. *Environmental Toxicology and Chemistry*, **28**, 2572–2579.

Withers P. J., Jordan P., May L., Jarvie H. P. and Deal N. E. (2013). Do septic tank systems pose a hidden threat to water quality? *Frontiers in Ecology and the Environment*, **12**, 123–130.

Wu M., Atchley D., Greer L., Janssen S., Rosenberg D. and Sass J. (2009). Dosed without prescription: preventing pharmaceutical contamination of our nation's drinking water. NRDC White Paper. Available from: http://docs.nrdc.org/health/files/hea_10012001a.pdf.

Yin X.-T., Zhou W.-D., Li J., Wang Q., Wu F.-Y., Dastan D., Wang D., Garmestani H., Wang X.-M. and Talu S. (2019). A highly sensitivity and selectivity Pt-SnO_2 nanoparticles for sensing applications at extremely low level hydrogen gas detection. *Journal of Alloys and Compounds*, **805**, 229–236.

Yoon Y., Westerhoff P., Snyder S. A., Wert E. C. and Yoon J. (2007). Removal of endocrine disrupting compounds and pharmaceuticals by nanofiltration and ultrafiltration membranes. *Desalination*, **202**(1–3), 16–23.

Yu J. T., Bouwer E. J. and Coelhan M. (2006). Occurrence and biodegradability studies of selected pharmaceuticals and personal care products in sewage effluent. *Agricultural Water Management*, **86**, 72–80.

Yuan F., Hu C., Hu X., Qu J. and Yang M. (2009). Degradation of selected pharmaceuticals in aqueous solution with UV and UV/H_2O_2. *Water Research*, **43**, 1766–1774.

Yuan Q., Snow D. D. and Bartelt-Hunt S. N. (2013). Potential water quality impacts originating from land burial of cattle carcasses. *Science of the Total Environment*, **456–457**, 246–253.

Zahirnejad M., Ziarati P. and Asgarpanah J. (2017). The Efficiency of Bio-adsorption of Heavy Metals from Pharmaceutical Effluent by Rumex crispus L. Seed. *Journal of Pharmaceutical and Health Sciences*, **5**(3), 231–243.

Zhao J. L., Ying G. G., Liu Y. S., Chen F., Yang J. F., Wang L., Yang X. B., Stauber J. L. and Warne M. S. (2010). Occurrence and a screening-level risk assessment of human pharmaceuticals in the Pearl River system, South China. *Environmental Toxicology and Chemistry*, **29**(6), 1377–1384.

Ziarati P., Mostafidi M., Shirkhan F. and Tamaskoni Zahedi M. (2018). Analysis of removal methods of toxic heavy metals using bio-absorbs. *Technogenic and Ecological Safety*, **4**(2), 62–76. DOI: 10.5281/zenodo.1402587.

Ziarati P., Vambol V. and Vambol S. (2020). Use of inductively coupled plasma optical emission spectrometry detection in determination of arsenic bioaccumulation in *Trifolium pratense* L. from contaminated soil. *Ecological Questions*, **31**(1), 15–22. DOI: 10.12775/EQ.2020.003.

Chapter 3

Prioritisation of pharmaceutical in effluent considering its biohazard

Aastha Dhingra[1], Sirajuddin Ahmed[1,2], Siddhartha Gautam[2], Tilemachos K. Koliopoulos[3], Ahmed Kadhim Hussein[4] and Fateh Mebarek-oudina[5]

[1]Civil Engineering Department, Jamia Millia Islamia, New Delhi, India
[2]Delhi Pollution Control Committee, New Delhi
[3]Department of Public Health, University of West Attica, Athens, Greece
[4]Department of Mechanical Engineering, University of Babylon, Al Ḫillah, Iraq
[5]Department of Physics, Faculty of Sciences, University of 20 août 1955- Skikda, Skikda, Algeria

ABSTRACT

There has been a substantial growth in the population and technological advancement. This has led to the increasing consumption of a variety of pharmaceuticals. The occurrence of drug residues and other chemicals has been confirmed by various research studies on aquatic environments throughout the world. Traces of pharmaceuticals are known to affect water quality, increase bacterial resistance, reduce self-purifying capacities of rivers, and adversely impact aquatic life. Many cytotoxic and antibiotic pharmaceuticals are known to be teratogenic, mutagenic, and even genotoxic. The eco-toxicity of pharmaceuticals in different organisms has been studied by various scientists. Despite the potential implications of drug residues, there are no discharge limits in various parts of the world including India. Pharmaceuticals have the potential

Editor(s): Nadeem A Khan, Sirajuddin Ahmed, Viola Vambol and Sergij Vambol
doi: 10.2166/9781789061338_0047

to bio-accumulate and persist in the ecosystem for years. This chapter will focus on the important criteria for identifying the drugs that could have a significant impact on the environment. The 34 most commonly used pharmaceuticals in in-patient departments of hospitals in Delhi were selected for the study. The following priority list has been formulated: high priority pharmaceuticals are those which are likely to have a detrimental impact on the aquatic environment and which are important to be analysed in the water discharged from hospitals, medical centres and industrial sources etc., before these are led into the water bodies. This is an important tool for addressing the risks of pharmaceutical residues affecting the aquatic environment.

Keywords: Aquatic, Contaminants, Environment, Hospitals, Pharmaceutical residues.

Contents

3.1 Introduction 49
3.2 Need for Prioritisation 51
3.3 Selection of Criteria 51
3.4 Prioritisation of Drugs and Rationale for Selection of Criteria 53
3.5 Selection of Pharmaceuticals 53
3.6 Results and Discussion 57
 3.6.1 Development of priority list 57
 3.6.2 High priority list 57
 3.6.3 Moderate priority list 64
 3.6.4 Low priority list 64
3.7 Conclusion 64
References 65

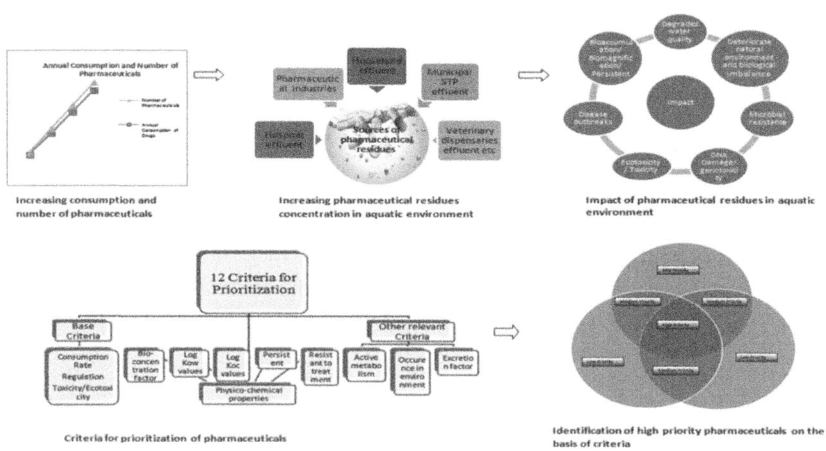

3.1 INTRODUCTION

Delhi, being the capital of India, is the main centre of health care facilities. As per the Delhi Pollution Control Committee, there are around 48,500 beds in the hospitals/nursing home/medical care centres of the city. The average number of beds per 1000 population is just 2.58. The number of beds and hospitals are following an upward trend to meet the growing demand for healthcare for local, national, as well as international patients. The average occupancy rate of hospitals is around 78–80% throughout the year. Healthcare is one of the major water consuming sectors. The water is majorly consumed in wards, kitchens, laundries, laboratories, operation theatres and haemodialysis units. As per a research study, the water consumption in a healthcare facility varies from 400–1200 L/bed/day depending on various factors. In America, the volume of wastewater generated from a healthcare facility is approximately 1000 L/person/day (Al-ajlouni *et al.*, 2010). A questionnaire survey was conducted for hospitals in Delhi and it was observed that the average water consumption is 550–600 L/bed/day. The wastewater generated from them constitutes domestic wastewater, as well as infectious and toxic wastewater, since it contains large amounts of microbes, pharmaceutical residues, chemicals, tissues, cultures, chemical toxins, metabolized radioactive medicines, traces of nuclear medicines, heavy metals, body fluids etc. The quantity of wastewater generated from this sector is huge and poses a harmful impact to the environment and hence human health. Hospital effluents, if discharged untreated or improperly treated, can lead to long term devastating impacts on the natural environment and cause a biological imbalance.

The management of solid hospital waste has always been a matter of concern. However, inadequate attention has been paid to liquid waste generated from hospitals and other healthcare facilities such as research and pathological labs etc. The hospitals widely use toxic substances, such as pharmaceutical residues, cytotoxic drugs, disinfectants, active substances, pigments, dyes and chemical reagents (Gautam *et al.*, 2007). Antibacterial and other disinfectants traces co-exist with the microbes responsible for biological processes involved in treatment of the effluent. Such wastewater could lead to various communicable epidemics such as dysentry, cholera and environment degradation if discharged untreated. It also has the potential to cause radioactive pollution. The hospital effluent needs to be treated at source before discharging it to the centralised water treatment facility to prevent contamination of the aquatic system (Singh *et al.*, 2013). The various chemicals and toxins present in the liquid discharge have the potential to cause skin diseases or enteric illness. These impacts are presented in Figure 3.1.

Hospital effluents should be treated prior to their disposal into the municipal sewage system because they contain many toxic contaminants. Though the hospital effluent is diluted to a higher degree, when it reaches the municipal sewage treatment plant there is still a significant possibility that it can cause a

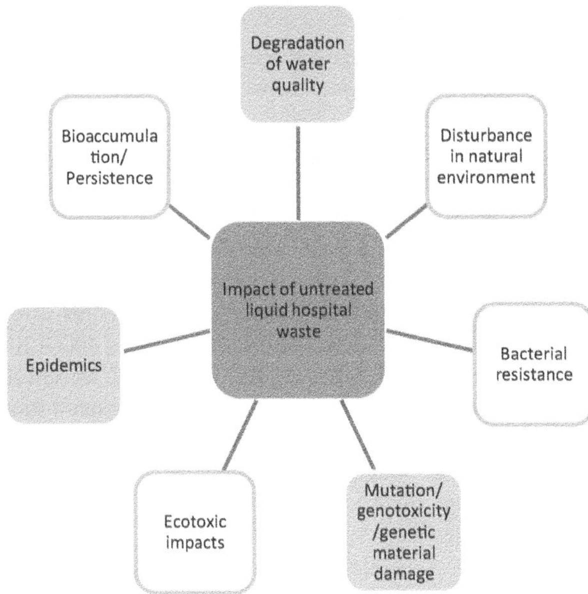

Figure 3.1 Impact of untreated liquid hospital waste (Dhingra *et al.*, 2017).

biological imbalance in the aquatic environment. To protect the aquatic system from toxic contaminants discharged from hospitals, it is important to provide adequate treatment before there are discharged into the municipal sewage system (Brandmayr *et al.*, 2015).

The occurrence of pharmaceutical residues in the aquatic environment is an emerging concern for environmentalists. They have been found in concentrations of μg/L to ng/L in surface water, groundwater and even in drinking water samples across the globe. Although the concentration of an individual pharmaceutical is quite low in the environment as compared to the dose administered to the patient, the combined concentration of the active pharmaceutical compounds is significant with respect to the aquatic environment. They may have synergistic effects (Kummerer, 2001) and are shown in Figure 3.1.

Pharmaceuticals can be present as parent compounds or its metabolites in the environment (Carlsson *et al.*, 2006). They enter the environment through the urine of treated patients, veterinary farms and healthcare and even the municipal sewage has traces of pharmaceuticals since the over the counter sale of many pharmaceuticals is high. There could also be direct release from the pharmaceutical industries (Ileri *et al.*, 2003). The ground water is generally contaminated by the leachates from landfills having residues of these pharmaceuticals or through soil because of the use of contaminated water for horticulture (Sui *et al.*, 2015).

Lin and Tsai (2009) reported the presence of many pharmaceutical classes like anti-inflammatory, antimicrobials, beta-blockers, estrogens in hospital wastewater in Taiwan. The study included eleven antimicrobials, three hormones, five anti-inflammatories, one beta blocker and one lipid regulator for analysis. The pharmaceutical residues were reported in μg/L in surface water.

3.2 NEED FOR PRIORITISATION

The occurrence and impacts of pharmaceutical residues are now known in all spheres of aquatic environment and soil. Even when the hospital effluent is known to degrade the environment and is likely to affect public health by causing bacterial resistance, there are no specific standards for discharging in India. Currently, the parameters to be tested for hospital effluent prescribed by the Central Pollution Control Board are pH, TSS, BOD, COD, O&G and bio-assay test for fishes (Pollution *et al.*, 2000). These standards are not sufficient considering the complexity of the hospital effluent. There is a vast difference between toxicity of hospital effluents and general industrial effluents in terms of their composition (pharmaceutical traces and pathogen counts). The standards prescribed by the World Health Organisation for hospital wastewater include pH, BOD_5, COD, TSS, O&G, Cd, Cr, Pb, Hg, Cl (total residue), phenols and faecal coli. However, there are still no specific standards prescribed by any regulatory body for hospital wastewater despite their occurrence in all spheres of the aquatic environment and its potential to cause environmental imbalance.

To move forward in this regard it is important to analyse pharmaceutical residues at the source of generation and this includes hospitals, pharmaceutical industries, research labs, pathological labs etc. There has been tremendous advancement in the analytical technologies and methods to confirm their presence, even in ng/L concentrations. However, since there are thousands of pharmaceuticals used in medical treatment, there is a dire need to devise a ranking system and prioritise the pharmaceuticals so that at least high priority pharmaceuticals could be analysed at the source according to their environmental relevance (e.g. their occurrence in the environment, consumption, its pharmacology, physico-chemical properties that decides its fate in the aquatic environment and their potential for ecotoxicological and other adverse effects).

3.3 SELECTION OF CRITERIA

Literature review and expert opinion has helped to select the criteria for prioritisation of the pharmaceuticals. This study included the opinions of 39 subject experts including 12 doctors, nine chemists, four pharmacologists, six professors and researchers, five experts from non-government organisations and four from local authorities, including one from the Indian Pharmacoepia

Commission, two from the Directorate of Health Services and one representative from the Delhi Pollution Control Committee. Initially, 20 criteria were screened from the literature followed by an opinion poll of these experts which provided 12 more relevant criteria. The approach for selecting these criteria is shown in Figure 3.2. All 12 selected criteria were given an

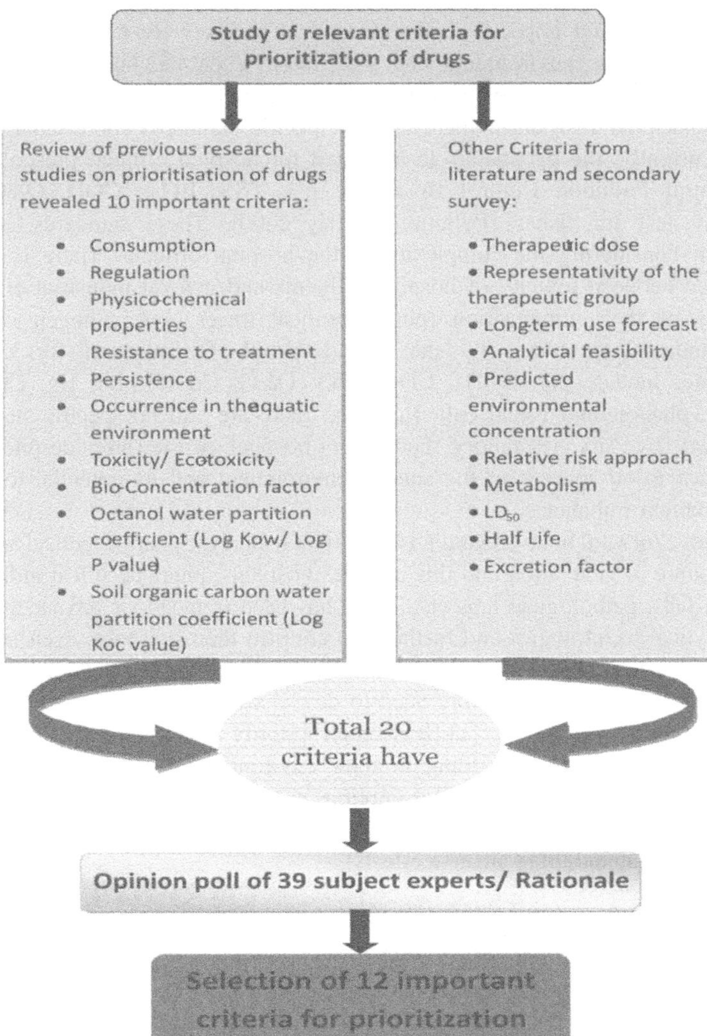

Figure 3.2 Flowchart showing selection of criteria (Dhingra *et al.*, 2017).

equal weightage and no further prioritisation was done. To start with, a questionnaire survey was carried out in the hospitals of Delhi to determine the consumption of pharmaceuticals in the inpatient department. This was carried out as a preliminary screening for selection of high priority pharmaceuticals and formed the basis of this study. As per the literature review, there are many existing prioritisation methods that have used consumption, as the basic criterion for selection of drugs (Carlsson *et al.*, 2006; Cooper *et al.*, 2008; Huschek *et al.*, 2004; Jean *et al.*, 2012; Jones *et al.*, 2002; Stuer-Lauridsen *et al.*, 2000). Some studies have also considered bio-accumulation potential as the base criterion since it is a fundamental risk factor for the aquatic environment (Jean *et al.*, 2012).

3.4 PRIORITISATION OF DRUGS AND RATIONALE FOR SELECTION OF CRITERIA

As discussed, the literature review initially came up with 20 criteria. To further curtail the number of criteria and to understand their relevance, an opinion poll was conducted. All the criteria hold equal weightage. To qualify onto the list of criteria to be considered for prioritization, more than 75% of the experts must agree about the importance of that criterion. This provided a list of 12 more relevant criteria based on scientific and logical basis. The selection of criteria is presented in Table 3.1. The percentage of responses of the experts in the favour of each criterion is shown in Figure 3.3.

3.5 SELECTION OF PHARMACEUTICALS

On the basis of literature available on the prioritisation of pharmaceuticals, 86 drugs were initially screened for the study. Following the meetings with stakeholders including doctors, pharmacists etc., the list was reduced to 43 pharmaceuticals for a questionnaire survey to be conducted in the multispecialty hospitals for collecting data on average consumption in the inpatient department. The basis of this list of pharmaceuticals was their eco-toxic and toxic properties; occurrence in aquatic environment and bioaccumulation.

On the basis of the questionnaire survey, it was noted that out of these only 19 were being extensively used, however, the questionnaire survey suggested 15 more pharmaceuticals including antibiotics and cytotoxics. Hence, a total of 34 pharmaceuticals was finalised for prioritisation.

Out of the 20 criteria mentioned above, the following 12 criteria were selected for prioritisation. The criteria considered in this study and the condition to fulfil these criteria is explained in Table 3.2.

Table 3.1 Rationale for selecting criteria.

Criterion	Criterion Selected	Rationale
Consumption (Carlsson *et al.*, 2006; Castiglioni *et al.*, 2004; Cooper *et al.*, 2008; Huschek *et al.*, 2004; Jean *et al.*, 2012; Jones *et al.*, 2002; Stuer-Lauridsen *et al.*, 2000)	Yes	The consumption questionnaire survey was carried out in six multispecialty hospitals of Delhi. This is the base criterion considered for this study. The more the consumption of the pharmaceutical/chemical, the more would the residues be expected in the effluents, depending on other factors such as metabolism
Regulation (Castiglioni *et al.*, 2004; Jean *et al.*, 2012; Voogt *et al.*, 2009; Zuccato *et al.*, 2005)	Yes	Every country's water supply body has to fulfil the regulation. The pharmaceuticals listed in any international regulation have been considered important. The information that the pharmaceutical is listed in any such regulation was confirmed from CAS database maintained by the Chemical Abstract Society. The CAS database was accessed using Sci Finder
Therapeutic dose (Huschek *et al.*, 2004; Voogt *et al.*, 2009)	No	This factor is related to the toxicity caused by the pharmaceutical residues. This can be to understand the concentration until it does not have hazardous impacts. Other relevant parameters have been considered in this study
Representativity of therapeutic group (Huschek *et al.*, 2004; Voogt *et al.*, 2009)	No	Being representative of any therapeutic group will not give an understanding of the properties of the pharmaceutical residues nor it is possible to consider each class due to lack of data
Long-term use forecast (Cooper *et al.*, 2008; Voogt *et al.*, 2009)	No	This factor itself depends on many factors and there are no stable models to predict the same. Thus, we have used consumption in the current scenario as a base criterion

Physico-chemical properties (Carlsson *et al.*, 2006); Castiglioni *et al.*, 2004; Voogt *et al.*, 2009)	Yes	This is a very important factor to be considered as it decides the mode of action and behaviour in the aquatic environment and the likely impacts it could have. The physico-chemical properties include its water solubility, absorption, etc.
Analytical feasibility (Voogt *et al.*, 2009)	No	This factor cannot be relied upon as we need to develop methods if they are not already there
Resistance to treatment (Cooper *et al.*, 2008; Huschek *et al.*, 2004; Voogt *et al.*, 2009)	Yes	The pharmaceuticals residues that cannot be degraded by effluent treatment plants should definitely be considered
Predicted environmental concentration (Voogt *et al.*, 2009)	No	The environmental concentration can be predicted by the consumption, occurrence and physico-chemical properties already considered in the study
Relative risk approach (Castiglioni *et al.*, 2004; Jones *et al.*, 2002; Voogt *et al.*, 2009; Zuccato *et al.*, 2005)	No	This is calculated as the ratio of the predicted environmental concentration and the predicted no effect concentration (PEC/PNEC)
Occurrence in surface waters, groundwater, drinking water, wastewater (Castiglioni *et al.*, 2004; Huschek *et al.*, 2004; Jones *et al.*, 2002; Voogt *et al.*, 2009)	Yes	This is one of the base criteria other than consumption. The literature review marks their presence in the aquatic environment and thus needs to be evaluated and monitored
Half life (Jean *et al.*, 2012; Jones *et al.*, 2001)	No	This factor is important from the point of view of pharmacology and to decide the therapeutic dosage. The excretion factor could be considered more important for this study
Bio-concentration factor (BFC) (Cooper *et al.*, 2008; Jean *et al.*, 2012)	Yes	The BCF suggests the transfer of pharmaceutical residue within the food chain. The BCF values at pH 1–10 have been taken from the CAS database. The pH in the natural environment varies from 6–9. Drugs having BCF greater than 1000 can be considered as potentially bio-accumulable

(*Continued*)

Table 3.1 Rationale for selecting criteria (*Continued*).

Criterion	Criterion Selected	Rationale
Log Kow value (Huschek *et al.*, 2004; Jean *et al.*, 2012)	Yes	This suggests the adsorption properties. It estimates molecular lipophilicity and gives an idea of the accumulation potential
Excretion factor (Jean *et al.*, 2012; Zuccato *et al.*, 2005)	Yes	The excretion factor is the fraction of unchanged pharmaceutical excreted from the patient in urine or faeces. This is important to assess the eco-toxicological risk
Log Koc value (organic carbon partition coefficient) (Jean *et al.*, 2012)	Yes	This gives an estimation of the mobility of residues in the sediments. A low value suggests it could contaminate the groundwater
Persistent (Voogt *et al.*, 2009)	Yes	Persistence can be estimated by Log Kow and Log Koc values. Pharmaceuticals that are persistent in the aquatic system should be considered important
Toxicity/ecotoxicity (Voogt *et al.*, 2009)	Yes	This is a very important factor to be considered as this suggests the implication of the residue in the environment and aquatic life. There is a need to undertake more toxicity studies as the data is limited
Metabolism (expert opinion)	Yes	However, the occurrence and consumption data are more relevant than this criterion, but there is a need to consider the active metabolite if excreted and released in the environment
Lethal dose 50 (LD_{50}) (opinion of some experts)	No	This was considered by some experts but did not qualify in the opinion poll as the pharmaceutical traces reported in the environment are too low if compared with LD_{50}

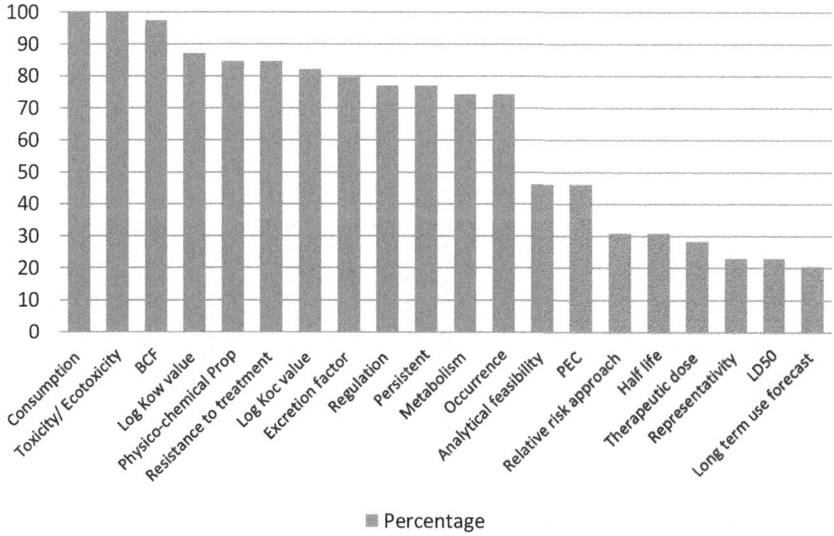

Figure 3.3 Percentage of experts considering the criteria to be important (Dhingra *et al.*, 2017).

3.6 RESULTS AND DISCUSSION

3.6.1 Development of priority list

A scaling weighing checklist method has been adopted to develop the prioritization list. For meeting the criteria, 1 point was given and 0 was given for not fulfilling the condition. Each criterion has been given equal weightage. The points for all the 12 criteria were added together to give a total score. On the basis of the total score earned, the 34 pharmaceuticals were classified into three categories, high priority (Class I); moderate priority (Class II) and low priority pharmaceuticals (Class III). Pharmaceuticals having a cumulative total of nine and above are considered as a high priority pharmaceutical. The pharmaceuticals having 6–8 criteria are considered in the moderate priority list and those having five or less criteria are in the list of low priority pharmaceuticals, as shown in Figure 3.4.

The characteristics of high priority pharmaceuticals, moderate priority pharmaceuticals and low priority pharmaceuticals are discussed below.

3.6.2 High priority list

Pharmaceuticals satisfying nine or more criteria are classified as high priority pharmaceuticals. There are nine pharmaceuticals in this category that must be analysed in the effluents before these are discharged into drains. All these pharmaceutical residues have been found to have maximum impact on the aquatic

Table 3.2 Criteria selected and its condition for fulfilment.

S. No	Criteria selected for prioritisation	Condition for the fulfilment of the criterion
1	Consumption in multi-specialty hospitals in Delhi	Highly consumed should be considered
2	Regulation	If the pharmaceutical is listed in international regulations
3	Metabolism	If they form active metabolites
4	Excretion factor (percentage of unchanged drug excreted)	If it exceeds 10%
5	Toxicity/eco-toxicity	If they have toxic and ecotoxic properties
6	Occurrence in surface water, groundwater, drinking water, wastewater	If they have been reported in the aquatic environment
7	Physico-chemical properties	If they are likely to get adsorbed in water and sediments of ecosystem
8	Bio-concentration factor (BCF)	BCF value of <100 – not expected to bio-accumulate BCF value of >100 but <1000 have the potential to bio-accumulate BCF value of >1000 has the potential to bio-accumulate significantly (Jean *et al.*, 2012)
9	Octanol water partition coefficient (Log Kow/Log P values)	According to Jones *et al.* (2002), the following is the criterion for deciding the potential of adsorption: Log Kow <2.5: low potential for adsorption $4.0<$ Log Kow >2.5: moderate potential Log Kow >4.0: high potential
10	Soil organic carbon water partition coefficient (Log Koc value)	If logKoc ≥ 3, it is taken as a trigger value for sediment effect assessment
11	Resistant to treatment	If they are reported resistant in the literature
12	Persistent	If they are persistent depending on BCF values and other literature support

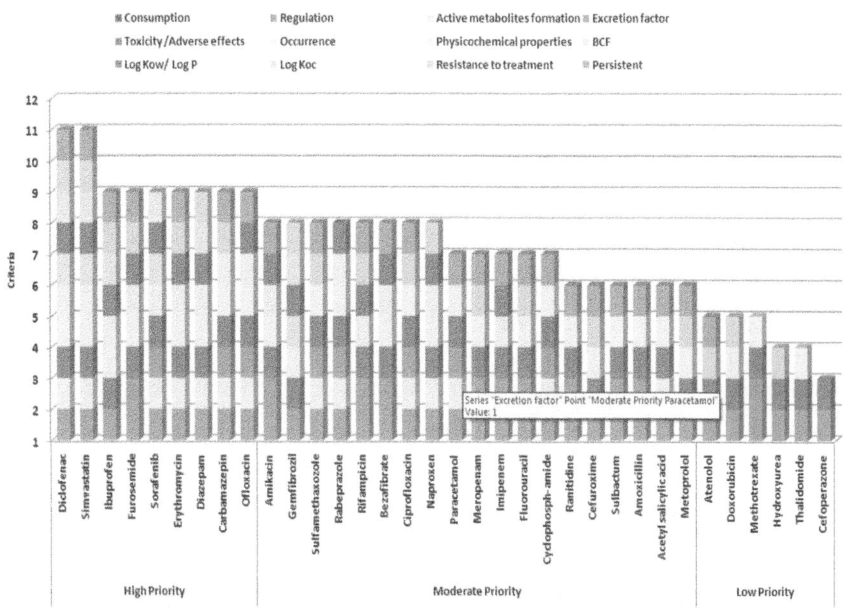

Figure 3.4 Prioritization on the basis of criteria for selection (Dhingra *et al.*, 2017).

ecology. These pharmaceuticals are considered as Class I. Following are the details of high priority pharmaceuticals:

(1) **Consumption**: The list of 34 pharmaceuticals has been discussed with the experts in the field to determine if these are commonly used. The high priority pharmaceuticals are among those selected by the experts. The annual consumption data was obtained from the Central Procurement Agency which provides pharmaceuticals to the hospitals under the aegis of the Delhi Government. Also, a questionnaire survey was undertaken in six multispecialty hospitals of Delhi. Sorafenib, diazepam and simvastatin pharmaceuticals are not used in huge quantities, but these have high eco-toxicity and bio-accumulation potential. The major pharmaceutical consumption in Government hospitals of Delhi is shown in Table 3.3. Apart from this, the Delhi Government hospitals also procure pharmaceuticals from vendors other than Central Procurement Agency and this has not been reflected in the table.

(2) **Regulation**: The chemicals/pharmaceuticals that need to be used judiciously are listed in international regulations. It should be made mandatory to analyse these pharmaceuticals in the effluents generated at the source, as well as the water bodies. The high priority pharmaceuticals are listed in various regulations which have been confirmed from the CAS database accessed through Sci finder (28) for

Table 3.3 Annual consumption of high priority pharmaceuticals.

Pharmaceutical	Annual consumption in Delhi Government hospitals from March, 2016 to March, 2017 (in grams)
Diclofenac	3,131,868
Ibuprofen	6,441,402
Sorafenib	9038
Carbamazepin	1,372,020
Diazepam	6270
Ofloxacin	159,176
Erythromycin	600,262
Simvastatin	3960
Furesmide	94,074

Source: Office of Central Procurement Agency.

example diclofenac is listed in the Australian Inventory of Chemical Substances, sorafenib in REACH etc.

(3) **Toxicity/eco-toxicity**: Toxicity assessment studies have been conducted by various researchers on these pharmaceuticals. Pharmaceutical residues are found to be toxic in fishes, daphnia and other aquatic organisms. Apart from this, a few pharmaceutical residues are genotoxic, mutagenic, and even teratogenic (Al-ajlouni *et al.*, 2010; Castiglioni *et al.*, 2004; Cooper *et al.*, 2008; González Alonso *et al.*, 2010; Heberer, 2002; Jones *et al.*, 2001; Lakshmi Prasanna *et al.*, 2015; Lin & Tsai 2009; Zhang *et al.*, 2008; Zuccato *et al.*, 2005).

(4) **Occurrence in aquatic environment**: There is ample literature available on the presence of these pharmaceuticals in different compartments of the aquatic environment. The occurrence of pharmaceutical residues has been reported worldwide. The concentration of pharmaceutical residues has been reported in the range of mg/L to ng/L in wastewater and the aquatic environment respectively (Al-ajlouni *et al.*, 2010; Castiglioni *et al.*, 2004; Cooper *et al.*, 2008; González Alonso *et al.*, 2010; Huschek *et al.*, 2004; Kummerer, 2001; Lakshmi Prasanna *et al.*, 2015; Lin and Tsai 2009; Voogt *et al.*, 2009; Zuccato *et al.*, 2005).

(5) **Bio-concentration factor**: The BCF values at pH 1–10 were taken from the CAS database (28). Pharmaceuticals having BCF greater than 1000 can be considered as potentially bio-accumulable.

BCF value of <100 – not expected to bio-accumulate (low potential)

BCF value of >100 but <1000 – has the potential to bio-accumulate (medium potential)

BCF value of >1000 – has the potential to bioaccumulate significantly (high potential) (Cooper *et al.*, 2008)

Table 3.4 Bio-concentration factors of high priority pharmaceuticals.

Pharmaceutical	pH									
	1	**2**	**3**	**4**	**5**	**6**	**7**	**8**	**9**	**10**
Diclofenac	1680	1670	1580	1010	220	25.2	2.82	1	1	1
Ibuprofen	270	269	260	195	55.4	6.83	1	1	1	1
Sorafenib	31	218	1240	2420	2670	2700	2700	2700	2680	2510
Carbamazepin	15.7	16.2	16.2	16.2	16.2	16.2	16.2	16.2	16.2	16.2
Ofloxacin	1	1	1	1	1	1	1	1	1	1
Erythromycin	1	1	1	1	1	1	1.34	7.73	14.9	16.4
Diazepam	1.1	3.81	23.2	63.5	77.3	79	79.1	79.2	79.2	79.2
Simvastatin	2290	2290	2290	2290	2290	2290	2290	2290	2290	2290
Furosemide	32.8	30.2	16.7	3.09	1	1	1	1	1	1

Source: CAS Database (28).

The values of bio-concentration factors are given in Table 3.4 above:

Table 3.4 determines the bio-accumulation potential of the high priority pharmaceuticals. Figure 3.5 depicts the pharmaceuticals belonging to categories of bio-accumulation potential in the pH range of the natural environment (low, medium and high). The pH range in the natural environment generally ranges from 5–9. Diclofenac has moderate potential in this range, however it has high potential in the

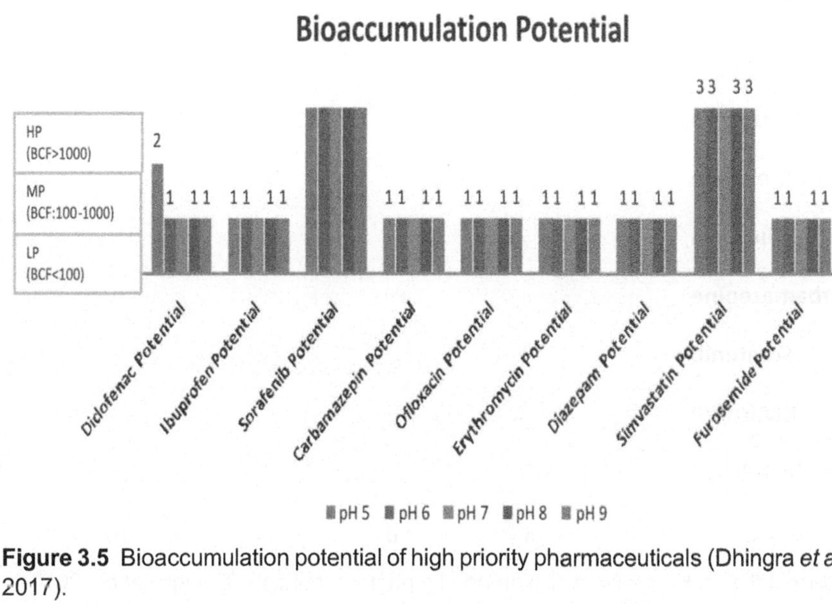

Figure 3.5 Bioaccumulation potential of high priority pharmaceuticals (Dhingra *et al.*, 2017).

low pH range. Ibuprofen has low potential in the low pH range, that is acidic medium. Simvastatin and sorafenib have higher potential for bio-accumulation.

(6) **Log Kow values**: The octanol water partition coefficient decides the potential of adsorption as well as bioaccumulation. The Log Kow values of these pharmaceuticals were taken from the CAS database.
Log Kow <2.5: low potential for adsorption
4.0< Log Kow >2.5: moderate potential
Log Kow >4.0: high potential (Cooper *et al.*, 2008; Voogt *et al.*, 2009; Zuccato *et al.*, 2005)
Pharmaceuticals having potential of adsorption as per their Log Kow values are depicted in Figure 3.6 below.

(7) **Log Koc values**: A logKoc ≥3 is taken as a trigger value for sediment effect assessment. Pharmaceuticals having such values are persistent in the sediment compartment. The Log Koc value of these pharmaceuticals has been taken from the CAS database (http://scifinder.cas.org/scifinder/ view/scifinder/scifinderExplore.jsf DOA 10/10/2017).
Pharmaceuticals having Log Koc ≥3 are depicted in Figure 3.7 below.

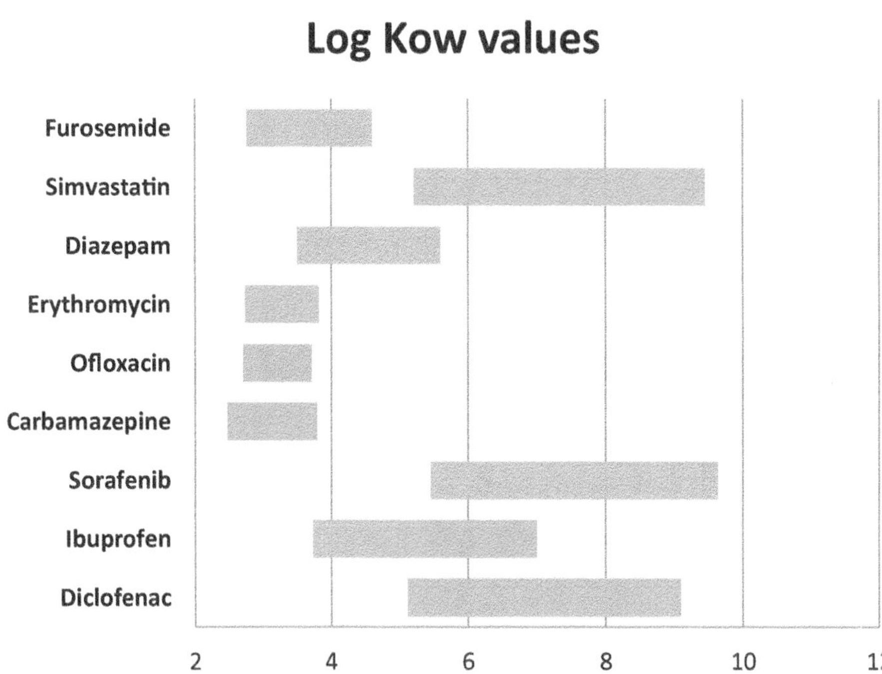

Figure 3.6 Log Kow values of high priority pharmaceuticals (Dhingra *et al.*, 2017).

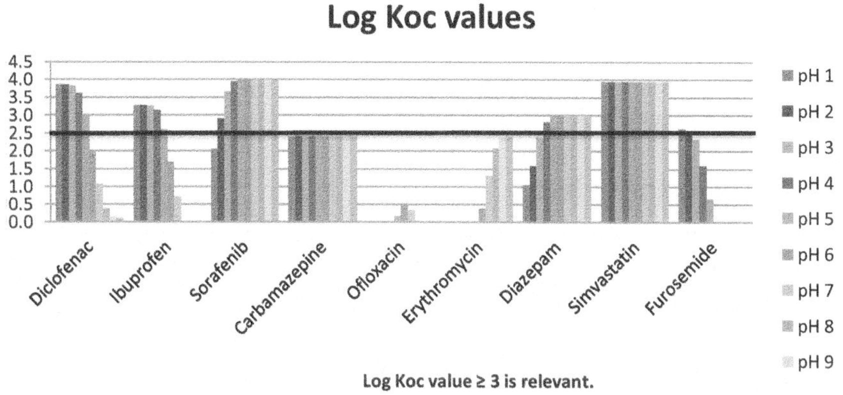

Figure 3.7 Log Koc values of high priority pharmaceuticals (Dhingra *et al.*, 2017).

(8) **Physico-chemical properties**: These pharmaceuticals have such physico-chemical properties that make them relevant from the environmental point of view. The physico-chemical properties are depicted by Log Kow values and Log Koc values. The physico-chemical properties of the pharmaceuticals determine the bio-accumulation potential, their persistence in the environment and their resistance to treatment.

(9) **Active metabolites**: The pharmaceuticals diclofenac, ibuprofen, sorafenib, erythromycin, diazepam, carbamazepine and ofloxacin form active metabolites. Pharmaceuticals having active metabolites are given an extra weightage as the metabolites formed are also relevant from the point of view of the environment.

(10) **Excretion factor**: An excretion factor higher than 10% constitutes an additional risk. Frusemide, sorafenib, amikacin, carbamazepine and ofloxacin have relevant excretion factors.

(11) **Resistant to effluent treatment**: The literature survey has revealed that diclofenac, ibuprofen, amikacin, erythromycin, frusemide, simvastatin, carbamazepine and diazepam are resistant to conventional wastewater treatment technologies. The pharmaceuticals resistant to effluent treatment tend to get slowly decayed and will remain in the ecosystem for a longer duration and thus impact the ecosystem (Gautam *et al.*, 2007; Xu *et al.*, 2007; Zhang *et al.*, 2008).

(12) **Persistence**: Persistence of a pharmaceutical is guided by bio-accumulation potential, adsorption potential and other physico-chemical properties. The resistance to treatment also leads to the persistence of the pharmaceutical residue in the aquatic ecosystem. The literature survey has revealed that diclofenac, ibuprofen, amikacin, erythromycin, frusemide, simvastatin, carbamazepine and ofloxacin are persistent in the aquatic environment (Jean *et al.*, 2012; Zuccato *et al.*, 2005).

3.6.3 Moderate priority list

These are class II pharmaceuticals. There are 19 pharmaceuticals in this category. The pharmaceuticals in these category complied with 6–8 criteria. These are also relevant pharmaceuticals for analysis.

3.6.4 Low priority list

There are six pharmaceuticals in this category which are relatively less relevant for different compartments of the aquatic ecosystem. These pharmaceuticals meet five or less than five criteria.

3.7 CONCLUSION

The occurrence of the pharmaceutical residues in all the compartments of the aquatic environment points towards the need to develop prioritisation of pharmaceutical residues and imparting treatment at the source with more stringent measures before discharging into the centralized water treatment facility. Prioritisation would help the authorities to monitor and keep a check on these residues in the aquatic environment as well as potable water supplies.

This priority list has been formed considering the limited set of selected pharmaceuticals from the literature and consumption data of in-patient departments of hospitals and the Central Procurement Agency. Twelve important criteria have been considered in this study for identifying the priority pharmaceuticals. The criteria included in the study were: consumption, toxicity/eco-toxicity, occurrence in surface water, bio-accumulation potential and other important physico-chemical properties. In this study, equal importance was imparted to all the selection criteria, however, if different weightage was given to each criterion, it may have led to some changes between the lists, however, these lists have given an idea about the relevance of prioritisation, some important pharmaceuticals and the need for analysing them and keeping a check on their concentrations in the aquatic environment. These criteria will remain the same locally, as well as globally. However, pharmaceuticals relevant for the study may vary, depending upon their consumption.

The high priority list has been developed including nine pharmaceuticals being most commonly consumed in hospitals in Delhi. The high priority pharmaceuticals belong to various classes including non-steroidal anti-inflammatory pharmaceuticals such as diclofenac and ibuprofen; antibiotics such as ofloxacin and erythromycin; anti-epileptic pharmaceuticals such as carbamazepine; anti-depressants such as diazepam; lipid regulators such as simvastatin and furosemide, which is a diuretic. This list of pharmaceuticals represents the minimum that should be considered while assessing the quality of hospital effluents being discharged into the drains which thus leads to the contamination of water bodies, thereby affecting their quality. The most commonly consumed pharmaceuticals have been incorporated. However, there is a possibility that other important pharmaceuticals may have not been incorporated due to constraints in collecting consumption data. Also, the literature on occurrence of various pharmaceuticals belonging to different therapeutic classes is limited. Three lists of pharmaceuticals (high, moderate and low priority) have been developed. With respect to its impact on the aquatic environment, this work would develop a protocol for the selection of priority pharmaceuticals. The approach for selecting high priority pharmaceuticals provides an efficient and practical base to manage the risks related to the discharge of pharmaceuticals into water bodies. This study would help the regulating agency to consider setting standards for important pharmaceutical residues in wastewater discharged from hospitals, veterinary farms, animal farms, pharmaceutical industries and other major sources of pharmaceutical consumption.

REFERENCES

Al-ajlouni K., Shakhatreh S., Ibraheem N. A. and Jawarneh M. (2010). Evaluation of wastewater discharge from hospitals in Amman – Jordan. *International Journal of Basic Applied Science. IJBAS-IJENS*, **13**(4), 44–50.

Brandmayr C., Kerber H., Winker M. and Schramm E. (2015). Impact assessment of emission management strategies of the pharmaceuticals Metformin and Metoprolol to the aquatic environment using Bayesian networks. *Science of the Total Environment*, **532**, 605–616. DOI: 10.1016/j.scitotenv.2015.05.074

Carlsson C., Johansson A. K., Alvan G., Bergman K. and Kühler T. (2006). Are pharmaceuticals potent environmental pollutants? Part I: environmental risk assessments of selected active pharmaceutical ingredients. *Science of the Total Environment*, **364**(1–3), 67–87. DOI: 10.1016/j.scitotenv.2005.06.035

Castiglioni S., Fanelli R., Calamari D., Bagnati R. and Zuccato E. (2004). Methodological approaches for studying pharmaceuticals in the environment by comparing predicted and measured concentrations in River Po, Italy. *Regulatory Toxicology and Pharmacology: RTP*, **39**(1), 25–32. DOI: 10.1016/j.yrtph.2003.10.002

Cooper E. R., Siewicki T. C. and Phillips K. (2008). Preliminary risk assessment database and risk ranking of pharmaceuticals in the environment. *Science of the Total Environment*, **398**(1–3), 26–33. DOI: 10.1016/j.scitotenv.2008.02.061

Dhingra A., Ahmed S., Siddiqui W. A., Gautam S. and Khan N. A. (2017). Identification of priority pharmaceutical residues in hospital effluent. *Global Journal of Engineering Science and Researches*, **4**, 28–34. DOI: 10.5281/zenodo.1320966

Gautam A. K., Kumar S. and Sabumon P. C. (2007). Preliminary study of physico-chemical treatment options for hospital wastewater. *Journal of Environmental Management*, **83**(3), 298–306. DOI: 10.1016/j.jenvman. (2006).03.009.

González Alonso S., Catalá M., Maroto R. R., Gil J. L. R., de Miguel Á. G. and Valcárcel Y. (2010). Pollution by psychoactive pharmaceuticals in the Rivers of Madrid metropolitan area (Spain). *Environment International*, **36**(2), 195–201. DOI: 10.1016/j.envint.2009.11.004

Heberer T. (2002). Tracking persistent pharmaceutical residues from municipal sewage to drinking water. *Journal of Hydrology*, **266**(3–4), 175–189. DOI: 10.1016/S0022-1694(02)00165-8

http://scifinder.cas.org/scifinder/view/scifinder/scifinderExplore.jsf DOA 10/10/2017.

Huschek G., Hansen P. D., Maurer H. H., Krengel D. and Kayser A. (2004). Environmental risk assesssment of medicinal products for human use according to European Commission recommendations. *Environmental Toxicology*, **19**(3), 226–240. DOI: 10.1002/tox.20015

Ileri R., Sengil I. A., Kulac S. and Damar Y. (2003). Treatment of mixed pharmaceutical industry and domestic wastewater by sequencing batch reactor. *Journal of Environmental Science and Health. Part A, Toxic/Hazardous Substances and Environmental Engineering*, **38**(10), 2101–2111. DOI: 10.1081/Ese-120023336

Jean J., Perrodin Y., Pivot C., Trepo D., Perraud M., Droguet J., Tissot-Guerraz F. and Locher F. (2012). Identification and prioritization of bioaccumulable pharmaceutical substances discharged in hospital effluents. *Journal of Environmental Management*, **103**, 113–121. DOI: 10.1016/j.jenvman.2012.03.005

Jones O. A. H., Voulvoulis N. and Lester J. N. (2001). Human pharmaceuticals in the aquatic environment – A review. Environmental Technology, (United Kingdom), **22**(12), 1383–1394. DOI: 10.1080/09593330.2001.11090873

Jones O. A. H., Voulvoulis N. and Lester J. N. (2002). Aquatic environmental assessment of the top 25 English prescription pharmaceuticals. *Water Research*, **36**(20), 5013–5022. DOI: 10.1016/S0043-1354(02)00227-0

Kummerer K. (2001). Drugs in the environment: Emission of drugs, diagnostic aids and disinfectants into wastewater by hospitals in relation to other sources – A review. *Chemosphere*, **45**(6–7), 957–969. DOI: 10.1016/S0045-6535(01)00144-8

Lakshmi Prasanna B., Padmini V. L., Navle K., Dometti H. S. and Moodu S. (2015). Analysis of drugs in aquatic environment. *Journal of Chemical and Pharmaceutical Research*, **7** (4), 71–79.

Lin A. Y. C. and Tsai Y. T. (2009). Occurrence of pharmaceuticals in Taiwan's surface waters: Impact of waste streams from hospitals and pharmaceutical production facilities. *Science of the Total Environment*, **407**(12), 3793–3802. DOI: 10.1016/j.scitotenv.2009.03.009

Pollution C., Board C. and Environment M. O. F. (2000). Environmental Standards. Annual Report, 2000. Central Pollution Control Board, Delhi.

Singh R., Mishra S. S. and Mishra A. (2013). Hospital solid waste and waste water analysis in Lucknow. *Research in Environment and Life Sciences*, **6**(1), 1–8.

Stuer-Lauridsen F., Birkved M., Hansen L. P., Holten Lützhøft H.-C. and Halling-Sørensen B. (2000). Environmental risk assessment of human pharmaceuticals in Denmark after normal therapeutic use. *Chemosphere*, **40**(7), 783–793. DOI: 10.1016/S0045-6535(99)00453-1

Sui Q., Cao X., Lu S., Zhao W., Qiu Z. and Yu G. (2015). Occurrence, sources and fate of pharmaceuticals and personal care products in the groundwater: a review. *Emerging Contaminants*, **1**(1), 14–24. DOI: 10.1016/j.emcon.2015.07.001

De Voogt P., Janex-Habibi M. L., Sacher F., Puijker L. and Mons M. (2009). Development of a common priority list of pharmaceuticals relevant for the water cycle. *Water Science and Technology*, **59**(1), 39–46. DOI: 10.2166/wst.2009.764

Xu W., Zhang G., Li X., Zou S., Lis P., Hu Z. and Li J. (2007). Occurrence and elimination of antibiotics at four sewage treatment plants in the Pearl River Delta (PRD), South China. *Water Research*, **41**(19), 4526–4534. DOI: 10.1016/j.watres.2007.06.023

Zhang Y., Geißen S. U. and Gal C. (2008). Carbamazepine and diclofenac: removal in wastewater treatment plants and occurrence in water bodies. *Chemosphere*, **73**(8), 1151–1161. DOI: 10.1016/j.chemosphere.2008.07.086

Zuccato E., Castiglioni S. and Fanelli R. (2005). Identification of the pharmaceuticals for human use contaminating the Italian aquatic environment. *Journal of Hazardous Materials*, **122**(3), 205–209. DOI: 10.1016/j.jhazmat.2005.03.00118

Chapter 4

Occurrences: pharmaceutical wastewater in environment

Humzah Jamshaid[1], Fakhar-ud-Din[1] and Izhar Ullah[2]

[1]Nanomedicine Research Group, Department of Pharmacy, Faculty of Biological Sciences, Quaid-e-Azam University, Islamabad, Pakistan

[2]Department of Pharmacy, University of the Poonch Rawalakot, AJK, Pakistan

ABSTRACT

The surge in water pollution is alarming due to massive industrialization all over the globe. Despite the importance of the pharmaceutical industry, in terms of a country's economy and provision of health care products, it is also an important source of pharmaceutical contaminants, not only for the fresh surface water bodies but also for the ground water reservoirs. Effluent wastewater from drug manufacturing units and hospitals has harmful impacts on the aquatic habitat of freshwater bodies. Most importantly, the primary producers of the aquatic food chain are under serious harm. However, pharmaceutically polluted ground water is a continuously imposing threat to human health as it is a principle source for drinking water. Another major occurrence of the unwanted pharmaceuticals and medicinal chemicals is in the soil; contaminated mainly through wastewater and municipal landfill leaches. The presence of pharmaceuticals in soil tends to have direct harm to plant growth. The contamination level in wastewater can be predicted, depending upon the type of pharmaceutical processing plant from which it originates. It is necessary to timely curb the growing source of pharmaceutical pollutants in water. This can only be achieved by making the water treatment plants more efficient and modernized, using more advanced wastewater treatment strategies.

 Pharmaceutical Wastewater Treatment Technologies: Concepts and Implementation Strategies
Editor(s): Nadeem A Khan, Sirajuddin Ahmed, Viola Vambol and Sergij Vambol
doi: 10.2166/9781789061338_0069

Keywords: Freshwater, Ground water reservoir, Pharmaceuticals, Primary producers, Wastewater, Wastewater treatment.

Contents

4.1 Introduction 70
4.1.1 Pharmaceutical industry – echelons in country's economy 71
4.1.2 Pharmaceutical pollution – an emerging crisis 71
4.2 Sources of Contamination 72
4.2.1 Point sources 72
4.2.1.1 Wastewater 72
4.2.1.2 Septic tanks 73
4.2.1.3 Municipal landfills leachates 73
4.2.2 Diffused (miscellaneous) sources 74
4.2.2.1 Sewage sludge 74
4.2.2.2 Groundwater-surface interface 74
4.3 Major Pharmaceutical (APIS) Pollutants 74
4.4 Pharmaceutical Wastewater 75
4.4.1 Types of wastewater effluent from the pharmaceutical industry 75
4.4.1.1 Basic (raw material) manufacturing unit wastewater 75
4.4.1.2 Formulation manufacturing unit wastewater 76
4.5 Occurrence of Pharmaceutical Wastewater in the Environment 76
4.5.1 Aquatic bodies 76
4.5.1.1 Occurrence in ground water 77
4.5.1.2 Occurrence in surface water 77
4.5.2 In soil 80
4.6 Egregious Influences of Pharmaceuticals Occurred in Aquatic Bodies and Soil 81
4.6.1 Microorganisms – the basics of an ecosystem chain 81
4.6.2 Algae – primary producers of aquatic environment 81
4.6.3 Fishes – an important nutrition source 83
4.6.4 Birds, mammals and humans 83
4.6.5 Phytotoxicity 83
4.7 Curbing Pharmaceutical Pollutants – Brief Discussion of Treatment Strategies 84
4.7.1 Recovery process 84
4.7.2 Physical–chemical methods 84
4.7.3 Biological methods 84
4.7.4 Activated carbon 87
4.7.5 Innovative (oxidation) treatment strategies 87
4.8 Concluding Remarks 87
References 88

4.1 INTRODUCTION

High quality water is the basic necessity of life, not only for drinking purposes but also for irrigation, livestock, irrigation and industrial purposes. Indeed, in the pharmaceutical industry it is employed as a critical raw material. Waters commonly utilized for this purpose are ground water and water from surface freshwater bodies (Gadipelly *et al.*, 2014). Rapid global industrialization by developed as well as developing countries, to stabilize their economies, is deemed to be the prominent reason for poor water environments which is not only harmful to humans but also to pythons, aquatic and marine ecosystems. In addition, drastic effects regarding the climate are also emerging (Inyinbor Adejumoke *et al.*, 2018).

4.1.1 Pharmaceutical industry – echelons in country's economy

Healthcare products, being manufactured by pharmaceutical industries, are among one of the basic needs of a country's population. In this modern era of science and technology, the collaboration between healthcare providers and the research and development (R&D) domain of pharmaceutical industries have evolved drastically. This is found to be fruitful in terms of advancement of new drug discoveries. In other words, pharmaceutical industries invest heavily in the development of new drug entities, not only for monetary benefits but also to revolutionize medical treatment and diagnostics (Taylor, 2015).

Today, the aim to set up pharmaceutical industries is not only to cater to the state's health needs but also to promote the country's economy. When we talk about world stabilized economies, that is USA, UK and China, their pharmaceutical manufacturers are playing a crucial role in trade, business and state's economy. Facts are quite supportive to these statements as in China there were approximately 7108 pharmaceutical manufacturing units as of 2014 (Guo *et al.*, 2017). Similarly, in the UK, the same enterprise numbers are over 3000 and these make up the second-most revenue generating industry of the country after the petroleum sector (Earl-Slater, 1998; Khaleeq *et al.*, 2019).

Despite all these benefits of the pharmaceutical sector, it has some limiting parameters as well. Starting from the implementation of strict rules and regulation, to the expulsive fallout of pharmaceutical waste water pollutants, are among the major problems of pharmaceutical industries (Guo *et al.*, 2017). Embellishing economies through this sector, policies should be devised and promulgated to curb emerging pharmaceutical based pollutions.

4.1.2 Pharmaceutical pollution – an emerging crisis

As mentioned earlier, with the advancement of science, the medical and pharmaceutical sector is overwhelmed. Every year, millions of dollars have been doled out for pharmaceutical product development and production, to be used for humans and animals. Even though we all are the beneficiaries of pharmaceuticals in one way or another, the evolution of pollution via pharmaceutical effluents in form of wastewater is quite alarming. Pharmaceutical industries work day and night for the synthesis of raw materials (active pharmaceutical ingredients (APIs) and excipients) as well as finished products, excreting effluents in the form of wastewater. That wastewater is considered to be the composite of multiple API and non-active moieties, possibly hostile to not only humans but also to the aquatic environment as well as the ecosystem (Guo *et al.*, 2017; Khaleeq *et al.*, 2019; Sim *et al.*, 2011).

4.2 SOURCES OF CONTAMINATION

In order to address this emerging form of pollution, its source should be identified for the purpose of impeding its environmentally hazardous impact. In general, pharmaceutical industries, hospitals, domestic (household) and animals waste are some prominent sources of possibly devastating pollution. The pharmaceutical pollution sources are usually classified as point (identifiable) sources and diffused sources (Li, 2014). The sources of pharmaceutical wastewater are illustrated in Figure 4.1.

4.2.1 Point sources

Point sources of pollution are the identifiable sources originating from separate and spatially distinct locations. The most prominent point sources of pharmaceutical pollution are the waste effluents from hospitals, pharmaceutical industries, septic tanks and municipal landfills leachates (Lapworth *et al.*, 2012).

4.2.1.1 Wastewater

One of the most important point sources, wastewater excreting from hospitals, pharmaceutical industries and domestic establishments, contains various

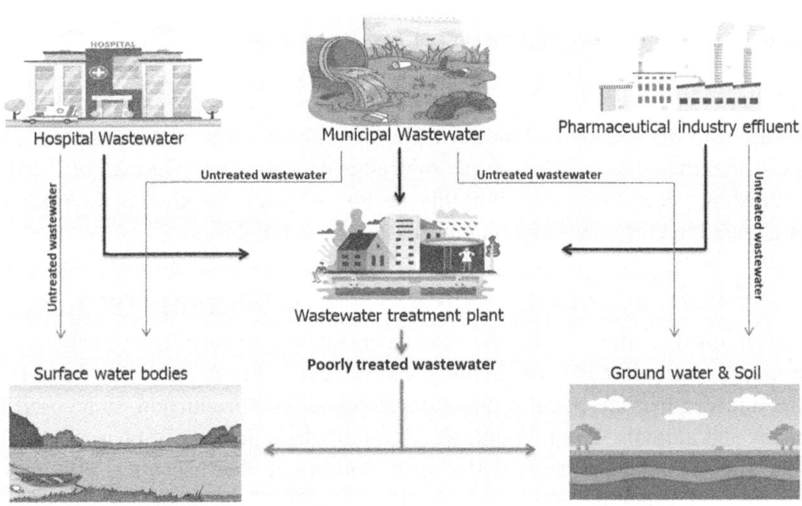

Figure 4.1 Sources and occurrences of pharmaceutical wastewater.

pharmaceutical candidates that have wide egregious events over the environment and ecosystems. These effluents not only pollute the ground water but also the soil, which could be harmful to botanical growth (Li, 2014). Sewage discharge water polluting the freshwater bodies was first reported in 1977, when the metabolites of clofibrate and aspirin were identified in the basin of the Missouri river (Hignite & Azarnoff 1977). This growing issue was not addressed at that time and, due to rapid industrialization, the toll of APIs in pharmaceutical industry sewage plant wastewater continued to increase. However, this remains limited not only to one or a few countries; rather this sort of pollution has emerged all around the globe including Europe and the American states. Adding insult to injury, the number of APIs identified from water bodies and soil is also increasing (Cardoso *et al.*, 2014). A study was carried out to estimate the occurrence of pharmaceuticals in various livestock, hospital, municipal and pharmaceutical manufacturing unit wastewater treatment plants. The results, however, showed the greatest incidence in livestock and pharmaceutical industry wastewater. The commonly identified APIs in that study were NSAIDs, antibiotics and anti-epileptics. For this purpose, wastewater should be required to process through several techniques to make it less harmful. However, in developing countries, due to lack of finances, wastewater is not treated in most cases (Sim *et al.*, 2011). The detail of occurrence and fallout of wastewater effluents is discussed in Section 4.4.

4.2.1.2 Septic tanks

Septic tanks collect household or domestic wastewater. They are also among one of the most common sources of pharmaceutical pollution. Humans use pharmaceuticals in the form of healthcare products which can become the part of domestic wastewater that is retained in the septic tanks. Unfortunately, leakage of the sewage system can result in the intermingling of wastewater with aquatic bodies. The most commonly appearing APIs in septic tanks are caffeine, anti-epileptics and antibiotics (Godfrey *et al.*, 2007).

4.2.1.3 Municipal landfills leachates

Municipal landfills possess certain chemical entities which can be harmful to humanity, wildlife and the ecosystem. In addition to organic waste, chlorinated alkylphosphates, perflorinated compounds and insect repellents, several pharmaceutical and personal care substances, including NSAIDs and antibiotics, were found to be a part of the leachate in municipal landfill waste (Eggen *et al.*, 2010). The leachate water also appeared to be the source of certain hormones and non-active pharmaceuticals. Later, this landfill leachate is believed to be part of ground water and surface water bodies (Barnes *et al.*, 2004).

4.2.2 Diffused (miscellaneous) sources

4.2.2.1 Sewage sludge

Sewage sledge, the wastewater treatment residue, is another source of contamination to ground water and the soil. In several countries it has been used as a soil quality modifier because it possesses a number of organic substances and nutrients. However, the problem is the remnants of some unwanted material; pharmaceuticals could be the reason for soil and ground water contamination (Lapworth *et al.*, 2012). According to a study carried out in 2008, it has been ascertained that the most commonly found pharmaceuticals in the sludge were thiabendazole caffeine and carbamazepine (Kinney *et al.*, 2008).

4.2.2.2 Groundwater-surface interface

An indirect route for the contamination of ground water is the ground water–surface water interface. Far too many mechanisms are involved in this indirect movement of contaminants, including downward migration in the soil (Lapworth *et al.*, 2012).

4.3 MAJOR PHARMACEUTICAL (APIS) POLLUTANTS

The identification of API in wastewater effluent depends upon a number of factors which can be a detection technique used to detect the type of wastewater identified the and use or manufacturing of APIs in that certain region where the study was conducted. The following enlisted pharmaceutical APIs have been detected in ground water, surface water, drinking water, soil and wastewater treatment plants (WWTPs) effluents of factories, domestic, hospitals and livestock farms, evidenced by a number of studies carried out worldwide (Aus der Beek *et al.*, 2016). Pharmaceutical pollutants are discussed in detail in Table 4.1.

Table 4.1 Common active pharmaceutical ingredients (APIs) in wastewater.

Pharmacological Class of APIs Detected	Commonly Detected APIs (Ashton *et al.*, 2004; Deblonde *et al.*, 2011; Koopaei & Abdollahi 2017)
Antibiotics	Clarithromycin, ciprofloxacin, doxycycline, erythromycin, metronidazole, ofloxacin, norfloxacin, sulfamethoxazole
Anti-epileptics	Carbamazepine, gabapentin
Anti-inflammatory and analgesics	Diclofenac, ibuprofen, indomethacin, naproxen
Anti-hyperlipidemias	Clofibric acid, fenofibrate, gemfibrozil, ezetimibe
Steroids	17α-ethinylestradiol and other androgenic drugs
Anti-hypertensive	Beta-blockers (atenolol and others), diuretics (hydrochlorothiazide and furosemide)
Miscellaneous	Caffeine, X-ray contrasts, cosmetics

4.4 PHARMACEUTICAL WASTEWATER

As discussed in the previous section on sources of contamination, wastewater containing pharmaceuticals originated mainly as wastage from pharmaceutical factories, hospitals, and livestock farms (Sim *et al.*, 2011). In this section the discussion is focused towards pharmaceutical wastewater – an effluent of pharmaceutical manufacturing units.

4.4.1 Types of wastewater effluent from the pharmaceutical industry

Not just for washing and cleansing, water is an important raw material in different processes. While discussing the use of water in the pharmaceutical industry, it is used in chemical manufacturing processes, fermentation processes for antibiotics production, material processing, extraction, purification, as well as a cooling agent. Its pharmaceutical benefits do not just end here; indeed, it is also used as a solvent, absorbent, as well as a suspending agent. After the deionization process, it is also used in injectables and large volume parenterals (LVPs) preparation as well (Gadipelly *et al.*, 2014; WHO, 2012).

4.4.1.1 Basic (raw material) manufacturing unit wastewater

Synthetic process wastewater. Pharmacologically active ingredients are usually fabricated by the utilization of basic raw materials – several organic and inorganic compounds. Furthermore, the synthesized active ingredients in the form of mixture with water were dried using separation techniques such as liquid–liquid extraction, solid–liquid extraction, filtration and deliquescence. Another critical role of water is the washing of cakes of synthesized products, through a crystallization process. In addition to all of this water can also be obtained as a by-product of the complex synthetic process. Evaluating the composition of manufacturing wastewater was complex and if it is expelled without treatment, it could cause serious environmental hazards. Synthetic process wastewater usually has high biological oxygen demand (BOD), chemical oxygen demand (COD), total suspended solids (TSS) and pH in the range of 1–11 (Gadipelly *et al.*, 2014; Pal, 2018). The composition of manufacturing pharmaceutical wastewater is stated in Table 4.2.

Table 4.2 Comparison of wastewater composition produced in the raw manufacturing pharmaceutical industry.

Basic Pharmaceutical Process	Biological Oxygen Demand (BOD)	Chemical Oxygen Demand (COD)	Total Suspended Solids (TSS)	pH
Synthesis	High	High	High	1–11
Fermentation	High	High	High	4–8
Extraction	Low	Low	Low	6–8

Fermentation process wastewater. To biologically synthesize several antibiotics, vitamins and anti-neoplastic agents, the fermentation process is commonly employed. In this, several bacteria, that is lactobacilli spp., yeast and growth media containing nutrients and electrolytes are used. Moreover, fermentation comprises of four basic steps including seed inoculum, preparation, fermentation and lastly, product recovery. The contaminants contained in fermentation wastewater usually depend upon the process. Broth wastewater contains organic nutrients, salts, nitrates and the phenolic anti-bacterial. The fermentation process usually takes place in an aqueous environment; thus, the dead bacterial cells or fungal cells could also be a part of fermentation wastewater. Such wastewater also has high BOD, COD and TSS (Gadipelly *et al.*, 2014; Lu *et al.*, 2000; Pal, 2018).

Extraction process wastewater. In the pharmaceutical industry, a number of medicinal entities are extracted through medicinal plants. This critically important step is carried out using an extraction or drug leaching process, comprising of solubility based separation from crude drugs (He *et al.*, 2016). A large volume of water, containing acids or bases to sustain particular pH, is deployed with a crude medicinal drug to remove lipophilic impurities and obtain hydrophilic, medicinally useful extracts. So, the fact is, extraction wastewater effluent of industry is less contaminated with curtailed levels of BOD, COD and TSS (Gadipelly *et al.*, 2014).

4.4.1.2 Formulation manufacturing unit wastewater

Formulation process wastewater. The APIs being produced are converted to the intended use dosage form in the formulation development wing of the pharmaceutical industry. A variety of processes and steps are involved in the dosage form development such as mixing, milling, grinding, crushing, and compression. Moreover, a highly clean environment is necessary for this type of pharmaceutical industry which is maintained by steam sterilization or the use of organic anti-bacterials. Non-active pharmaceutical ingredients such as fillers, binders, diluents etc. are also employed for dosage form development. To surmise, wastewater of dosage form manufacturing units may result in these substances polluting ground water and freshwater bodies (Bottoni *et al.*, 2010; Gadipelly *et al.*, 2014).

Use of water in solvent recycling. Benzene, toluene, and other non-conventional solvents, that is acetone, ethanol, and ethyl acetate, are employed in the industry for the removal of by-products, gaseous, liquid and solid impurities (Perez-Vega *et al.*, 2013). The presence of organic solvents in industrial wastewater would result in devastating effects over the environment and the ecosystem (Gadipelly *et al.*, 2014).

4.5 OCCURRENCE OF PHARMACEUTICAL WASTEWATER IN THE ENVIRONMENT

4.5.1 Aquatic bodies

In 1970, a number of pharmaceuticals were reported for the first time in aquatic bodies by a study carried out in the United States. Vast research has been carried out to qualify and quantify pharmaceutical pollutants in surface water bodies, ground water and WWTPs effluents. A few studies have claimed that in European countries and United States, more than 100 pharmaceuticals have been detected in aquatic water bodies, regardless of groundwater or surface water (Aus der Beek *et al.*, 2016). Owing to several naturally occurring processes, such as volatilization, biotransformation and photolysis, the concentration of APIs in ground water and surface water have reduced to a greater extent as compared to the concentration in WWTPs effluents (Gurr & Reinhard 2006). In addition to the aquatic bodies, soil also gets contaminated, as illustrated in Figure 4.1.

4.5.1.1 Occurrence in ground water

Ground water, the most neglected form of natural water resource, is being increasingly polluted via on-site sewage effluent, seepage through septic tanks, landfills and veterinary manure (Godfrey *et al.*, 2007). Based on critical evaluation of the available literature, it has been identified that the occurrence of pharmaceuticals varies significantly by region. In the United States, antibiotics mainly identified in ground water were sulfamethoxazole, trimethoprim and ciprofloxacin. Almost all WWTP effluent in Canada detected the presence of macrolide antibiotics, erythromycin and roxithromyin (Miao *et al.*, 2004). Analgesics identified through various techniques were acetaminophen and ibuprofen. The concentration of acetaminophen identified was extremely troublesome. In addition to antibiotics and analgesics, carbamazepine and caffeine were also quantified in US regions, that is in California and Canada. These drugs are not only qualified but also quantified, as stated in Table 4.3 (Fram & Belitz 2011; Van Stempvoort *et al.*, 2013). Studies in Europe have clearly shown the presence of sulfamethoxazole, ibuprofen, diclofenac, ketoprofen and higher concentrations of carbamazepine, along with stimulant-caffeine (Loos *et al.*, 2010). The quantified values of those pharmaceutical contaminants are listed in ng/L in the Table 4.3.

4.5.1.2 Occurrence in surface water

Fresh water sources, that is rivers, lakes, and ponds, are also the common fate of pharmaceuticals in the form of wastewater excreting from industries, hospitals and municipal WWTPs. Contrary to ground water, surface water can pollute directly from the effluents. More extensive data is available regarding the occurrence of pharmaceuticals in surface aquatic bodies globally, which is summarized in Table 4.3. In the United States and in its peripheries, a critically higher concentration of antibiotics was observed in fresh aquatic bodies because of the absence of natural mechanisms observed for ground water. Another unusual observation was the under detection of caffeine in freshwater bodies of US regions (Kleywegt *et al.*, 2011). Furthermore, in European regions, among detected antibiotics, the concentration of ciprofloxacin is exceptionally lower – the reason for which has not been quantified. The number of identified analgesics is also

Table 4.3 Occurrence and detected concentration of most common APIs in aquatic bodies.

Wastewater Occurrence	Country/Region	Pharmaceuticals Identified	Detected Concentration (ng/L)	References
Ground water	United States/Canada	Sulfamethoxazole	458	Fram and Belitz (2011); Miao *et al.* (2004); Van Stempvoort *et al.* (2013)
		Ciprofloxacin	0.28	
		Trimethoprim	18	
		Ibuprofen	3.97	
		Acetaminophen	1890	
		Carbamazepine	420	
		Caffeine	290	
		Erythromycin-Roxithromycin	—	
	European states	Sulfamethoxazole	38	Loos *et al.* (2010)
		Ibuprofen	395	
		Ketoprofen	2886	
		Diclofenac	24	
		Carbamazepine	3600	
		Caffeine	4500	
Surface water (lakes, rivers)	United States/Canada	Trimethoprim	145	Kleywegt *et al.* (2011)
		Sulfamethoxazole	170	
		Ciprofloxacin	77	
		Naproxen	555	

	Ibuprofen	203	
	Ketoprofen	16	
	Carbamazepine	735	
European states	Trimethoprim	60	López-Serna *et al.* (2012); WHO (2012)
	Sulfamethoxazole	33	
	Ciprofloxacin	Not quantified	
	Naproxen	80.5	
	Ibuprofen	468	
	Acetaminophen	40	
	Ketoprofen	293	
	Diclofenac	794	
	Salicylic acid	184	
	Carbamazepine	366	
	Bezafibrate	350	
	Lopromide	Not quantified	
Asia	Trimethoprim	1808	Hoa *et al.* (2011)
	Sulfamethoxazole	4330	
	Ibuprofen	30	
	Ketoprofen	620	
	Diclofenac	62	
	Carbamazepine	120	

high, imaging the overuse of analgesics in European regions and the detected analgesics include naproxen, ibuprofen, ketoprofen, diclofenac, acetaminophen and salicylic acid. Few drugs were only above the detection limit but less than the quantification limit, so were not quantified (López-Serna *et al.*, 2012; WHO, 2012). Studies performed by Japanese researchers showed the presence of numerous pharmaceuticals in Asian territories as well. Elevated levels of antibiotics in Asian freshwater bodies are alarming as they can be the reason of microbial resistance to certain antibiotics. In addition, they are also harmful to certain crops and plants because surface water is the principle irrigation source (Hoa *et al.*, 2011).

4.5.2 In soil

Soil, being the nutrition media for kingdom Plantae, is mainly polluted through one of the direct sources, that is veterinary pharmaceuticals that appear in animal manure (Gadipelly *et al.*, 2014; Khetan & Collins 2007). However, wastewater effluents are also the second common source for soil pollution. The impact of antibiotics and anti-virals in soil not only threaten the plant habitat but are also the reason for resistance emergence in bacteria and viruses. A study conducted in Spain, however, claimed the emergence of diclofenac, acetaminophen, caffeine, hydrochlorothiazide, mefenamic acid, flumequine, carbamazipine, trimethoprim and sulfamethoxazole in the soil (Biel-Maeso *et al.*, 2018). The most commonly identified pharmaceuticals residing in the soil are sulfadiazine, carbamazepine, triclosan, ibuprofen, diclofenac and trimethoprim. Table 4.4 provides a list of the

Table 4.4 Occurrence and detected concentration of most common APIs in soil.

Compound Identified	Country/Region	Concentration (ng/g)	Reference
Trimethoprim	China	0.64–2.15	Chen *et al.* (2011)
	USA	n.d.–0.64	Kinney *et al.* (2008)
Triclosan	Mexico	n.d.–16.7	Gibson *et al.* (2010)
Ibuprofen	China	1.51–5.03	Chen *et al.* (2011)
Diclofenac	Spain	n.d.–5.06	Biel-Maeso *et al.* (2018)
	China	0.35–1.16	Chen *et al.* (2011)
Carbamazepine	Mexico	2.6–7.5	Gibson *et al.* (2010)
	USA	0.7–1.4	Wu *et al.* (2010)
Ketoprofen	China	3.19–10.64	Chen *et al.* (2011)
Norfloxacin	China	4.2–14	Chen *et al.* (2011)
Acetaminophen	Spain	n.d.–5.95	Biel-Maeso *et al.* (2018)

evidence-based occurrence of the most commonly found pharmaceuticals in soil among various regions of the globe (Chen *et al.*, 2011; Gibson *et al.*, 2010; Kinney *et al.*, 2008).

4.6 EGREGIOUS INFLUENCES OF PHARMACEUTICALS OCCURRED IN AQUATIC BODIES AND SOIL

The occurrence of human and veterinary pharmaceuticals waste in soil, surface water bodies and ground water have emerged with eco-toxic influences in every part of the ecosystem. Most probably, it is due to the aqueous solubility of most of the pharmaceuticals that can easily get absorbed by the cells (Bottoni *et al.*, 2010; Li, 2014]. In this section, the evidence based convulsive events of pharmaceutical pollution are discussed in detail, with the enlistment of major pharmaceutical contaminants and their associated harmful implications shown in Table 4.5.

4.6.1 Microorganisms – the basics of an ecosystem chain

Antibiotics in surface wastewater bodies have harmful effects in the form of impeded reproduction of bacteria, that is alteration in normal flora. A study found the reduced reproduction rate of bacteria of genus *Vibrio* after being exposed to several antibiotics (Backhaus & Grimme 1999; Froehner *et al.*, 2000). Antibiotic exposure; due to farms, hospital and industries wastewater, is not only devastating to the aquatic ecosystem but could also be the reason for resistance emergence against imipenem, ceftazidime and cefotaxime, among the several microorganisms (Ash *et al.*, 2002). However, there is no important impact of resistance emergence in aquatic microbes but it has quite important implications on human health (Hoa *et al.*, 2011).

4.6.2 Algae – primary producers of aquatic environment

Merely a single cellular to basic multicellular structured organism, algae are very essential aquatic ecosystem members and the only primary producers in the aquatic food chain. Moreover, not limited to that, the algae are also the major source of oxygen production along with the higher plants (Chapman, 2013). As a core component of aquatic ecosystems, when algae is exposed to various pharmaceuticals originating from waste water, it could also experience harmful degenerative changes (Crane *et al.*, 2006). In 2011, a study was conducted demonstrating that the concentration of several pharmaceuticals, including carbamazepine and diclofenac, has crucial effects upon the algal chloroplast, the photosynthesis factory of the cell. In addition, they are also sensitive to several antibiotic exposures. Chloroplast damage would ultimately result in the reduction of total algae burden in the aquatic food system a threat to the entire ecosystem balance (Vannini *et al.*, 2011).

Table 4.5 Harmful implications of pharmaceutical wastewater.

Pharmaceutical Contaminant	Major Ecosystem Player Effected	Type of Risks Involved	Reference
Antibiotics; imipenem, ceftazidime, cefotaxime	Aquatic microorganisms	Hindrance in reproduction Resistance emergence	Ash *et al.* (2002); Backhaus and Grimme (1999); Hoa *et al.* (2011)
Carbamazepine and Diclofenac Sulfamethoxazole Ibuprofen	Unicellular algae	Proteomic analysis yield Damage to Chloroplast and photosynthetic assembly	Liu *et al.* (2011); Vannini *et al.* (2011)
Diclofenac 17α-ethinylestradiol Gemfibrozil Ibuprofen	Fishes, alligators and other aquatic organisms of kingdom Animalia (Hydra)	Cause anatomical destruction of kidneys and intestine Alter genetic and metabolic makeup Endocrine abnormalities Reduction in hydra attachment and hydranth number	Lyssimachou and Arukwe (2007); Mehinto *et al.* (2010); Pawlowski *et al.* (2004); Quinn *et al.* (2008)
Diclofenac 17α-ethinylestradiol	Birds and Mammals	Cause anatomical destruction of kidneys and intestine Gonadal defects	Gawad *et al.* (2016); Li (2014)
Fermentation contaminants Microbiologically contaminated hospital wastewater	Humans	Risk of infections	Emmanuel *et al.* (2009); Koopaei and Abdollahi (2017); Kumaraswamy *et al.* (2014)
Antibiotics – metronidazole, Chloroquine	Plants (legumes)	Growth retardation and death	Jjemba (2002); Yakubu (2017)

4.6.3 Fishes – an important nutrition source

The aquatic vertebrates belonging to the kingdom Animalia, the fishes, also appear to be critically vulnerable to pharmaceutical wastewater by its union with surface water bodies. Diclofenac and 17α-ethinylestradiol, a steroidal contraceptive, appeared to alter the kidney and intestine structure of this important nutrition source for humans and birds, and modified the genetic makeup of fishes which ultimately led to alterations in body metabolic activities (Lyssimachou & Arukwe 2007; Mehinto *et al.*, 2010). Moreover, the higher concentration of the steroidal contraceptive was also found to cause impaired development of secondary male characteristic and complete feminization of the fish population by causing endocrine abnormalities in response to estrogenic effect (Pawlowski *et al.*, 2004).

4.6.4 Birds, mammals and humans

Several birds and mammals also appeared to be affected by the estradiol's estrogenic mechanism, even in low concentrations, because they are biologically very active (Santos *et al.*, 2010). Furthermore, vultures and cattle are vulnerable to diclofenac and its metabolites (Gawad *et al.*, 2016). Although there is probably less direct exposure of pharmaceutical wastewater to land vertebrates, some of the evidence claims its toxicity in larger animals owing to the contamination of soil and ground water. While focusing our discussion on humans, the drinking water derived from contaminated ground water accompanied with improperly treated fermentation and hospital wastewater could possibly cause a high risk of multiple infections such as entero-pathogenic infections, helminthiasis and entero-virus infections (Emmanuel *et al.*, 2009; Kumaraswamy *et al.*, 2014). Hormonal disturbing agents, which cause complete feminization of fishes, are also known to have alarming effects on the human population (Anderson, 2005). Another study claimed the direct transfer of several drugs (carbamazepine, diclofenac and gabapentin) and X-ray contrast media in untreated irrigated wastewater, which could easily be transferred from plants to the humans. Through a deep literature review it has been concluded that no remarkable human toxicity from pharmaceutical wastewater has yet been reported yet (Koopaei & Abdollahi 2017).

4.6.5 Phytotoxicity

Toxic implications of pharmaceutical wastewater are not only limited to the kingdoms Protista or Animalia, as it also adversely affects the plants in terms of their growth retardation. Plant growth is highly vulnerable to the level of antibiotic incident in soil or ground water. Metronidazole was found to significantly retard the growth rate of leguminous plants and some studies also revealed antibiotic associated plant death (Jjemba, 2002; Yakubu, 2017).

4.7 CURBING PHARMACEUTICAL POLLUTANTS – BRIEF DISCUSSION OF TREATMENT STRATEGIES

To curtail this emerging type of aquatic pollution, there should be requirements to adopt worthwhile and cost-effective methodology by the WWTPs (Li & Li 2015). Details of pharmaceutical wastewater strategies are discussed in Chapter 15. As mentioned earlier regarding the variant composition of wastewater produce by different sources, thus the treatment strategy should be opted after the detailed observation of the source of pharmaceutical wastewater. A number of methodologies currently being applied (Table 4.6) to impede the rising concentration of pharmaceutical wastewater include:

(1) recovery process;
(2) physical-chemical methods; coagulation, sedimentation and floating;
(3) aerobic-anaerobic biological treatment;
(4) innovative treatment technologies.

4.7.1 Recovery process

The most suitable wastewater treatment strategy for pharmaceutical plants is the recovery process. This is due to its multiple advantages in terms of cost-effectiveness and recovery of the numerous pharmaceuticals (APIs), raw material, and culture media. Moreover, it also reduces the water treatment cost. Superior to both the above stated reasons, this strategy is feasible to adopt as there are certain substances which are non-biodegradable. The recovery can be achieved by several means comprising of reverse osmosis, nano-filtration, ultrafiltration, and electrochemical processes, though its utilization would depend upon the physical size and nature of the contaminant (Bezawada *et al.*, 2011; Gadipelly *et al.*, 2014; Shahtalebi *et al.*, 2011).

4.7.2 Physical–chemical methods

Coagulation involves the precipitation of pharmaceutical impurities followed by the addition of chemical agents to the wastewater secreted from many sources. Those chemical agents are most probably inorganic metallic salts and polymers, capable of removing toxic organic impurities dissolved in wastewater. Contrary to that, sedimentation is the separation phenomenon under the influence of gravitational force (Guo *et al.*, 2017).

Floating, a physical method, using air to form a tiny floating floccule of suspended particles, can be utilized to remove impurities from wastewater, except those which are highly soluble (Guo *et al.*, 2017).

4.7.3 Biological methods

Aerobic and anaerobic methods are usually employed for the treatment of effective neutralization of fermentation wastewater. The major aim of this treatment strategy

Table 4.6 Wastewater treatment strategies.

Wastewater Treatment Strategy	Beneficial Aspect	Techniques	Components Removed Significantly	Example	Reference
Recovery process	Cost effective and economical	Re-use of wastewater Reverse osmosis Membrane/nano-filtration Electrochemical process to remove metal impurities Ultrafiltration for organic impurities	Antibiotics and drugs having MW > 250 Da	Use of nano-filtration to isolate amoxicillin	Bezawada *et al.* (2011); Gadipelly *et al.* (2014); Shahtalebi *et al.* (2011); Sun *et al.* (2000)
Physical–chemical methods	Easy operation and economical	Coagulation Sedimentation Floating	Organic impurities from wastewater	—	Guo *et al.* (2017)
Biological methods (aerobic)	Low cost	Activated sludge method	Fermentation wastewater impurities, that is COD Pharmaceuticals	Removal of 90% COD and 90.4% BOD from fermentation wastewater	Noble (2006); Radjenovic *et al.* (2007)
	High sludge retention time	Membrane Bioreactor	TSS, COD, BOD and Pharmaceuticals		
Biological methods (anaerobic)	Recovery of by-product *Biomethane*	Anaerobic hybrid reactor	Pharmaceuticals Organic impurities COD and BOD removal	90% tylosin removal and 75% COD removal from antibiotic waste effluent	Chelliapan *et al.* (2006); Deegan *et al.* (2011)
	Removal of high pharmaceutical concentration	Up-flow anaerobic batch reactor			

(*Continued*)

Table 4.6 Wastewater treatment strategies (*Continued*).

Wastewater Treatment Strategy	Beneficial Aspect	Techniques	Components Removed Significantly	Example	Reference
Activated carbon	Cheap techniques due to low costs of raw materials	—	For effective removal of organic matter. Can be used as a pre-treatment with other techniques	Removal of ibuprofen through AC	Mestre *et al.* (2007); Stoquart *et al.* (2012); Ternes *et al.* (2002)
Advance (oxidation) strategies	For low biodegradable pollutants	Ozone/H_2O_2 treatment	Removal of antibiotics accept those containing amide group	Removal of penicillin from fermentation wastewater	Arslan-Alaton and Dogruel (2004); Dantas *et al.* (2008)
		Fenton's oxidation treatment	Removal of antibiotics, BOD and COD	95% COD removal from waste water containing antibiotics	Alaton *et al.* (2004); Badawy *et al.* (2009)
		Photocatalysis	Removal of non-biodegradable effluents	—	Deegan *et al.* (2011); Gadipelly *et al.* (2014)

is the removal of TSS, BOD and COD from the fermentation wastewater. However, the pharmaceuticals also experience a good removal rate. Among a few of the aerobic strategies, the activated sludge method was traditionally used but, nowadays, membrane bioreactors have gained more attention because of their low operational technicality and high sludge retention time. Furthermore, this method has more than 90% removal rate of BOD, COD, TSS and several pharmaceuticals (Noble, 2006; Radjenovic *et al.*, 2007). Focusing towards anaerobic treatment strategies, they include up-flow anaerobic batch reactors, stirred tank reactors, fluidized bed reactors and anaerobic hybrid reactors. It can be employed for the effective removal of pharmaceuticals in addition to BOD and COD (Chelliapan *et al.*, 2006; Deegan *et al.*, 2011).

4.7.4 Activated carbon

The appropriate technique for the removal of organic impurities present in wastewater is the activated carbon technique, particularly because of the large absorptive surface area of activated carbon. The low production cost of carbon is the advantage of this technique. The activated carbon occurs in two forms, powdered activated carbon (PAC) and granular activated carbon (GAC). Few studies have been conducted that suggest employing this technique in combination with the rest of the techniques (Stoquart *et al.*, 2012; Ternes *et al.*, 2002).

4.7.5 Innovative (oxidation) treatment strategies

Advance oxidation strategies usually comprise the utilization of such agents that oxidize the electrophiles present in wastewater effluents. Antibiotics are usually treated using ozone gas, a very strong oxidizing agent, or hydroxyl ion – an even stronger oxidizing agent than ozone. However, certain amide groups containing compounds are resistant to this mode of treatment (Dantas *et al.*, 2008). Fenton's oxidation utilized Fenton's reagent to carry out the oxidation of pharmaceutical effluents. Moreover, it can effectively reduce COD and BOD. However, it is more suitable to apply this technique in combination and prior to the biological treatment of wastewater (Alaton *et al.*, 2004). Photocatalysis, an advance strategy, is basically a photochemical reaction governed by a catalyst such as Fenton's reagent/TiO_2. It is considered to be the most effective means for the removal of COD and BOD to convert non-biodegradable contaminants to be suitable for biodegradation (Deegan *et al.*, 2011).

4.8 CONCLUDING REMARKS

Despite the higher significance of pharmaceutical industries in healthcare systems and the country's economy, the fledgling pharmaceutical pollution caused by them are in fact troublesome. The waste excretions of domestic, hospitals and industries are continually polluting aquatic surface bodies, ground water and soil. The occurrence of pharmaceuticals in these water bodies not only causes problems for the aquatic ecosystem, but also persistently contaminate drinking water. In addition, natural food chains and food webs are also disrupted, possibly due to the assassination of microbes and primary producers – the algae. Devising essential strategies to impede the growing pharmaceutical contaminants in the environment and ecosystem is, however, the necessity of time.

REFERENCES

Alaton I. A., Dogruel S., Baykal E. and Gerone G. (2004). Combined chemical and biological oxidation of penicillin formulation effluent. *Journal of Environmental Management*, **73**, 155–163. http://doi.org/10.1016/j.jenvman.2004.06.007

Anderson P. D. (2005). Technical Brief: Endocrine Disrupting Compounds and Implications for Wastewater Treatment. Water Environment Research Foundation, London, UK.

Arslan-Alaton I. and Dogruel S. (2004). Pre-treatment of penicillin formulation effluent by advanced oxidation processes. *Journal of Hazardous Materials*, **112**, 105–113.

Ash R. J., Mauck B. and Morgan M. (2002). Antibiotic resistance of gram-negative bacteria in rivers, United States. *Emerging Infectious Diseases*, **8**, 713–716. http://dx.doi.org/10.3201%2Feid0807.010264

Ashton D., Hilton M. and Thomas K. V. (2004). Investigating the environmental transport of human pharmaceuticals to streams in the United Kingdom. *Science of the Total Environment*, **333**, 167–184. http://doi.org/10.1016/j.scitotenv.2004.04.062

Aus der Beek T., Weber F. A., Bergmann A., Hickmann S., Ebert I., Hein A. and Küster A. (2016). Pharmaceuticals in the environment – Global occurrences and perspectives. *Environmental Toxicology and Chemistry*, **35**, 823–835. http://doi.org/10.1002/etc.3339

Backhaus T. and Grimme L. H. (1999). The toxicity of antibiotic agents to the luminescent bacterium Vibrio fischeri. *Chemosphere*, **38**, 3291–3301. http://doi.org/10.1016/S0045-6535(98)00560-8

Badawy M. I., Wahaab R. A. and El-Kalliny A. S. (2009). Fenton-biological treatment processes for the removal of some pharmaceuticals from industrial wastewater. *Journal of Hazardous Materials*, **167**, 567–574. http://doi.org/10.1016/j.jhazmat.2009.01.023

Barnes K. K., Christenson S. C., Kolpin D. W., Focazio M. J., Furlong E. T., Zaugg S. D., Meyer M. T. and Barber L. B. (2004). Pharmaceuticals and other organic waste water contaminants within a leachate plume downgradient of a municipal landfill. *Groundwater Monitoring and Remediation*, **24**, 119–126. http://doi.org/10.1111/j.1745-6592.2004.tb00720.x

Bezawada J., Yan S., John R. P., Tyagi R. D. and Surampalli R. Y. (2011). Recovery of Bacillus licheniformis alkaline protease from supernatant of fermented wastewater sludge using ultrafiltration and its characterization. *Biotechnology Research International*, **2011**, 1–11. http://doi.org/10.4061/2011/238549

Biel-Maeso M., Corada-Fernández C. and Lara-Martin P. A. (2018). Monitoring the occurrence of pharmaceuticals in soils irrigated with reclaimed wastewater. *Environmental Pollution*, **235**, 312–321. http://doi.org/10.1016/j.envpol.2017.12.085

Bottoni P., Caroli S. and Caracciolo A. B. (2010). Pharmaceuticals as priority water contaminants. *Toxicological and Environmental Chemistry*, **92**, 549–565. http://doi.org/10.1080/02772241003614320

Cardoso O., Porcher J.-M. and Sanchez W. (2014). Factory-discharged pharmaceuticals could be a relevant source of aquatic environment contamination: review of evidence and need for knowledge. *Chemosphere*, **115**, 20–30. http://doi.org/10.1016/j.chemosphere.2014.02.004

Chapman R. L. (2013). Algae: the world's most important 'plants' – an introduction. *Mitigation and Adaptation Strategies for Global Change*, **18**, 5–12. http://doi.org/10.1007/s11027-010-9255-9

Chelliapan S., Wilby T. and Sallis P. J. (2006). Performance of an up-flow anaerobic stage reactor (UASR) in the treatment of pharmaceutical wastewater containing macrolide antibiotics. *Water Research*, **40**, 507–516. http://doi.org/10.1016/j.watres.2005.11.020

Chen F., Ying G.-G., Kong L.-X., Wang L., Zhao J.-L., Zhou L.-J. and Zhang L.-J. (2011). Distribution and accumulation of endocrine-disrupting chemicals and pharmaceuticals in wastewater irrigated soils in Hebei, China. *Environmental Pollution*, **159**, 1490–1498. http://doi.org/10.1016/j.envpol.2011.03.016

Crane M., Watts C. and Boucard T. (2006). Chronic aquatic environmental risks from exposure to human pharmaceuticals. *Science of the Total Environment*, **367**, 23–41. http://doi.org/10.1016/j.scitotenv.2006.04.010

Dantas R. F., Contreras S., Sans C. and Esplugas S. (2008). Sulfamethoxazole abatement by means of ozonation. *Journal of Hazardous Materials*, **150**, 790–794. http://doi.org/10.1016/j.jhazmat.2007.05.034

Deblonde T., Cossu-Leguille C. and Hartemann P. (2011). Emerging pollutants in wastewater: a review of the literature. *International Journal of Hygiene and Environmental Health*, **214**, 442–448. http://doi.org/10.1016/j.ijheh.2011.08.002

Deegan A. M., Shaik B., Nolan K., Urell K., Oelgemöller M., Tobin J. and Morrissey A. (2011). Treatment options for wastewater effluents from pharmaceutical companies. *International Journal of Environmental Science and Technology*, **8**, 649–666. http://doi.org/10.1007/BF03326250

Earl-Slater A. (1998). The importance of the pharmaceutical industry to the UK economy. *Journal of Management in Medicine*, **12**, 5–20. http://dx.doi.org/10.1108/02689239810225210

Eggen T., Moeder M. and Arukwe A. (2010). Municipal landfill leachates: a significant source for new and emerging pollutants. *Science of the Total Environment*, **408**, 5147–5157. http://doi.org/10.1016/j.scitotenv.2010.07.049

Emmanuel E., Pierre M. G. and Perrodin Y. (2009). Groundwater contamination by microbiological and chemical substances released from hospital wastewater: Health risk assessment for drinking water consumers. *Environment International*, **35**, 718–726. http://doi.org/10.1016/j.envint.2009.01.011

Fram M. S. and Belitz K. (2011). Occurrence and concentrations of pharmaceutical compounds in groundwater used for public drinking-water supply in California. *Science of the Total Environment*, **409**, 3409–3417. http://doi.org/10.1016/j.scitotenv.2011.05.053

Froehner K., Backhaus T. and Grimme L. H. (2000). Bioassays with Vibrio fischeri for the assessment of delayed toxicity. *Chemosphere*, **40**, 821–828. http://doi.org/10.1016/S0045-6535(99)00274-X

Gadipelly C., Pérez-González A., Yadav G. D., Ortiz I., Ibáñez R., Rathod V. K. and Marathe K. V. (2014). Pharmaceutical industry wastewater: review of the technologies for water treatment and reuse. *Industrial and Engineering Chemistry Research*, **53**, 11571–11592. http://doi.org/10.1021/ie501210j

Gawad J. B., Deole M. K., Chavan B. A. and Tauro S. J. (2016). Environment and pharmaceutical industry in 21st century: an update. *Journal of Environmental Science, Computer Science and Engineering & Technology*, **5**, 230–234.

Gibson R., Durán-Álvarez J. C., Estrada K. L., Chávez A. and Cisneros B. J. (2010). Accumulation and leaching potential of some pharmaceuticals and potential endocrine disruptors in soils irrigated with wastewater in the Tula Valley, Mexico. *Chemosphere*, **81**, 1437–1445. http://doi.org/10.1016/j.chemosphere.2010.09.006

Godfrey E., Woessner W. W. and Benotti M. J. (2007). Pharmaceuticals in on-site sewage effluent and ground water, western Montana. *Groundwater*, **45**, 263–271

Guo Y., Qi P. S. and Liu Y. Z. (2017). A review on advanced treatment of pharmaceutical wastewater. IOP Conference Series: Earth and Environmental Science 63, 012025, IOP Publishing, UK.

Gurr C. J. and Reinhard M. (2006). Harnessing Natural Attenuation of Pharmaceuticals and Hormones in Rivers. ACS Publications, Washington, DC, USA.

He D., Wang Z., Yang L., Liu T., Yao Y. and Mao Z. (2016). Modeling and optimization of the drug extraction production process. *Scientific Programming*, **2016**, 1–15. http://doi.org/10.1155/2016/3279423

Hignite C. and Azarnoff D. L. (1977). Drugs and drug metabolites as environmental contaminants: chlorophenoxyisobutyrate and salicylic acid in sewage water effluent. *Life Sciences*, **20**, 337–341. http://doi.org/10.1016/0024-3205(77)90329-0

Hoa P. T. P., Managaki S., Nakada N., Takada H., Shimizu A., Anh D. H., Viet P. H. and Suzuki S. (2011). Antibiotic contamination and occurrence of antibiotic-resistant bacteria in aquatic environments of northern Vietnam. *Science of the Total Environment*, **409**, 2894–2901. http://doi.org/10.1016/j.scitotenv.2011.04.030

Inyinbor Adejumoke A., Adebesin Babatunde O., Oluyori Abimbola P., Adelani Akande Tabitha A., Dada Adewumi O. and Oreofe Toyin A. (2018). Water pollution: effects, prevention, and climatic impact. *Water Challenges of an Urbanizing World*, **33**, 33–47. http://doi.org/10.5772/intechopen.72018

Jjemba P. K. (2002). The effect of chloroquine, quinacrine, and metronidazole on both soybean plants and soil microbiota. *Chemosphere*, **46**, 1019–1025. http://doi.org/10.1016/S0045-6535(01)00139-4

Khaleeq N., Rabia S., Khan G. M. and Din F. U. (2019). Pharmaceutical Pollutants and Bioremediation, Chapter 5, pp. 10–135. Cambridge Scholars Publishing, Newcastle upon Tyne, UK. Available from: www.cambridgescholars.com/environmental-contamination-and-remediation; www.cambridgescholars.com/download/sample/64816

Khetan S. K. and Collins T. J. (2007). Human pharmaceuticals in the aquatic environment: a challenge to green chemistry. *Chemical Reviews*, **107**, 2319–2364. http://doi.org/10.1021/cr020441w

Kinney C. A., Furlong E. T., Kolpin D. W., Burkhardt M. R., Zaugg S. D., Werner S. L., Bossio J. P. and Benotti M. J. (2008). Bioaccumulation of pharmaceuticals and other anthropogenic waste indicators in earthworms from agricultural soil amended with biosolid or swine manure. *Environmental Science and Technology*, **42**, 1863–1870. http://doi.org/10.1021/es702304c

Kleywegt S., Pileggi V., Yang P., Hao C., Zhao X., Rocks C., Thach S., Cheung P. and Whitehead B. (2011). Pharmaceuticals, hormones and bisphenol A in untreated source and finished drinking water in Ontario, Canada – occurrence and treatment efficiency. *Science of the Total Environment*, **409**, 1481–1488. http://doi.org/10.1016/j.scitotenv.2011.01.010

Koopaei N. N. and Abdollahi M. (2017). Health Risks Associated with the Pharmaceuticals in Wastewater. Springer, BioMed Central, London, UK.

Kumaraswamy R., Amha Y. M., Anwar M. Z., Henschel A., Rodríguez J. and Ahmad F. (2014). Molecular analysis for screening human bacterial pathogens in municipal wastewater treatment and reuse. *Environmental Science and Technology*, **48**, 11610–11619. http://doi.org/10.1021/es502546t

Lapworth D. J., Baran N., Stuart M. E. and Ward R. S. (2012). Emerging organic contaminants in groundwater: a review of sources, fate and occurrence. *Environmental Pollution*, **163**, 287–303. http://doi.org/10.1016/j.envpol.2011.12.034

Li W. C. (2014). Occurrence, sources, and fate of pharmaceuticals in aquatic environment and soil. *Environmental Pollution*, **187**, 193–201. http://doi.org.10.1016/j.envpol.2014.01.015

Li X. and Li G. (2015). A Review: Pharmaceutical Wastewater Treatment Technology and Research in China. Atlantis Press, Paris, France. http://dx.doi.org/10.2991/ap3er-15.2015.81

Liu B.-Y., Nie X.-P., Liu W.-Q., Snoeijs P., Guan C. and Tsui M. T. K. (2011). Toxic effects of erythromycin, ciprofloxacin and sulfamethoxazole on photosynthetic apparatus in Selenastrum capricornutum. *Ecotoxicology and Environmental Safety*, **74**, 1027–1035. http://doi.org/10.1016/j.ecoenv.2011.01.022

Loos R., Locoro G., Comero S., Contini S., Schwesig D., Werres F., Balsaa P., Gans O., Weiss S. and Blaha L. (2010). Pan-European survey on the occurrence of selected polar organic persistent pollutants in ground water. *Water Research*, **44**, 4115–4126. http://doi.org/10.1016/j.watres.2010.05.032

López-Serna R., Petrović M. and Barceló D. (2012). Occurrence and distribution of multi-class pharmaceuticals and their active metabolites and transformation products in the Ebro River basin (NE Spain). *Science of the Total Environment*, **440**, 280–289. http://doi.org/10.1016/j.scitotenv.2012.06.027

Lu S. G., Imai T., Ukita M., Sekine M., Fukagawa M. and Nakanishi H. (2000). The performance of fermentation wastewater treatment in ultrafiltration membrane bioreactor by contiuous and intermittent aeration processes. *Water Science and Technology*, **42**, 323–329. http://doi.org/10.2166/wst.2000.0398

Lyssimachou A. and Arukwe A. (2007). Alteration of brain and interrenal StAR protein, P450 scc, and Cyp11β mRNA levels in Atlantic Salmon after nominal waterborne exposure to the synthetic pharmaceutical estrogen ethynylestradiol. *Journal of Toxicology and Environmental Health, Part A*, **70**, 606–613. http://doi.org/10.1080/10937400600882905

Mehinto A. C., Hill E. M. and Tyler C. R. (2010). Uptake and biological effects of environmentally relevant concentrations of the nonsteroidal anti-inflammatory pharmaceutical diclofenac in rainbow trout (Oncorhynchus mykiss). *Environmental Science and Technology*, **44**, 2176–2182. http://doi.org/10.1021/es903702m

Mestre A. S., Pires J., Nogueira J. M. F. and Carvalho A. P. (2007). Activated carbons for the adsorption of ibuprofen. *Carbon*, **45**, 1979–1988. http://doi.org/10.1016/j.carbon.2007.06.005

Miao X.-S., Bishay F., Chen M. and Metcalfe C. D. (2004). Occurrence of antimicrobials in the final effluents of wastewater treatment plants in Canada. *Environmental Science and Technology*, **38**, 3533–3541. http://doi.org/10.1021/es030653q

Noble J. (2006). GE ZeeWeed MBR technology for pharmaceutical wastewater treatment. *Membrane Technology*, **2006**, 7–9. http://doi.org/10.1016/S0958-2118(06)70790-X

Pal P. (2018). Treatment and disposal of pharmaceutical wastewater: toward the sustainable strategy. *Separation and Purification Reviews*, **47**, 179–198. http://dpi.org/10.1080/15422119.2017.1354888

Pawlowski S., Van Aerle R., Tyler C. R. and Braunbeck T. (2004). Effects of 17α-ethinylestradiol in a fathead minnow (Pimephales promelas) gonadal recrudescence assay. *Ecotoxicology and Environmental Safety*, **57**, 330–345. http://doi.org/10.1016/j.ecoenv.2003.07.019

Perez-Vega S., Ortega-Rivas E., Salmeron-Ochoa I. and Sharratt P. N. (2013). A system view of solvent selection in the pharmaceutical industry: towards a sustainable choice. *Environment, Development and Sustainability*, **15**, 1–21. http://doi.org/10.1007/s10668-012-9365-5

Quinn B., Gagné F. and Blaise C. (2008). An investigation into the acute and chronic toxicity of eleven pharmaceuticals (and their solvents) found in wastewater effluent on the cnidarian, Hydra attenuata. *Science of the Total Environment*, **389**, 306–314. http://doi.org/10.1016/j.scitotenv.2007.08.038

Radjenovic J., Petrovic M. and Barceló D. (2007). Analysis of pharmaceuticals in wastewater and removal using a membrane bioreactor. *Analytical and Bioanalytical Chemistry*, **387**, 1365–1377. http://doi.org/10.1007/s00216-006-0883-6

Santos L. H., Araújo A. N., Fachini A., Pena A., Delerue-Matos C. and Montenegro M. (2010). Ecotoxicological aspects related to the presence of pharmaceuticals in the aquatic environment. *Journal of Hazardous Materials*, **175**, 45–95. http://doi.org/10.1016/j.jhazmat.2009.10.100

Shahtalebi A., Sarrafzadeh M. H. and Montazer R. M. (2011). Application of nanofiltration membrane in the separation of amoxicillin from pharmaceutical wastewater. *Iranian Journal of Environmental Health Science and Engineering (IJEHSE)*, **8**, 109–116.

Sim W.-J., Lee J.-W., Lee E.-S., Shin S.-K., Hwang S.-R. and Oh J.-E. (2011). Occurrence and distribution of pharmaceuticals in wastewater from households, livestock farms, hospitals and pharmaceutical manufactures. *Chemosphere*, **82**, 179–186. http://doi.org/10.1016/j.chemosphere.2010.10.026

Stoquart C., Servais P., Bérubé P. R. and Barbeau B. (2012). Hybrid membrane processes using activated carbon treatment for drinking water: a review. *Journal of Membrane Science*, **411**, 1–12. http://doi.org/10.1016/j.memsci.2012.04.012

Sun M., Gan S. X., Yin D. F., Liu H. Y. and Yang W. D. (2000). Application of nanofiltration membrane in the purification process of tylosin. *Chinese Journal of Antibiotics*, **25**, 172–174.

Taylor D. (2015). The Pharmaceutical Industry and the Future of Drug Development. Pharmaceuticals in the Environment. Royal Society of Chemistry, London, UK, pp. 1–33. https://doi.org/10.1039/9781782622345-00001

Ternes T. A., Meisenheimer M., McDowell D., Sacher F., Brauch H.-J., Haist-Gulde B., Preuss G., Wilme U. and Zulei-Seibert N. (2002). Removal of pharmaceuticals during drinking water treatment. *Environmental Science and Technology*, **36**, 3855–3863. http://doi.org/10.1021/es015757k

Van Stempvoort D. R., Roy J. W., Grabuski J., Brown S. J., Bickerton G. and Sverko E. (2013). An artificial sweetener and pharmaceutical compounds as co-tracers of urban wastewater in groundwater. *Science of the Total Environment*, **461**, 348–359. http://doi.org/10.1016/j.scitotenv.2013.05.001

Vannini C., Domingo G., Marsoni M., De Mattia F., Labra M., Castiglioni S. and Bracale M. (2011). Effects of a complex mixture of therapeutic drugs on unicellular algae Pseudokirchneriella subcapitata. *Aquatic Toxicology*, **101**, 459–465. http://doi.org/10.1016/j.aquatox.2010.10.011

WHO. (2012). Pharmaceuticals in Drinking-Water. World Health Organization, Geneva, Switzerland. Available from: https://apps.who.int/iris/bitstream/handle/10665/44630/9789241502085_eng.pdf;jsessionid=372EC7E46F44DB2B1DB8ED86D4FB65ED?sequence=1 (accessed May 17, 2020).

WHO. (2012). WHO Good Manufacturing Practices: Water for Pharmaceutical use. World Health Organization, Geneva, Switzerland. Available from: http://www.who.int/medicines/areas/quality_safety/quality_assurance/GMPWatePharmaceuticalUseTRS970Annex2.pdf?ua=1 (accessed May 5, 2020).

Wu C., Spongberg A. L., Witter J. D., Fang M. and Czajkowski K. P. (2010). Uptake of pharmaceutical and personal care products by soybean plants from soils applied with biosolids and irrigated with contaminated water. *Environmental Science and Technology*, **44**, 6157–6161. http://doi.org/10.1021/es1011115

Yakubu O. H. (2017). Pharmaceutical wastewater effluent – source of contaminants of emerging concern: phytotoxicity of metronidazole to soybean (Glycine max). *Toxics*, **5**, 10. http://doi.org/10.3390/toxics5020010

Chapter 5

Pharmaceutical presence in effluent detected till date

Aastha Dhingra[1], Sirajuddin Ahmed[1,2], Siddhartha Gautam[2], T. Koliopoulos[3], A. K. Hussein[4] and Fateh Mebarek-oudina[5]

[1]Civil Engineering Department, Jamia Millia Islamia, New Delhi, India
[2]Delhi Pollution Control Committee, New Delhi
[3]Department of Public Health, University of West Attica, Athens, Greece
[4]Department of Mechanical Engineering, University of Babylon, Al Ḫillah, Iraq
[5]Department of Physics, Faculty of Sciences, University of 20 août 1955- Skikda, Skikda, Algeria

ABSTRACT

The effluent which is ejected from the Health Care Facilities (HCFs) can pose a threat to both humans and the environment, due to the presence of pathogens, pharmaceutical compounds, harmful chemicals and products generated in laboratories and research. This chapter will explain the characterisation of the hospital effluents with respect to physio-chemical, biological parameters, selected chemicals and nine high priority pharmaceutical residues. The samples for analysis were taken from the seven multi-specialty renowned hospitals of Delhi, selected on the basis of their sewage treatment technology. The study revealed that the characterisation of the hospital effluents is a hybrid of domestic and industrial effluents, due to the presence of environmentally relevant chemicals and pharmaceutical residues. The concentration of pharmaceuticals was found in the range of μg/L to ng/L. Ofloxacin and diclofenac were found to be in highest concentration of 64,040 and 11,613 ng/L,

 Pharmaceutical Wastewater Treatment Technologies: Concepts and Implementation Strategies
Editor(s): Nadeem A Khan, Sirajuddin Ahmed, Viola Vambol and Sergij Vambol
doi: 10.2166/9781789061338_0095

respectively. This was followed by diazepam and simvastatin, found to be as high as 7405 and 6779 ng/L, respectively. The BOD and COD signify the organic and chemical load in the effluents which have been found as high as 270 and 1142 mg/L, respectively. Phosphate, nitrates, lead, anionic detergents, phenolic compounds, and even harmful compounds including formaldehyde and gluteraldehyde, have also been found in the effluent samples from hospitals. Thus, hospital effluents should be given special attention with respect to the regulation of discharge of these contaminants. This is only possible if the effluents are treated with suitable treatment technologies.

Keywords: Concentration, Contaminants, Pharmaceutical residues, Presence.

Contents

5.1 Introduction 96
5.1.1 Characteristics of hospital wastewater and subsequent risks 97
5.1.2 Biological characteristics and risks 98
5.1.3 Occurrence of pharmaceutical residues in hospital effluents in the aquatic environment 99
5.1.4 Impact of HWW on the aquatic ecosystem 99
5.2 Material and Methods 100
5.3 Analytical Method for Pharmaceuticals on LCMS 101
5.3.1 Preparation of standard solutions 101
5.3.2 Procedure of sample preparation 101
5.3.3 UPLC conditions 101
5.3.4 Mass spectrometry 102
5.4 Results and Discussion 103
5.5 Comparative Analysis of the Values of Parameters Found in Hospital Effluent Globally and in this Study 107
5.6 Conclusion 108
References 109

5.1 INTRODUCTION

Hospitals generate mixed effluents, domestic as well as industrial, including those generated from medical and research activities (Emmanuel *et al.*, 2005). Hospital effluents contain various hazardous substances including:

(1) microbial pathogens like enteric pathogens (bacteria, viruses and helminths) that can be easily transmitted through water and thus cause water-borne diseases;
(2) toxic chemicals, sourced from several cleaning and disinfection operations executed in the facility, to maintain optimum or sterile conditions;
(3) pharmaceuticals administered to the patients in wards, or sold in the pharmacies; and

(4) radioactive isotopes, from the oncology department as a part of radiation therapy for the treatment of cancer.

These compounds are present in traces in the excretory products of patients as non-metabolised pharmaceuticals which then reach the sewage system (Carraro *et al.*, 2016).

There are prescribed discharge standards and norms for the treatment and disposal of industrial effluents in the natural system across the world. However, the scenario for the management of hospital effluents is not very encouraging. Only a few countries have prescribed guidelines and standards for the treatment of hospital wastewater (HWW). Since the majority of countries consider hospital generated wastewater on similar lines as domestic wastewater, HWW is generally being directly disposed into public sewers. Some countries are considering the hospital effluents equivalent to industrial and, therefore, they pre-treat the effluents before disposing them into drains.

Only conventional parameters such as BOD, COD, pH, temperature and TSS have been legislated for the disposal of wastewater in public sewers. For industries or hospitals, a few specific micropollutants such as anionic detergents, adsorbable organic halogens (AOX), nitrates, total nitrogen, sulphates, total and free chlorine, disinfectants, tensio-actives, cyanides, organophosphates, heavy metals and biological parameters (*E. coli* and faecal coli, etc.) are also to be quantified (Carraro *et al.*, 2016; Emmanuel *et al.*, 2002). However, the probable emerging contaminants, such as pharmaceutical residues, and chemical residues such as hypochlorite and glutaraldehyde which are used in huge quantities in hospitals, antibiotic resistant strains, and radioactive elements in hospital effluents, are not being regulated. Their impact on the environment and human health has been underestimated (Arnold *et al.*, 2013). Therefore, proper disposal of hospital sludge is a principal zone of focus to decrease the concerns pertaining to public health.

5.1.1 Characteristics of hospital wastewater and subsequent risks

The characteristics of the HWW depends on a myriad of determinants, such as the number of beds, average occupancy, footfall, number and types of wards, types of treatment, facilities provided and the economic status of the country (Verlicchi *et al.*, 2012). According to the World Health Organization (WHO) in 2013, wastewater is generated from all the activities in a hospital; ranging from medical activities such as diagnosis, treatment, operation, first-aid and radiation to non-medical activities (toilet, kitchen and laundry activities.

In 2013, the WHO identified pharmaceuticals belonging to various therapeutic classes such as antibiotics, analgesics, lipid regulators, psychiatric drugs, cytotoxics and chemicals like disinfectants, detergents, chemicals from laboratories and treatment facilities and X-ray contrast media. These traces reach the effluents as parent compounds, metabolites, or conjugates with other compounds.

Some of these pharmaceutical compounds can be detrimental for aquatic ecosystems, even in very low concentrations. Thus, it becomes imperative to characterise their contribution to the level of toxicity of the HWW, so that these can be ranked. This will further help the hospital management to identify the major risk compounds. As a result, it would be possible to take steps to decrease their release into the environment.

Studies have been centred on the impact of pharmaceuticals present in effluent upon human health, the ecosystem and endocrinology and sexual behaviour of aquatic organisms. Although there are no discharge standards with respect to pharmaceuticals, considering their consequences, these have been listed by the European Union (2009) in their priority list, which includes diclofenac, 17b-oestradiol, and 17a-ethinylestradiol (EU, 2013/39, about water policy). Similarly, the US contaminant candidate list includes antibiotics like erythromycin and hormones such as 17a-ethinylestradiol, equisetin, oestrone, mestranol and norethindrone.

As per the 2013 report by the WHO, hospitals are responsible for up to 5% of the release of mercury into the aquatic environment. When the UK is considered, approximately 50% of total mercury emissions in the wastewater are from the healthcare sector. The discharge of amalgam in wastewater should be prohibited since it contains metals like mercury, silver, tin, zinc and copper. In 2013, the WHO also reported that the fate of chemicals like X-ray contrast media, disinfectants, halogens, surfactants, formaldehyde, and glutaraldehyde, which are largely used in hospitals, is not well known and these are being discharged into the surface water. These compounds are ecotoxic and are persistent in the environment because of their bio-accumulation potential. There is also the issue of the release of radioactive and nuclear medicines from the hospitals. These pose a maximum risk to the aquatic environment and have very high bio-accumulation potential.

5.1.2 Biological characteristics and risks

Hospital effluents can be host to a wide diversity of pathogenic microbes, including bacteria, virus, and protozoa, which are primarily present in the faeces of infected humans. Researchers have always been attracted towards larger problems such as microbial resistance, the presence of pathogenic microbes in the wastewater, or the aquatic environment, even though not much data is available on the quantification of these microbes in hospital wastewater. (Emmanuel *et al.*, 2005) reported the concentration of entero-pathogenic viruses like rotavirus and adenovirus to be 2–3 times more in hospital effluents. Apprehension over microbial resistance has been persistent since the last decade and has garnered worldwide concern because of the excessive use of antimicrobials. In sewage treatment plants, the potential menace is caused because of the antibiotic resistant bacteria (ARB) released from patients or equipment and their behaviour is also unknown. The antimicrobials in the wastewater make microbes more resistant. In Europe, the consumption of antimicrobials is high and 20–30% patients are prescribed antibiotics. Antibiotic resistant genes are also transferred to other community members, thus increasing the gravity of the problem (Bound & Voulvoulis, 2004). Beta lactams are the most widely used group of antibiotic classes for patients, followed by fluoro-quinolones. However, since β-lactams degrade by hydrolysis, these are rarely detected in wastewater, whereas fluoroquinolones are ubiquitous (Lakshmi Prasanna *et al.*, 2015). Almost similar is the case with macrolides and sulphonamides. Since the potential of the antibiotics is reduced on ARBs, the risk to humans and animals increase.

5.1.3 Occurrence of pharmaceutical residues in hospital effluents in the aquatic environment

The drug consumption pattern is variable in many of the healthcare and hospital facilities worldwide (Lee *et al.*, 2016). A case study of Germany regarding the drugs consumption pattern in different healthcare facilities found variations up to a maximum of 1263 kg/year for general hospitals, with a maximum annual average individual use of 1000 g/bed. In another case study carried out in Delhi (India), it was found that 7101 enrolled patients (out of 17,995) visited private pharmacies for the purchase of antibiotics, while 3615 patients (out of 9205) visited public healthcare facilities. This variable consumption pattern resulted in the uneven discharge of pharmaceutical pollutants into the acquatic environment (Kotwani & Holloway, 2011).

After such variable consumption, antibiotics were discharged either through urine (50–80%) or faeces (5–30%) as drugs residue, metabolites or inactivating substance (Carraro *et al.*, 2016; Dimian & Bildea, 2008). The concentration of drugs residues in the hospital effluents is mainly due to three factors: (1) excretion, (2) administered quantity, and (3) laboratory waste (stable and biodegradable to some extent) (Verlicchi *et al.*, 2010). The drugs residue and its load varies from one geographic location to another, and can be as high as 5 mg/L. Cytotoxic drugs are used in chemotherapy and the treatment of cancer patients. These usually interfere in DNA synthesis disrupting the cellular proliferation, and are thus proven to be genotoxic, mutagenic, carcinogenic and as endocrine disruptors in several organisms. There are many reports about the scarce occurrence of cytostatic drug compounds in the ecosystem, but these are present in aquatic life (Escher *et al.*, 2010; Fent *et al.*, 2006; Isidori *et al.*, 2016).

5.1.4 Impact of HWW on the aquatic ecosystem

HWW is the source of a multitude of pollutants such as silver and mercury from X-rays, mammograms and some dental procedures, sewage having high BOD, kitchen waste high in oil and grease, detergents from laundries, and pharmaceutical residues through the urine of patients. The pollution load has already surpassed the removal efficiency of the municipal sewage treatment plants and thus these can be commonly found in the aquatic environment. Since the hospital wastewater is a cocktail of chemicals, even lower concentrations of the pollutants should not be underestimated since these may have additive effects. Thus, the cumulative effect of these can have a major impact on the ecosystem (Emmanuel *et al.*, 2005). Hospital effluents also contain heavy metals which again have detrimental effects on the ecosystem. In cases where the environmental conditions are not suitable for degradation of these pollutants it represents a short- to long-term risk to the organisms living there because of their high bio-accumulation potential and bio-magnification. Because of its complex composition, hospital effluent poses a physical, chemical, as well as biological, risk to organisms in the aquatic ecosystem (Emmanuel *et al.*, 2002). Many compounds, such as pharmaceutical residues and phenolic compounds, are known to cause acute or chronic toxicity, changed behaviour, reduced reproduction potential and stunted growth in the lower organisms. The level of risk can be determined by eco-toxicological tests. Unfortunately, these are rarely carried out (Pinto & Garcı, 2014).

5.2 MATERIAL AND METHODS

According to the literature review on regulations with respect to hospital wastewater, it has been found that physio-chemical, biological and specific chemicals, including drug residues, formaldehyde and glutaraldehyde, should be analysed since hospital effluents pose a physical, chemical and biological risk to the aquatic environment and human health. Effluent samples were collected from seven multi-specialty hospitals having high bed strength, different treatment technologies and belonging to different sectors such as private, government and charitable. Composite sampling was carried out every 4 hours for 24 hours. The list of hospitals selected for the study are given in Table 5.1.

Table 5.1 List of hospitals selected for study.

Hospital No.	Name of the Hospital	Sector
1	Indraprastha Apollo	Private
2	Batra	Private
3	AIIMS	Government
4	GTB	Government
5	Holy family	Charitable
6	Baba Saheb Ambedkar	Government
7	Maharaja Agrasen	Private

Characterization with respect to various parameters has been carried out as per standard protocols. The physio-chemical parameters analysed in the effluent are: pH value, total suspended solids, BOD, COD, phosphate, lead, anionic detergent, nitrate, phenolic compounds, free residual chlorine, alkalinity and hypochlorite. Biological parameters such as faecal coli and *Escherichia coli* (*E. coli*) have also been analysed since hospital effluents are reservoirs of pathogens and are mainly responsible for microbial resistance. The most commonly consumed chemicals such as formaldehyde, glutaraldehyde and hypochlorite have also been analysed. Nine high priority pharmaceuticals which include analgesics have been identified: diclofenac and ibuprofen, antibiotics such as erythromycin, ofloxacin and the diuretic frusemide, antidepressants such as diazepam and carbamazepine, the cytotoxic drug sorafenib and the lipid regulator simvastatin. These are known to have potential implications on the aquatic environment and are most commonly consumed in the hospitals. Therefore, these were analysed using an LCMS instrument in the hospital effluent.

5.3 ANALYTICAL METHOD FOR PHARMACEUTICALS ON LCMS

5.3.1 Preparation of standard solutions

The stock standard used in this study was 1 mg/mL in methanol. The standard was then diluted in diluents (mixture of 0.3% formic acid in methanol and water, 50:50). Linearity was achieved from 2.5 to 300 ng/L with a five-point calibration curve with r^2 more than 0.99 for all the analytes.

5.3.2 Procedure of sample preparation

Approximately 150 mL of the sample was taken. The sample was filtered with Whatman filter paper No. 1. The 100 mL filtrate for sample was used in the preparation. An HLB Oasis Extraction Cartridge (200 mg) was used for solid phase extraction. The cartridge was activated with 6 mL methanol and 6 mL purified water, loaded with 100 mL sample filtrate and then washed with 5 mL purified water. Drying was conducted by a vacuum pump. The dilution procedure was carried out with 5 ml methanol. The diluted solution was evaporated with a nitrogen evaporator at 40°C. Reconstitution of the sample was taken with 1 ml of solution (a mixture of 0.3% formic acid in methanol and water, 50:50). The sample was filtered with a 0.45 micron nylon filter and injected into the LCMS MS system (de Almeida *et al.*, 2015).

5.3.3 UPLC conditions

The analytical determinations of carbamazepine, diazepam, ofloxacin, simvastatin, sorafenib, erythromycin, ibuprofen, diclofenac and furosemide were analysed by the given method. An automated method based on liquid chromatography-MS/MS was developed for this study. The operational conditions for the analysis of these pharmaceutical residues are as follows: the symmetry shiel RT 18, 5u, 4.6 × 150 mm chromatographic column, a mobile phase of 10 mM ammonium format in water (A) and 10 mm ammonium formate in methanol (B), both containing 0.2% formic acid. An isocratic programme was used, with 20% mobile phase A and 80% mobile phase B. The applied flow rate was 0.8 mL per minute and the injection volume was 25 μL. The temperature of the column oven was 40°C. Water Mass Lynx 4.1 software was employed for the control of the instrument and data acquisition (Chang *et al.*, 2009).

5.3.4 Mass spectrometry

Analysis of the pharmaceutical concentration was performed using liquid chromatography with UPLC H-Class and Waters Xevo TQ-S Micro (Tandem Quadrupole Mass Spectrometer). The UPLC identifies and quantifies the molecule with the help of an electrospray ionization source (ESI), operating in a positive mode. The dependent parameters were optimized directly infusing 10 μL min^{-1} of work solution of all analytes. The next step was the selection of the ionisation and the precursor ion modes. Thereafter, the automated quantitative optimisation method was used to determine the product ions, as well as their corresponding capillary voltage (4.5 kv), cone voltage (33 V), source temperature (150°C), desolvation temperature (450°C), cone gas flow (50 L/Hr), desolvation gas flow (850 L/h), collision gas flow (50 mL/min) (de Almeida *et al.*, 2015). At the same time, the quantification and confirmation transitions were monitored for the analytes. The quantification transition was selected by the precursor ion and the product ions. This gave the most abundant response and the highest sensitivity. Mass transitions are given in Table 5.2.

Table 5.2 Mass transitions of the selected pharmaceuticals for analysis.

Sr. No.	Compound Name	Precursor Ion	Product Ion
1	Carbamazepin	237.03	179
2	Diazepam	284.84	192,985
3	Ofloxacin	361.904	260,965
4	Simvastatin	419.032	243,009
5	Sorafenib	464.84	251,933
6	Erythromycin	7734.204	82,967
7	Ibuprofen	205.1	161.1
8	Diclofenac	294	250
9	Furosemide	328.777	204,804

5.4 RESULTS AND DISCUSSION

Only basic physio-chemical parameters have been mandated by the Central Pollution Control Board to be analysed in the hospital effluents. These include BOD, COD, pH, TSS and bioassay test. However, EPA and European Directives have suggested a few other parameters also. However, to date, there are no discharge standards for the emerging contaminants like pharmaceutical residues and chemicals such as formaldehyde and glutaraldehyde, which can have hazardous impacts on the environment and thus public health (Carraro *et al.*, 2016). BOD and COD signify the organic and chemical load in the effluent, which has been found as high as 270 and 1142 mg/L respectively. This is because the hospital effluents contain both domestic sewage and trade effluents. Phosphate, nitrate, lead, anionic detergents, phenolic compounds and even harmful compounds, including formaldehyde and glutaraldehyde, have also been found in the effluent samples from the hospitals. The phosphate present in the hospital wastewater when discharged to receiving water bodies may accelerate the eutrophication process. Nitrates have been detected in the hospital effluents because of the ammonia present in the urine. Thus, the hospital effluents should be given special attention with respect to regulation of these contaminants in compliance with the

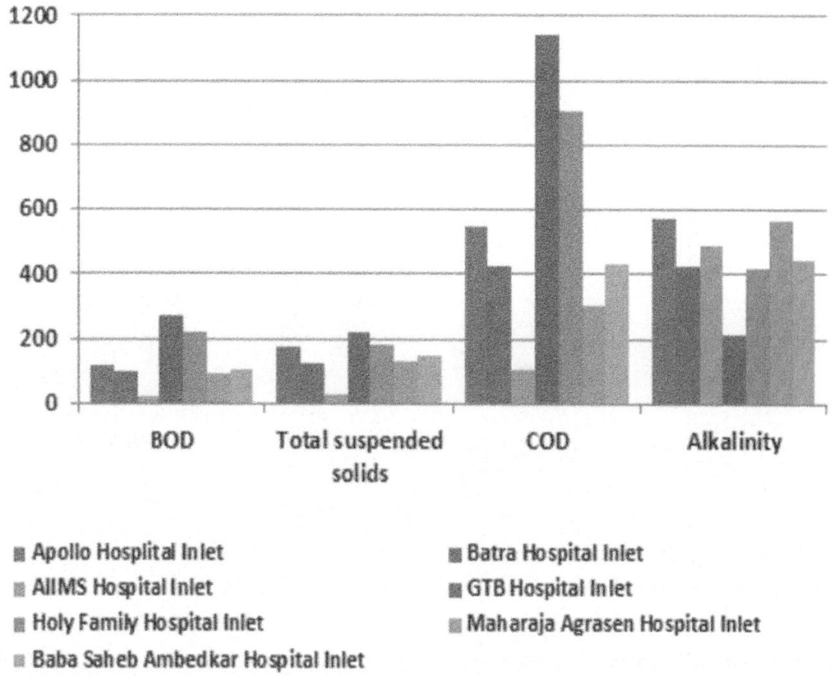

Figure 5.1 Physiochemical characterization of hospital influent (Source: Khan *et al.* 2020).

discharge standards. This would be possible only when the effluents are subjected to suitable treatment technologies. The physio- chemical parameters have been analysed in all the seven samples and the results are shown in Figure 5.1.

The biological parameters *E. coli* and faecal coli were analysed in the hospital effluents and it was found that *E. coli* was present in all the samples. However, it was not quantified and the results of faecal coli are presented in Figure 5.2. Faecal coli was found to be as high as 80,000 in the effluents of the HolyFamily Hospital and was least in the effluents of the AIIMS hospital.

Out of nine high priority pharmaceuticals identified for analysis in the hospital effluents, eight were found in all the seven hospitals. Only carbamazepine was not detected in hospital no. 2. Sorafenib, a cytotoxic pharmaceutical, was not measured in the inlet of all the seven hospitals. The concentration of pharmaceuticals was found in the range of μg/L to ng/L. Ofloxacin and diclofenac were found to be in the highest concentration of 64,040 and 11,613 in hospital nos. 4 and 3 respectively. This was followed by diazepam and simvastatin which were found to be as high as 7405 and 6779 ng/L in hospital nos. 4 and 2 respectively. These concentrations were calculated from the chromatograms on LCMS and are presented in Tables 5.2 and 5.3.

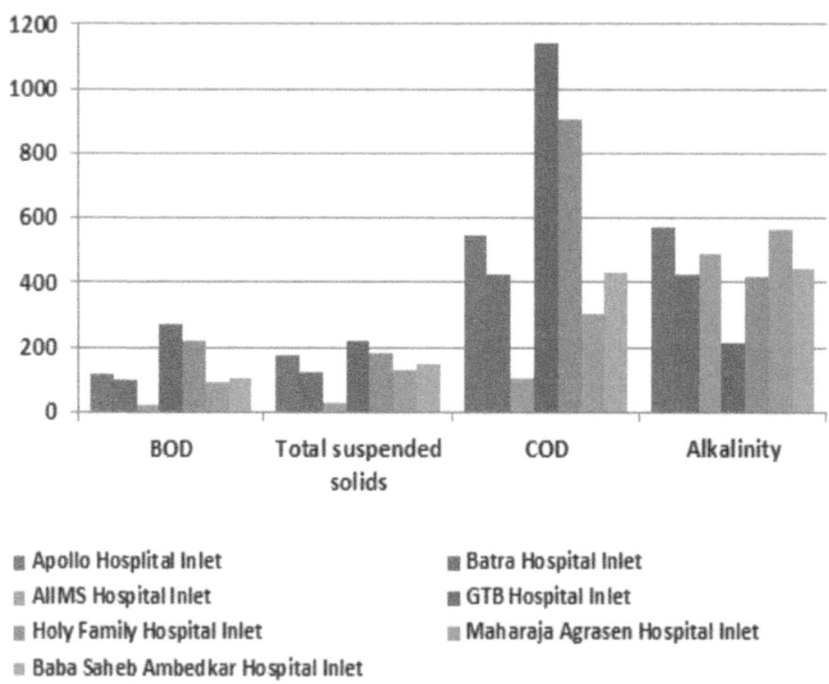

Figure 5.2 Biological characterisation of hospital influents (Source: Khan *et al.* 2020).

Table 5.3 Comparative analysis of physiochemical, biological characterization and pharmaceutical residues in hospital wastewater worldwide.

Test Parameters	RP1	RP2	RP3	RP4	RP5	RP6	RP7	RP8	RP9	RP10	RP11	RP12	RP13	RP14	RP15	RP16	RP17	RP18	RP19	RP20	Results Inlet (min–max)
pH value	7.97–8.35	—	6–9	—	7.63	7.6–8.2	7.41	6.94	6.26–8.52	8.53	—	6–8	—	—	—	6.6–7.1	—	—	—	7	6.6–7.2
Total suspended solids (mg/L)	42.70–83.33	225	16–326		402.29	157–421	42.66	183	155.16–297.6	74.7	225					61.1–305	—	—	—	30	31–221
BOD (mg/L)	86.96–238.98	603	16–2575		482.08	199–545	187.66		251–1559		603	270				68–437	—	—	—	30	24–270
COD (mg/L)	109.82–350.88	855	39–7764		996.98	456–918	300.33		604–2590	619	855	450				221.3–907.7	—	—	—	80	108–1142
Phosphate (mg/L)	2.31–6.25	8.8	6–19		13.07	1–3.4	7.79			2.50	8.80					4–9.6					3.1–5.7
Lead (mg/L)	0–0.025		0.003–0.019																	0.01	0.0–0.2
Anionic detergent	0.61–0.65						17.60									4.9–63				0.5	0.2–0.8
Nitrate (mg/L)			1–2		6.67					4.5											2.1–8.6
Phenolic compounds (mg/L)	0.01–0.04																			0.01	0.1–0.6
Free residual chlorine (mg/L)	1.04–1.63																			0.5	0.0–0.0
Alkalinity (mg/L)					113.66																211–537
Hypochlorite (mg/L)																					0.0–0.0
MLSS (mg/L)																					53–267
Formaldehyde (mg/L)																					0.0–0.3
																					1.0–0.0

(Continued)

Table 5.3 Comparative analysis of physiochemical, biological characterization and pharmaceutical residues in hospital wastewater worldwide (*Continued*).

Test Parameters	RP1	RP2	RP3	RP4	RP5	RP6	RP7	RP8	RP9	RP10	RP11	RP12	RP13	RP14	RP15	RP16	RP17	RP18	RP19	RP20	Results Inlet (min–max)
Gluteraldehyde (mg/L)																					
Faecal coliform (CFU/100 mL)	677–2172						7333.33					>1600				220–8 × 10 × 5				40,000	1700–80,000
E. coli						5.4 × 10,000,0000	44,000					4 × 1000				10 × 5–5.4 × 10×6					Present
Diclofenac (ng/L)			240–1500	300–2600									780–1630		300–2800	58–1920	600–1900	980–2500			4754–11613.0
Ibuprofen (ng/L)			70–4300	700–10,200									420–990		100–2600	813–4964	1500–151,000	27–1793		8500	924.9–3637.1
Carbamazepine (ng/L)			540–2000	1700–10,600									6300	344–479	700–3400	3–956	40	942–160		1300	0.0–1288.6
Diazepam (ng/L)														433.1–595.9							3230–7404.9
Erythromycin (ng/L)			27,000–83,000	200–900												127–575	10–30			5500	569.9–3198.9
Ofloxacin (ng/L)			350–35,000													218–20,032			25–500		5991–64,040.0
Sorafenib																					0.0–0.0
Frusemide (ng/L)				4500–13,700											150–350						577.9–1635.6
Simvastatin (ng/L)				200											2–4						1747–6779.4

Rp1: January *et al.* (2012); Rp2: Emmanuel *et al.* (2002); Rp3: Verlicchi (2018); Rp4: Le Corre *et al.* (2012); Rp5: Al-ajlouni *et al.* (2011); Rp6: Sciences *et al.* (2015); Rp7: Rezaee *et al.* (2005); Rp8: Mesdaghinia *et al.* (2015); Rp9: Mesdaghinia *et al.* (2015); Rp10: Mesdaghinia *et al.* (2015); Rp11: Gautam *et al.* (2007); Rp12: Rezaee *et al.* (2005); Rp13: Tran *et al.* (2014); Rp14: de Almeida *et al.* (2015); Rp15: Langford and Thomas (2009); Rp16: Carraro *et al.* (2016); Rp17: Gómez *et al.* (2006); Rp18: Kosma *et al.* (2010); Rp19: Brown *et al.* (2006); Rp20: January *et al.* (2012).

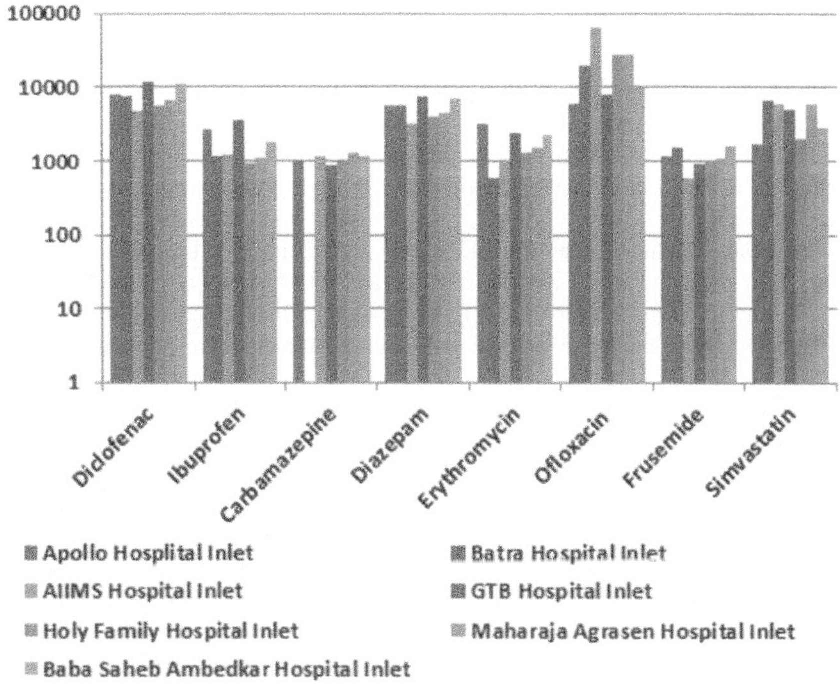

Figure 5.3 Pharmaceutical residues detected in hospital influents (Source: Khan *et al.* 2020).

The values of the selected pharmaceutical residues are presented in Figure 5.3. These drugs were found to be toxic to the aquatic environment and are persistent in nature. Thus, their unchecked discharge is very harmful and should be controlled.

5.5 COMPARATIVE ANALYSIS OF THE VALUES OF PARAMETERS FOUND IN HOSPITAL EFFLUENT GLOBALLY AND IN THIS STUDY

In this section of the chapter, the literature on physical, chemical and biological characterisation of the hospital effluents and occurrence of pharmaceutical residues in the effluents were studied. The same have been compared with the characterisation results of the study. It was found that almost all the parameters more or less have a similar range as reported in international studies. Thus, it can be concluded that the consequences caused by the hospital effluents on the aquatic environment and human health is the same worldwide and this is presented in Table 5.3.

5.6 CONCLUSION

Hospital effluents are still not being understood in the context of the presence of hazardous substances. These have been treated as industrial effluents by the Central Pollution Control Board and they are loaded with pathogens, infected urine, blood, traces of high priority pharmaceuticals and chemicals. Thus, it is important to have an in-depth knowledge of the physio-chemical and biological characterisation of the effluents, along with the analysis of probable contaminants, including harmful chemicals and pharmaceutical residues. This study has attempted to do the same. BOD and COD have been found as high as 270 and 1142 mg/L, respectively. In hospital effluents, the physio-chemical parameters depict moderate organic pollution. However, there are a number of specific pollutants that have been found in the effluents from hospitals and these include: phosphates, nitrate, anionic detergents, phenols, gluteraldehyde and formaldehyde. Pharmaceutical residues have been found to be the main emerging contaminants found in the hospital effluents. The high priority pharmaceuticals were analysed in the effluents using an LCMS instrument. These are relevant from the point of view of the environment considering their physio-chemical properties, occurrence in the effluents and environment, consumption, bio-accumulation potential, resistance to treatment, ecotoxicity and persistence. It has been noticed that the high priority compounds do not belong to one or two therapeutic classes. Therefore, it is important that in future, a treatment scheme is proposed that can work on it cumulatively considering the interaction between various molecules, should be determined. Eight out of nine pharmaceuticals were found in all the hospitals. Only carbamazepine was not found in the effluents of one out of the seven hospitals. Ofloxacin (antibiotic) and diclofenac (analgesic) were found to be in the highest concentrations of 64,040 and 11,613 in hospitals nos. 4 and 3 respectively. This was followed by diazepam (anti-depressant) and simvastatin (lipid regulator), found as high as 7405 and 6779 ng/L in hospitals nos. 4 and 2 respectively. Thus, this study provides an in-depth knowledge of the characterisation of the hospital effluents and stresses the need to focus on the suitable treatment for such effluents, in order to save the environment and public health. The characterisation highlights the importance of a separate treatment of hospital effluents. Stringent measures for regulation of these contaminants should be urgently taken, considering the potential implications on the environment. Also, the authorities should lay down specific discharge standards with respect to high priority pharmaceuticals and chemicals that have been found in the effluents. Further research should be carried out on the interaction of various chemical and pharmaceutical molecules and their degradation in the effluents. There is a need to develop a treatment scheme that would work well for the degradation of all these molecules.

REFERENCES

Al-ajlouni K., Shakhatreh S., Ibraheem N. A. and Jawarneh M. (2011). Evaluation of wastewater discharge from hospitals in Amman – Jordan. *International Journal of Basic Applied Science, IJBAS-IJENS*, **13**(4), 44–49.

Arnold K. E., Alistair B. A., Boxall A., Brown R., Cuthbert R. J., Gaw S., Hutchinson T. H., Jobling S., Madden J. C., Metcalfe C. D., Naidoo V., Shore R. F., Smits J. E., Taggart M. A. and Thompson H. M. (2013). Assessing the exposure risk and impacts of pharmaceuticals in the environment on individuals and ecosystems Assessing the exposure risk and impacts of pharmaceuticals in the environment on individuals and ecosystems. *Biology Letters*, **9**(4), 9–14. doi: 10.1098/rsbl.2013.0492.

Bound J. P. and Voulvoulis N. (2004). Pharmaceuticals in the aquatic environment – A comparison of risk assessment strategies. *Chemosphere*, **56**(11), 1143–1155. doi: 10.1016/j.chemosphere.2004.05.010.

Brown K. D., Kulis J., Thomson B., Chapman T. H. and Mawhinney D. B. (2006). Occurrence of antibiotics in hospital, residential, and dairy effluent, municipal wastewater, and the Rio Grande in New Mexico. *Science of the Total Environment*, **366**(2–3) 772–783. doi: 10.1016/j.scitotenv.2005.10.007.

Carraro E., Bonetta S., Bertino C., Lorenzi E., Bonetta S. and Gilli G. (2016). Hospital effluents management: Chemical, physical, microbiological risks and legislation in different countries. *Journal of Environmental Management*, **168**, 185–199, doi: 10.1016/j.jenvman.2015.11.021.

Chang H. S., Choo K. H., Lee B. and Choi S. J. (2009). The methods of identification, analysis and removal of endocrine disrupting compounds (EDCs) in water. *Journal of Hazardous Materials*, **172**(1), 1–12 doi: 10.1016/j.jhazmat.2009.06.135.

de Almeida C. A. A., Oliveira M. S., Mallmann C. A. and Martins A. F. (2015). Determination of the psychoactive drugs carbamazepine and diazepam in hospital effluent and identification of their metabolites. *Environmental Science and Pollution Research International*, **22**(21), 17192–17201. doi: 10.1007/s11356-015-4948-y.

Dimian A. C. and Bildea C. S. (2008). Bioethanol Manufacturing, *Wiley Online Library*, **28**(12), 429–460. https://doi.org/10.1002/9783527621583.ch15.

Emmanuel E., Perrodin Y., Keck G., Blanchard J. M. and Vermande P. (2002). Effects of hospital wastewater on aquatic ecosystem. International Conference XXVIII Congreso Interamericano de Ingeniería Sanitaria y Ambiental Cancún, México, 27–31st October, 2002.

Emmanuel E., Perrodin Y., Keck G., Blanchard J. M. and Vermande P. (2005). Ecotoxicological risk assessment of hospital wastewater: A proposed framework for raw effluents discharging into urban sewer network. *Journal of Hazardous Materials*, **117**(1), 1–11. doi: 10.1016/j.jhazmat.2004.08.032.

Escher B. I., Baumgartner R., Koller M., Treyer K., Lienert J. and Mcardell C. S. (2010). Environmental toxicology and risk assessment of pharmaceuticals from hospital wastewater. *Water Research*, **45**(1), 75–92. doi: 10.1016/j.watres.2010.08.019.

European Union (2009). https://eur-lex.europa.eu/LexUriServ/LexUriServ.do?uri=OJ:L:2013:226:0001:0017:EN:PDF ĐDIRECTIVESÝIRECTIVE 2013/39/ EU OF THE EUROPEAN PARLIAMENT AND OF THE COUNCILĐf 12 August 2013Ĺmending Directives 2000/60/EC and 2008/105/EC as regards priority substances in the field ofĐater policy.

Fent K., Weston A. A. and Caminada D. (2006). Ecotoxicology of human pharmaceuticals. *Aquatic Toxicology*, **76**(2), 122–159. doi: 10.1016/j.aquatox.2005.09.009.

Gautam A. K., Kumar S. and Sabumon P. C. (2007). Preliminary study of physico-chemical treatment options for hospital wastewater. *Journal of Environmental Management*, **83**(3) 298–306. doi: 10.1016/j.jenvman.2006.03.009.

Gómez M. J., Petrović M., Fernández-Alba A. R. and Barceló D. (2006). Determination of pharmaceuticals of various therapeutic classes by solid-phase extraction and liquid chromatography-tandem mass spectrometry analysis in hospital effluent wastewaters. *Journal of Chromatography*, **1114**(2), 224–233. doi: 10.1016/j.chroma.2006.02.038.

Isidori M., Lavorgna M., Russo C., Kundi M., Žegura B., Novak M., Filipič M., Mišík M., Knasmueller S., de Alda M. L., Barceló D., Žonja B., Česen M., Ščančar J., Kosjek T. and Heath E. (2016). Chemical and toxicological characterisation of anticancer drugs in hospital and municipal wastewaters from Slovenia and Spain. *Environmental Pollution*, **219**. doi: 10.1016/j.envpol.2016.10.039.

January I., Kusuma Z., Yanuwiadi B. and Laksmono R. W. (2012). Study of hospital wastewater characteristic in Malang City. *International Journal of Engineering and Science*, **2**(2) 13–16.

Khan A. H., Khan N. A., Ahmad S., Dhingra A., Singh C. P., Khan S. U., Mohammadi A. A., Changani F., Yousefi M., Shamshad A., Vambol S., Vambol V., Khursheed A. and Ali I. 2020. Application of advanced oxidation processes followed by different treatment technologies for hospital wastewater treatment. *Journal of Cleaner Production*, **269**, 122411. doi: 10.1016/j.jclepro.2020.122411.

Kosma C. I., Lambropoulou D. A. and Albanis T. A. (2010). Occurrence and removal of PPCPs in municipal and hospital wastewaters in Greece. *Journal of Hazardous Materials*, **179**(1–3) 804–817. doi: 10.1016/j.jhazmat.2010.03.075.

Kotwani A. and Holloway K. (2011). Trends in antibiotic use among outpatients in New Delhi, India. *BMC Infectious Diseases*, **11**, 1–9. doi: 10.1186/1471-2334-11-99.

Lakshmi Prasanna B., Padmini V. L., Navle K., Dometti H. S. and Moodu S. (2015). Analysis of drugs in aquatic environment. *Journal of Chemical and Pharmaceutical Research*, **7**(4) 71–79, [Online]. Available from: www.jocpr.com.

Langford K. H. and Thomas K. V. (2009). Determination of pharmaceutical compounds in hospital effluents and their contribution to wastewater treatment works. *Environment International*, **35**(5), 766–770. doi: 10.1016/j.envint.2009.02.007.

Le Corre K., Katouli M., Stratton H., Ort C. and Keller J. (2012). Hospital wastewater. *Urban Water Security Research Alliance Technical Report No*. **76**(76), 229.

Lee Y., Kovalova L., McArdell C. S. and von Gunten U. (2016). Prediction of micropollutant elimination during ozonation of a hospital wastewater effluent. *Water Research*, **64**, 134–148. doi: 10.1016/j.watres.2014.06.027.

Mesdaghinia A. R., Naddafi K., Nabizadeh R., Saeedi R. and Zamanzadeh M. (2015). wastewater characteristics and appropriate method for wastewater management in the hospitals. *Iranian Journal of Public Health*, **38**(1), 34–40.

Pinto P. and Garcı P. A. (2014). Ecotoxicity and environmental risk assessment of pharmaceuticals and personal care products in aquatic environments and wastewater treatment plants. *Ecotoxicology*, **23**, 1517–1533. doi: 10.1007/s10646-014-1293-8.

Rezaee A., Ansari M., Khavanin A., Sabzali A. and Aryan M. M. (2005). Hospital wastewater treatment using an integrated anaerobic aerobic fixed film bioreactor. *American Journal of Environmental Sciences*, **1**(4) 259–263.

Sciences M., Amouei A., Asgharnia H., Fallah H. and Faraji H. (2015). Characteristics of effluent wastewater in hospitals of Babol University of Iran. *Health Scope*, **4**(2), 4–7.

Tran N. H., Urase T. and Ta T. T. (2014). A preliminary study on the occurrence of pharmaceutically active compounds in hospital wastewater and surface water in Hanoi, Vietnam. *Clean – Soil, Air, Water*, **42**(3) 267–275. doi: 10.1002/clen.201300021.

Verlicchi P. (2018). Hospital Wastewaters. Springer Nature, Switzerland.

Verlicchi P., Galletti A., Petrovic M. and Barceló D. (2010). Hospital effluents as a source of emerging pollutants: An overview of micropollutants and sustainable treatment options. *Journal of Hydrology*, **389**(3–4), 416–428. doi: 10.1016/j.jhydrol.2010.06.005.

Verlicchi P., Al Aukidy M., Galletti A., Petrovic M. and Barceló D. (2012). Hospital effluent: Investigation of the concentrations and distribution of pharmaceuticals and environmental risk assessment. *Science of the Total Enviroment*, **430**, 109–118. doi: 10.1016/j.scitotenv.2012.04.055.

Chapter 6

Legislative and criminal law aspects of water protection and prevention of pharmaceuticals accumulation in nature

Viola Vambol[1], Andrii Shulha[2], Sergij Vambol[3], Parveen Sihag[4], Vitalii Pavlykivskyi[5], Niloofar Mozaffari[6], Nastaran Mozaffari[7], Nadeem A. Khan[8] and Barbara Sawicka[9]

[1]*Educational and Scientific Department of Occupational Safety and Health, National Scientific and Research Institute of Industrial Safety and Occupational Safety and Health, Vavilovykh str., 13, Kyyiv, 04060, Ukraine*
[2]*Kharkiv National University of Internal Affairs, L. Landau Avenue, 27, Kharkiv, 61080, Ukraine*
[3]*Life Safety and Law Department, Kharkiv Petro Vasylenko National Technical University of Agriculture, Alchevskyh str., 44, Kharkiv, 61002, Ukraine*
[4]*Department of Civil Engineering, Shoolini University, Solan, Himachal Pradesh, India*
[5]*Kharkov University, pr. Gagarina, 187, 61080, Kharkiv, Ukraine*
[6]*Department of Physics, Faculty of Sciences, Science and Research Branch, Islamic Azad University, Tehran, Iran*
[7]*Department of Environmental Engineering, Faculty of Natural Resources and Environment, Science and Research Branch, Islamic Azad University, Tehran, Iran*
[8]*Department of Civil Engineering, Jamia Millia Islamia, New Delhi*
[9]*University of Life Sciences in Lublin, Lublin, Poland*

 Pharmaceutical Wastewater Treatment Technologies: Concepts and Implementation Strategies
Editor(s): Nadeem A Khan, Sirajuddin Ahmed, Viola Vambol and Sergij Vambol
doi: 10.2166/9781789061338_0113

ABSTRACT

The rational use of water resources, the harmonization of the relationship between man and nature, and environmental protection are significant issues because the future of all mankind depends on their solution. Accordingly, the problem of proper protection of this natural site requires a comprehensive solution. One of the directions should be the provision of legal mechanisms to counteract illegal encroachment on water resources. Criminal legal aspects have a special place in relation to causing harm to water bodies. This chapter analyses the rules for protecting surface waters from pollution by wastewater, wastewater quality standards, as well as the dynamics of crimes against water resources, using Ukraine as an example. Also, the world experience in the field of the criminal legal protection of water bodies and the prevention of the accumulation of active pharmaceutical ingredients is presented in detail.

Keywords: Dynamics of crimes against water resources, Legislative and criminal law mechanisms, Pharmaceutical ingredients, Polluted water, Water protection.

Contents

6.1 Introduction 114
6.2 Regulations for the Conservation of Surface of Waters from Contamination by Waste-Water. Quality Standards of Ukraine 118
6.3 Crimes Against Water Resources in Ukraine 120
- 6.3.1 Statistical data 121
- 6.3.2 Dynamics indicators 124
- 6.3.3 The series average characteristics calculation 127

6.4 World Experience in the Criminal Law Protection of Water Bodies and in the Prevention of Active Pharmaceutical Ingredients 130
6.5 Conclusions 137
References 139

6.1 INTRODUCTION

The water resources of any state are one of the most significant biosphere components (Getirana *et al.*, 2017; Mazhar *et al.*, 2020; Vadez *et al.*, 2013). Water, as the principle of living in the world, is one of the substantial natural resources, subject to special protection by the state.

If we talk about water resources in a general sense, then water resources cover all the waters of our planet that are in a free, chemically unbound state. In a broad sense, they consist of the ocean waters and surface of the waters (rivers, ponds and reservoirs) and underground water, as well as waters concentrated in glaciers, marshy and waterlogged soils, and the atmosphere. In a narrow sense, the concept of water resources includes surface and groundwater of a specific territory and water area that is suitable for use in agricultural and industrial production and to satisfy the household needs of the public.

Of the total surface of the planet, whose area is 510 million km^2, water occupies 70.8%. However, more than 98% of all the planet's water resources are water with increased salinity, which is not cost-effective for utilising. The share of suitable freshwater for household consumption is only 4 200 000 km^2 (0.3% of the total hydrosphere); so, most of the global communities experience a lack of water or its unsatisfactory quality.

International experts of the World Health Organization (WHO) found that more than 60% of the world's diseases are caused by the use of low-quality water; So today, water is regarded not only as a natural resource, it has a pronounced social significance. In support of this, the Ministerial Declaration of the World Water Forum in The Hague in 2000, by the International Conference on Fresh Water (Bonn, December 2001), water quality is recognised as the main criterion of the balanced development of society, its safety and existence in general.

Therefore, the rational use of natural resources (in our case, water), harmonisation of the relationship between man and nature, and environmental protection are the most pressing problems, and the future of mankind depends on their solution. Accordingly, the proper protection of this natural site requires a comprehensive solution. One of the directions should be the provision of legal mechanisms to counteract illegal encroachment on water resources. A special place is taken through criminal law's influence on persons guilty of harm to these environmental objects.

The main regulatory document establishing the Community's framework for water policy in the EU is Directive 2000/60/EC as amended by Decision No. 2455/2001/EC and Directive 2008/32/EC of the European Parliament and Council of 11.03.2008 (Directive 2000/60/EC). The introduction of this document was intended to improve the surface water quality of the community, as well as protecting groundwater from contamination generated by hazardous materials as part of the general strategy of fresh-water conservation and monitoring. Paragraph (22) of the Directive states that freshwater protection policies should promote the gradual decrease of the release of perilous materials into the water. Paragraph (27) indicates the latest goal of this Directive, namely the final removal of perilous materials, the fight against pollution which requires priority measures, and the promotion of the fact that in the marine environment, substances naturally have concentrations near to the value of background concentrations.

In paragraph 30 of this Directive, the term 'Substances whose pollution control requires priority measures' means substances defined in accordance with Article 16 (2). The committee should suggest a plan containing a list of materials whose pollution control requires priority measures that have been selected among those substances that pose a notable hazard to the aquatic ecosystem or through it to other environmental objects. Substances are distributed according to their action based on risk for the aquatic environment or through it for other environmental objects, which is determined by: (A) an evaluation of risk accomplished in accordance with Council Regulation (EEC) N 793/93(1), Council Directive 91/414/EEC(2), and Directive 98/8/EC of the European Parliament and the Council (3), (B) a targeted risk-based evaluation (under the strategy of Regulation (EEC) N 793/93), which focuses only on aquatic eco-toxicity and toxicity to humans through the marine ecosystem. Among these substances, TV is "hazardous substances whose pollution control requires priority measures", which is defined under Article 16(3) and (6), and measures should be taken in agreement with Article 16(1) and (8) (Directive 2000/60/EC).

Annex X of the Directive contains a list of substances for which pollution control requires priority measures. According to the latest version of the publication and amendments to the Directive dated January 31, 2012, it has 48 priority pollutants for surface and ground waters. This list is perpetually expanded and supplemented by new hazardous substances.

Pharmaceutical substances are classified as 'hazardous substances that their pollution control requires priority measures', since the degree of their effect on the water depends on the physicochemical properties of these pharmaceutical substances.

Considering the existing problem, the European Commission on the Water Framework Directive has included three pharmaceutical substances in the list of preference materials, the ejection of which toward the environment is controlled (Dhingra *et al.*, 2020; Directive 2000/60/EC; Falås *et al.*, 2012): 17 beta-estradiol, diclofenac, 17 alpha-ethinyl estradiol.

According to the current legislation of Ukraine, the general necessities for the compound and quality of water in reservoirs of various categories are determined by two documents: sanitarian laws and principles for the conservation of surface of the waters from contamination SanPiN 4630–88 (SanPiN 4630–88) and legislations for the conservation of surface waters, 1991 (USSR State Committee for Nature Protection). Appendix 2 of the specified SanPiN (SanPiN 4630–88) contains a list of sanitary norms for the maximum allowable density of harmful matters in the drinking, cultural, and domestic water use (MPC – maximum permissible concentrations, and OSU – approximate permissible levels), which consists of 1345 substances. This list needs to be updated according to the appendices of the Water Framework Directive 2000/60/EC. Also, the terminology of the Water Code of Ukraine must be brought into line with the provisions of the guidance. The sections 'Surface waters' and 'Groundwater' of the State Water Cadastre of Ukraine should be harmonised with the legislative acts of the European Union.

Regarding the rules for the acceptance of wastewater of enterprises into urban and departmental sewer systems of settlements of Ukraine, confirmed by request of the State Committee for Construction, Architecture and Housing strategy of Ukraine on February 19, 2002 No. 37 in paragraph 4.3, it is strictly forbidden to discharge into the city sewer network: wastewater containing radioactive, toxic substances, salts of heavy metals and bacterial contaminants, including sewage of infectious medical institutions and departments; these wastewaters before being discharged into the sewers of a settlement should be neutralised at local treatment facilities with the obligatory disposal or burial of formed sediments (State Budget of Ukraine). However, if the wastewater is not sufficiently treated, then the sludge burial from these wastewater at landfills will result in leaching of pharmaceutical compounds, heavy metals and other toxic substances and its infiltration into groundwater (Vambol *et al.*, 2018, 2019; Ziarati *et al.*, 2020). That is, the wastewater of hospitals and other medical institutions that contain dangerous bacterial contaminants need additional local treatment before dumping them into the sewer system. Given this fact, the development of additional joint methods for the purification and disposal of wastewater at the local level is relevant. Appendix 1 of these rules provides requirements for the composition and properties of wastewater enterprises for their safe drainage network. This appendix gives more general indicators of wastewater quality control and permissible values such as temperature, pH, sulphates, sulphides, chlorides, etc. Appendix 2 shows the allowable values of wastewater and water quality indicators. This list contains 163 substances of both organic and inorganic origin, which indicate the maximum allowable densities of hazardous materials in the water and pollution in wastewater entering the treatment plant for biological treatment. Appendix 3 of the allowable heavy metal content in sewage sludge and the removal efficiency of heavy metals in urban wastewater treatment plants lists 11 types of them, including metals such as strontium, lead, and mercury.

Directive 91/271/EEC on urban wastewater remedy (supplemented by Directive 98/15/EC and Regulation (EC) 1882/2003 and 1137/2008) is the European Union regarding urban wastewater treatment. Appendix 1 in Table 1 shows the requirements for wastewater discharged from urban sewage treatment facilities (biochemical oxygen utilization (BOD), chemical oxygen request (COD)), which are regulated by Articles 4 and 5 of this Directive. Table 2 shows the justification for the need for wastewater evacuations from treatment facilities into vulnerable regions, including permissible concentrations of total phosphorus and nitrogen (Council Directive 91/271/EEC).

According to this document, the State should amend the rules for wastewater reception and provide all cities and towns with sewage systems; create a list of vulnerable areas in agreement with the criteria established in Appendix II to the Directive and introduce the principle of additional wastewater treatment before evacuation in such areas; the establishment of a ban on the discharge of sediment into surface water from ships, pipelines or in any other way; and determine the procedure for handling industrial wastewater from food industry enterprises. Today in Ukraine there are 16 cities and 375 urban-type settlements, that is, 97.5% of rural settlements are not equipped with centralised sewage systems, and in 187 urban settlements, sewage treatment plants are inefficient – more than 154 000 m^3 of untreated wastewater and insufficiently treated wastewater which are discharged daily into water bodies. Also, there are practically no treatment facilities in Ukraine with a completed technological process for processing and utilizing sludge that would meet the requirements of the Directive.

Also, the main regulatory legal acts regulating legal relations similar to those regulated in Directive 91/271/EEC are:

- Criminal Code of Ukraine (Article 242 'Violation of the rules for the protection of waters', Article 243 'Pollution of the seawater', Article 244 'Violation of the law on the continental shelf of Ukraine') (Criminal Code of Ukraine of 13.08.2020 2001);
- The Law of Ukraine 'On Housing and Communal Services' (Law of Ukraine 2018);
- The Law of Ukraine 'On Drinking Water and Drinking Water Supply' (Law of Ukraine 2002);
- The Law of Ukraine 'On Environmental Protection' (Law of Ukraine 1991);
- Decree of the Cabinet of Ministers of Ukraine dated July 21, 05 No. 630 'On approval of the Rules for the provision of centralized heating, cold and hot water supply and drainage and a standard contract for the provision of centralised heating, cold and hot water supply and drainage' (Cabinet of Ministers of Ukraine);
- Rules for the use of centralised universal water supply and improvement of systems in the settlements of Ukraine dated June 27, 2008 No. 190 (Ministry of Housing and Communal Services of Ukraine);
- DBN V.2.5–75: 2013 'Sewerage. External networks and facilities. The main provisions of the design' (Ministry of Regional Development, Construction and Housing of Ukraine).

To date, in Ukraine, the problem of pollution of natural waters with pharmaceutical compounds has not been considered. Accordingly, in the country, there are no regulatory documents that would regulate the discharge of these substances into water bodies. According to the Proposal for the Basic Plan for Adaptation of National Ecological Legislation with the Laws of the European Union, it requires an immediate solution to the issue of developing and implementing scientifically sound standards for the maximum allowable discharge of pharmaceutical compounds into surface waters.

6.2 REGULATIONS FOR THE CONSERVATION OF SURFACE OF WATERS FROM CONTAMINATION BY WASTE-WATER. QUALITY STANDARDS OF UKRAINE

The rules for wastewater intake into wastewater systems, according to Ukrainian standards, stipulate that wastewater that is taken to wastewater systems should not:

(1) Contain explosive impurities and soluble gaseous matters which are able to form combustible compounds;

(2) Contain substances that are capable of cluttering pipes, wells, grids or deposited on their surfaces (debris, soil, abrasive powders, and other coarse suspensions, gypsum, lime, sand, metal and plastic chips, fats, resins, fuel oil, beer grains, bread yeast, etc.);
(3) Contain only inorganic substances that are not amenable to biological degradation;
(4) Contain substances for which maximum permissible concentrations (MPC) are not established for water in reservoirs, or toxic substances that impede the biological treatment of wastewater as well as substances for the determination of which analytical control methods have not been developed;
(5) Contain dangerous bacterial, viral, toxic, and radioactive contaminants;
(6) Contain biologically rigid synthetic surfactants, the level of primary biological decomposition of which is less than 80%;
(7) Have a temperature above 40°C;
(8) Have a pH below 6.5 or above 9.0;
(9) Have a chemical oxygen consumption (COD) higher than the biochemical oxygen consumption over 5 days (hereinafter BOD_5) more than 2.5 times;
(10) Have a BSC that exceeds that specified in the draft WWTP of the relevant settlement;
(11) Create conditions for causing harm to the health of personnel servicing wastewater systems;
(12) To exclude the disposal of sewage sludge using methods that are safe for the environment;
(13) Contain pollutants above permissible concentrations established by these rules.

It is implied that consumers are legal entities (regardless of ownership and departmental affiliation) and individuals (entrepreneurs, those who carry out independent professional activities) who dump wastewater into wastewater systems or directly into sewage treatment plants.

Local treatment facilities of the consumer must meet the requirements of technical conditions. It is forbidden to discharge into the system of centralised drainage without prior disposal and decontamination at local treatment plants with obligatory disposal or disposal of formed sewage containing pollutants, which are listed (Ministry of Regional Development, Construction and Housing of Ukraine). Among the prohibited substances are: substances that can form explosive, toxic and (or) combustible gases, organic solvents, combustible and explosive substances (oil, gasoline, kerosene, acetone, etc.) in the sewage system; solutions of acids with pH <5.0 and alkalis with pH >10.0; bad smelling and other volatile substances; radioactive substances; substances that have increased toxicity, the ability to accumulate in the human body, including mono- and polycyclic organochlorine, organophosphorus, organo-nitrogen and organosulphur substances, biologically hard surfactants, toxic chemicals, potent toxic substances, medical waste, epidemiologically dangerous bacterial and viral contaminants; galvanic solutions (electrolytes) both output and spent, precipitation (sludge) of local treatment facilities, sedimentation of sedimentation tanks, chemical reagents and reagents; any solid waste from slaughterhouses and meat processing, whole blood, waste from leather processing, waste from animal husbandry and poultry farming, including faeces; municipal solid waste and trash; any fibrous materials; biomass of food, pharmaceutical industries and other biotechnological processes.

If the quantitative and qualitative indicators of the consumer's wastewater vary significantly during the day, and the concentration of pollutants exceeds the allowable concentration, the consumer should install special averaging tanks and devices that ensure uniform discharge of wastewater throughout the day. When the consumer cannot ensure compliance with the requirements, then he must contact the manufacturer with a statement and justification for the acceptance of wastewater contaminated with excess and indicate the concentration of pollution.

6.3 CRIMES AGAINST WATER RESOURCES IN UKRAINE

As stated in Article 174 of the Treaty establishing the European Community (No. 994_017 dated 01.01.2005) (European Community 2005), the Community's environmental policy has a contribution to the implementation of the tasks of conserving and amending the environmental quality, in the careful and logical use of natural resources. This policy is based on the principle of prudence, on the principle of eliminating harm to the environment, and on the principle that the polluter must pay.

The classic approach to solving environmental problems is, as a rule, to implement the achievements of environmental engineering. However, environmental engineering is not the only tool for effectively solving environmental problems. Non-standard tools for ensuring environmental safety should include a legal approach, such as the prevention of criminal acts (misuse, spoilage, pollution) against natural resources (through the adoption of laws on the conservation of natural resources and the establishment of a ban on criminal attacks on natural resources).

In this study, we focus on the relationship between crime (as a socially negative phenomenon) and environmental factors (as one of the factors that determine the level of crime).

The environmental factor of crime means that criminal manifestations are possible in the field of environmental relations, in particular in the field of conservation, logical use, and reproduction of natural resources, for example, the deterioration of the quality of individual elements of the environment, their destruction or damage, illegal seizure, and the like.

6.3.1 Statistical data

The current Ukrainian criminal law provides for liability for violation of water protection rules (Article 242), sea pollution (Article 243), and violation of the legislation on the continental shelf of Ukraine (Article 244). In turn, administrative responsibility for violations in the matter of water consumption is characterised by specific features that allow distinguishing this type of responsibility from other types, and are as follows. The basis of liability is the commission of an administrative offence in the water consumption. A special object of encroachment is the legal regime of water resources of Ukraine as a common object, the legal regime of a water body, water users' rights, and other persons, etc. as a specific object. According to Article 1 of the Water Code of Ukraine, water resources are the volumes of surface, underground, and sea waters of the corresponding territory (Water Code of Ukraine 1995). Administrative liability arises for such types of violations (according to the Code of Ukraine on Administrative Offences (Code of Ukraine on Administrative Offenses 1984): violation of the state ownership of water (Article 48), violation of the rules for the conservation of water resources (Article 59); violation of the requirements for the preservation of territorial and inland sea waters from contamination and clogging (Article 59-1); violation of the rules for water use (Article 60); damage to water facilities and devices, violation of the rules for their operation (Article 61); failure to comply with the obligations to register operations with harmful substances and mixtures in ship documents (Article 62).

According to Article 1 of the Water Code of Ukraine dated June 6, 1995 (No. 213/95-BP), with amendments and additions, water resources are the volumes of surface, ground, and sea waters of the corresponding territory.

It should be noted that Ukraine belongs to countries with insufficient water resources. It has the least water supply in Europe. The water bodies of Ukraine (rivers, lakes, reservoirs, ponds, canals) cover 24 200 km^2, which is 4.0% of its total territory (603 700 km^2). The territory of Ukraine does not have a very dense river network. Marshes, which were a natural regulator of river water content, are now half-drained. So, the water natural resources of Ukraine are, first of all, local and transit river flows, water reserves of lakes, artificial reservoirs, and underground horizons. The water supply in Ukraine is presented in Table 6.1 (Shapovalova (2007).

According to law enforcement agencies and the court, the percentage of established crimes against water bodies in Ukraine relative to other crimes against the environment is quite small. This is, first of all, because water bodies in the country are perceived as a natural object. Most accounted for are the crimes provided for in article 242 'Violation of the rules for the protection of waters' of the Criminal codex of Ukraine (Figure 6.1). It should be noted that the distribution of recorded crimes of this group and convictions are significantly distinguished, in that there are virtually no convictions (Table 6.2). Analyzing the statistics on criminal attacks on water resources, we note no observable steady

Table 6.1 Ukraine' surface water resources (Shapovalova 2007).

River Basin	Total Water Resources, km^3	
	P-50%	P-95%
Dnepr River	53.4	32.2
Dniester River	10.9	5.9
Prut River	1.3	0.6
Tisa River	4.2	2.1
Western Bug River	1.8	0.8
Rivers of the Black Sea	0.2	—

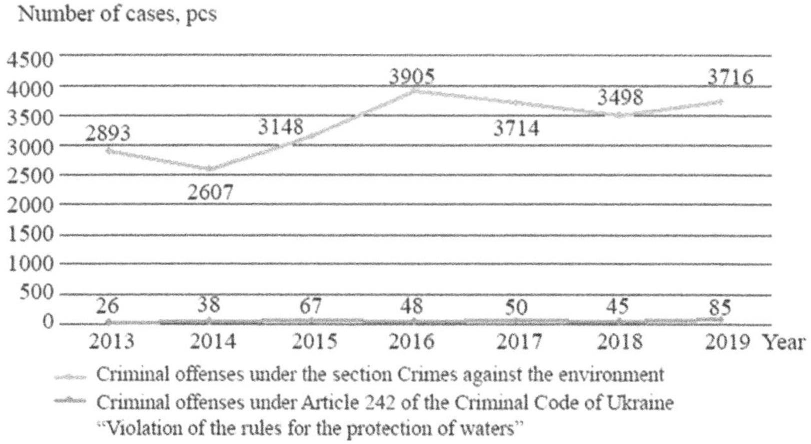

Figure 6.1 Dynamics of criminal offences against the environment and water resources which is compiled on the basis of the data of the National Police of Ukraine.

increase in such crimes. During the last seven years, for example, enormous violations of water protection rules were recorded in 2019 (85 cases) and marine pollution in 2015 (eight cases) (Table 6.2).

Thus, we can draw some conclusions. First, the bulk of crimes against water resources are crimes related to violation of water protection rules. Second, there is a discrepancy between the recorded criminal offences in the field of water protection and convictions (Figures 6.2 and 6.3), that is, there are significantly fewer accusatory sentences. Third, there is tendency to enhance the number of criminal offences in regard to preservation of the waters. This means that these objects of nature do need to be protected by legal measures. In any case, legal protection in general and criminal law in particular is a weak protection from committing criminal offences against the water resources of our state by potential offenders.

Table 6.2 Statistics of law enforcement agencies on crimes against water resources.

Offences Name	Year						
	2013	**2014**	**2015**	**2016**	**2017**	**2018**	**2019**
Criminal offences under the section VIII 'Crimes against the environment' of Criminal codex of Ukraine	2893	2607	3148	3905	3714	3498	3716
Article 242 of the Criminal Code of the of Ukraine 'Violation of the rules for the protection of waters'							
Criminal offences under Article 242 of the Criminal Code of the of Ukraine 'Violation of the rules for the protection of waters'	26	38	67	48	50	45	85
Criminal offences for which a bill of indictment has been sent to court	3	1	0	0	0	0	1
Criminal offences in which proceedings are closed	61	28	37	29	32	27	38
Article 243 of the Criminal Code of the of Ukraine 'Pollution of the sea'							
Criminal offences under Article 243 of the Criminal Code of Ukraine 'Pollution of the sea'	5	1	8	5	6	6	1
Criminal offences for which a bill of indictment has been sent to court	0	0	0	0	0	0	0
Criminal offences in which proceedings are closed	5	0	0	0	3	3	0
Incurred responsibility							
Number of persons suspected of environmental crimes	1215	1007	823	671	890	894	713
Number of persons suspected under Article 242 of the Criminal Code of Ukraine	3	1	0	0	0	0	1

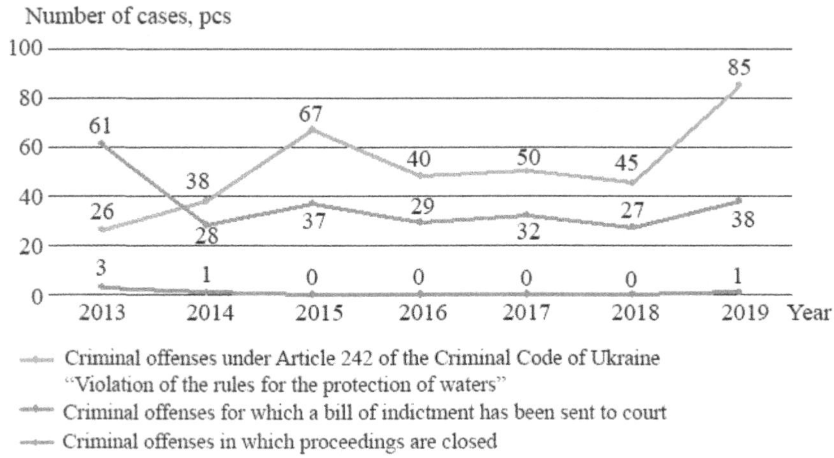

Figure 6.2 Results of consideration of criminal offences against water resources in Ukraine which is compiled on the basis of the data of the National Police of Ukraine.

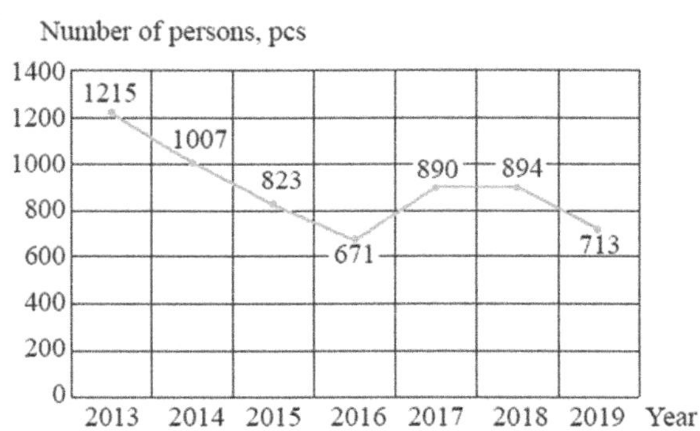

Figure 6.3 Number of persons suspected of environmental crimes.

6.3.2 Dynamics indicators

To calculate the dynamics indicators constantly, each level of the series is compared with the same base level. Such indicators are basic. To calculate the dynamics indicators on a variable base, each subsequent level of the series is compared with the previous one. Such dynamics indicators are chain. The most important statistical indicators of dynamics are:

(1) Absolute increase (it is determined in the different comparison of two levels of some dynamics in units of measurement of the initial information):

$$\Delta y_{ci} = y_i - y_{i-1} \tag{6.1}$$

$$\Delta y_{bi} = y_i - y_1 \tag{6.2}$$

where Δy_{ci} = chain increase; Δy_{bi} = basic increase.

(2) Absolute growth pace (it characterises the absolute increase in relative values (%) and shows how many percent the compared level has changed with the level taken as the basis of comparison)

$$T_{\text{Incr}_{ci}} = \Delta y_{ci}/y_{i-1} \tag{6.3}$$

$$T_{\text{Incr}_{bi}} = \Delta y_{bi}/y_1 \tag{6.4}$$

where $T_{\text{Incr}_{ci}}$ = chain; $T_{\text{Incr}_{bi}}$ = basic.

(3) Growth pace (it characterises the ratio of two levels of the series and can be expressed as a coefficient or as a percentage):

$$T_{\text{Gro}_{ci}} = y_{ci}/y_{i-1} \tag{6.5}$$

$$T_{\text{Gro}_{bi}} = y_{bi}/y_1 \tag{6.6}$$

where $T_{\text{Gro}_{ci}}$ = chain; $T_{\text{Gro}_{bi}}$ = basic.

(4) Absolute value of 1% increase

1% increase = $y_{i-1}/100\%$ – chain;
1% increase = $y_b/100\%$ – basic.

The calculation outcomes are demonstrated in Table 6.3.

Statistics for the entire study period indicate that in 2019 in comparison to 2013, the rate of criminal offences in agreement with section VIII "Crimes against the environment" of Criminal codex of Ukraine increased by 823 units (28.5%), and compared to the previous year by 218 units (6.2%). The maximum increase in criminal offences under Section VIII 'Crimes against the environment' of the Criminal codex of Ukraine is observed in 2016 and amounts to 757 units, and the minimum increase in 2014 amounts to 286 units. It should be emphasised that in 2019, the number of persons who received a notice of suspicion of crimes against the environment decreased by 502 persons (41.3%) and 181 persons (20.3%) compared to 2013 and 2018, respectively.

Meanwhile, there is an increment in the rate of criminal offences under Article 242 'Violation of the rules for the protection of water': by 59 units (226.9%) and 40 units (88.9%) in 2019 compared with 2013 and 2018, respectively. At that time, the number of criminal proceedings under Article 242 was decreased in 2019 compared to 2013, which were closed to 23 units (37.7%), and increased up to 11 units (40.7%) compared to 2018.

The growth pace shows an increasing trend for both the opening (account) of criminal offences and for closing criminal proceedings for various reasons under Section VIII 'Crimes against the environment', Article 242. This trend indicates an acceleration in the increase in the number of criminal proceedings.

Table 6.3 Offences dynamics' statistical indicators against water resources.

Period	Criminal Cases	Indicators of the Dynamics			
		Absolute Increase	**Absolute Growth Pace**	**Growth Pace**	**Absolute Value of 1% Increase**
Criminal offences under the section VIII 'Crimes against the environment' of Criminal codex of Ukraine					
2013	2893	—	—	100	—
2014	2607	–286	–9.89	90.11	28.93
2015	3148	541	20.75	120.75	26.07
2016	3905	757	24.05	124.05	31.48
2017	3714	–191	–4.89	95.11	39.05
2018	3498	–216	–5.82	94.18	37.14
2019	3716	218	6.23	106.23	34.98
Criminal offences in agreement with Article 242 of the Criminal Code of the of Ukraine 'Violation of the rules for the protection of waters'					
2013	26	—	—	100	—
2014	38	12	46.15	146.15	0.26
2015	67	29	76.32	176.32	0.38
2016	48	–19	–28.36	71.64	0.67
2017	50	2	4.17	104.17	0.48
2018	45	–5	–10	90	0.5
2019	85	40	88.89	188.89	0.45
Criminal offences in which proceedings are closed					
2013	61	—	—	100	—
2014	28	–33	–54.1	45.9	0.61
2015	37	9	32.14	132.14	0.28
2016	29	–8	–21.62	78.38	0.37
2017	32	3	10.34	110.34	0.29
2018	27	–5	–15.63	84.38	0.32
2019	38	11	40.74	140.74	0.27
Number of persons who received a notice of suspicion					
2013	1215	—	—	100	—
2014	1007	–208	–17.12	82.88	12.15
2015	823	–184	–18.27	81.73	10.07
2016	671	–152	–18.47	81.53	8.23
2017	890	219	32.64	132.64	6.71
2018	894	4	0.45	100.45	8.9
2019	713	–181	–20.25	79.75	8.94

6.3.3 The series average characteristics calculation

The average growth pace and average absolute growth pace were determined respectively by the formulas:

$$\bar{T}_{\text{Gro}} = \sqrt[n-1]{\frac{y_n}{y_1}} \tag{6.7}$$

$$\bar{T}_{\text{Incr}} = \bar{T}_{\text{Gro}} - 1 \tag{6.8}$$

The average absolute increase is a generalised characteristic of the individual absolute increase of a dynamics' series. The average absolute increase was determined by the formula:

$$\bar{d}_y = \frac{y_n - y_1}{n - 1} \tag{6.9}$$

The calculations result show that on average with each period (i.e. annually):

(1) The number of criminal offences under section VIII 'Crimes against the environment' increased by 137 units (4.3%);
(2) The number of criminal offences under Article 242 'Violation of the rules for the protection of waters' increased by 9.83 units (21.8%);
(3) The number of criminal offences in which proceedings are closed declined by 3.83 units (7.6%);
(4) The number of persons who received a notice of suspicion declined by 83.67 persons (8.5%).

The forecast for the next three years was made using the absolute increase indicator and is presented in Table 6.4.

While maintaining these trends, a negative forecast can be seen in the development of the situation connected with the criminal legal defence of the environment generally, and water protection particularly. As the forecast shows,

Table 6.4 Forecast of the criminal offences dynamics for the next three years.

Year	Criminal Offences Under the Section VIII 'Crimes Against the Environment'	Criminal Offences Under Article 242 'Violation of the Rules for the Protection of Waters'	Criminal offences in which Proceedings are Closed	Number of Persons who Received a Notice of Suspicion
2020	3853	95	34	629
2021	3990	105	30	546
2022	4128	115	27	462

the number of recorded crimes against the environment will grow (by at least 4.3%), and against water resources by almost a quarter a year (21.8%). It should be noted that this forecast is shown only for recorded offences. However, criminological studies show that latent crime (hidden, unregistered) is usually 2–3 times higher. This means that in general, the forecast in Ukraine confirms the dangerous situation in the field of environmental protection, namely its most dangerous manifestations associated with criminal offences. At the same time, the number of dismissed cases due to various circumstances and persons suspected will steadily decrease (by 7.6 and 8.5% annually, respectively). The aforesaid speaks about the ever-smaller influence of the law on environmental protection. The forecast shows that the number of registered violations is steadily increasing, and at the same time, fewer people are being prosecuted for their violation. Also, many criminal proceedings were closed on rehabilitating and non-rehabilitating grounds concerning persons who committed these offences. Consequently, a decrease in the number of persons who received a notice of suspicion indicates a lack of work in searching for and identifying the perpetrators of crimes of this category. As a rule, many do not give a consideration to the catastrophic consequences of pollution of environmental resources, especially water.

A similar situation with environmental violations was observed in the Republic of Moldova (Belecciu 2017). Analysis of the facts about water pollution allows us to conclude that law enforcement agencies are not in haste or do not want to initiate criminal proceedings on water pollution, based on existing criminal law. They do not work well enough to prevent and detect violations against water resources. Exemption from criminal liability in many cases is a consequence of the fact that it is difficult to demonstrate a psychological attitude (subjective aspect) to a crime against the environment. The subjective aspect is one of four key elements in shaping the offence corpus related to water pollution; its precise definition will be important for qualifying the crime. The guilt of a crime specified in Article 229 of the Criminal Code of the Republic of Moldova (Barbăneagră 2003) is expressed in both forms provided for by criminal law, both intentionally and recklessly, but this is not explicitly stated in the content of the article (Belecciu 2017). It is essential to mention that for the offence referred to Article 229 of the Criminal Code of the Republic of Moldova, the motive and purpose are not set forth as a mandatory sign of the subjective aspect. Therefore, it is often believed that the violation of water pollution was committed without special motives and goals. Therefore, avoiding the determination of motives and goals in the case of an offence related to water pollution does not affect the process of its qualification or individualisation of the applicable punishment (Belecciu 2017).

Besides, emphasis should be placed on a very important point in offences against water bodies, namely legal liability. Hence, the given statistics do not take into account the accompanying actions of interested parties that are not criminally punishable. An example of this is the shredding Lake Svityaz (Ukraine) for tens of meters due to a lack of rain. This deepest lake in Ukraine has become catastrophically shallow (Mel'nychok 2013). The water receded from the shore tens of meters. In the past few weeks, the lake has been declining. Environmentalists attribute this to the lack of rainfall, and this, in turn, is a direct consequence of global warming, which develops as a result of excessive human economic activity.

Another example of the indirect criminal influence on the country's water-bodies is the abuse of official position in the interests of third parties. This ultimately leads to deterioration in the condition of a particular water body. So, recently there has been a sharp drop in the water level in Shatsky Lakes, which are located in the Volyn region of Ukraine (Yanko *et al.*, 2014). This was the result of the fact that officials of the state environmental inspection, contrary to the requirements of the Law 'On Environmental Impact Assessment', assess the influence on the state of the lakes that were made by the illegal business entities operations. One of the enterprises built artificial water bodies near the indicated lakes on a leased area of 230 hectares without the relevant documents and carries out water intake for irrigation of blueberry seedlings on an area of over 150 hectares. This provides severe consequences for the ecological system and causes economic damage to this region of Ukraine.

This situation has arisen due to the lack of strict guidelines, an insufficiently effective judicial and executive system, as well as a lack of rigor in the laws (there is reason to believe that the law does not always indicate the specific manifestations of criminal behaviour aimed at worsening the state water bodies in crime). In this regard, there is a need to develop strict standards and laws for the administration of aquatic resources generally and wastewater particularly. This can be justified by the fact that uncontrolled discharge of untreated sewage and the waste ejection in water bodies are the most significant crimes against water resources and the environment. Moreover, much attention in standards and laws should be given to hospital wastewater and other wastewater containing active pharmaceutical ingredients, since the content of these harmful compounds in high concentrations contributes to the development of resistant genes in the environment. Standards and laws must necessarily include subsections that oblige the most effective treatment methods to be used for wastewater containing pharmaceutical compounds. In addition, laws should specify the consumption of cross border water channels and international springs, because at present the concept of 'water resource' stands out as a separate object of water relations and has an important economic and environmental significance. Thus, not strict regulations, an imperfect legislative and judicial system of one administrative unit (country, region, state) can contribute to the development of a water environmental crisis in a neighbouring administrative unit. One example is the situation in the United States (Franz 2011), where Under the Federal Water Pollution Control Act 1977, downstream regions are pre-empted from challenging upstream contamination source state permit issuance, Milwaukee v. Illinois (Milwaukee v. Illinois); and International Paper v. Ouellete (International Paper v. Ouellete). It is unclear whether the state can bring criminal charges against polluters upstream from the state, but at present, no state considers this treatment effective (Franz 2011). International agreements can be multilateral (e.g., the 1994 Convention on Cooperation for the Protection and Sustainable Development of the Danube River) or bilateral (e.g., the 1909 Border Waters Agreement between the United States and Canada). Relating to the violation of both conditions or any other conventions, a certain type of legal liability arises, including criminal. The Rules for the Water Consumption of International Rivers, signed in Helsinki in 1966, defined the use of water from the international drainage basin. Particular importance was given to the term 'water pollution', which alludes to any detrimental variation in the natural composition, content, and quality of the waters in the international watershed as a result of human behaviour.

It should be noted that criminal liability issues usually arise for pollution of oil in the sea and navigable waters (Donner and Johansson 2014; Greife and Stretesky 2013; Kopela 2011), while the active pharmaceutical ingredients' penetration contained in wastewater does no less harm, and in some cases (at high concentrations or very high toxicity of these ingredients) even more damage.

6.4 WORLD EXPERIENCE IN THE CRIMINAL LAW PROTECTION OF WATER BODIES AND IN THE PREVENTION OF ACTIVE PHARMACEUTICAL INGREDIENTS

In September 2002, at the World Summit on Sustainable Development in Johannesburg, the European Union (EU) officially announced the launch of the Global Water Initiative: 'Water for life – health, well-being, economic development, and security'. The Commission of the European Union entirely approved this innovation.

Innovation' political support is underpinned by commitments to achieve main objectives. The EU confirmed its obligation to contribute to the preparation of unified plans to administrate the water resources, and plans for the effective use of water supplies in all countries. In addition, the EU reaffirmed its commitment to help halve the ratio of public who do not have access to fresh water by 2015.

The Water Initiative involves the preparation and assistance in the establishment of legislation related to water resources (including the strengthening the system of water consumption licenses as well as permits for the discharge of sewage), and support of international agreements connected with administration of water supplies. Meanwhile, in the Water Initiative, the international legal regulation of water relations is underdeveloped. They do not refer to the Convention on the Law of the Non-Navigational Uses of International Watercourses, accepted in New York in 1997, which is the only legal instrument providing a comprehensive legal framework for cross border water management. Unfortunately, many initiatives related to international water issues do not provide for the participation of lawyers in the conceptual steps of these plans.

With the adoption of the Contract on the Preservation and Consumption of Cross Border Watercourses and International Lakes (Water Convention) (Helsinki, 1992) (Convention on the Protection and Use of Transboundary Watercourses and International Lakes 2013) and the Contract on the Law of the Non-Navigational Uses of International Watercourses (New York, 1997), the theory was introduced 'transboundary watercourses' and then 'transboundary waters'. Moreover, the concept of 'watercourse' means a system of surface and groundwater, which, due to their physical interconnection, constitute a single whole and usually have a common river mouth; and an international watercourse means a waterway, parts of which are located in various countries. In 1999, the UN adopted the Protocol on Water and Health to UNECE Water Convention (Economic and Social Council 1992). It says that water is necessary to maintain life, and it must be in such quantities and quality that are sufficient to meet basic human needs. According to this Protocol, this statement is one of the prerequisites for both improving people's health and sustainable development. This statement recognised the merits to public health that clean water with a good quality brings, as well as a harmonious and properly functioning aquatic environment. Of particular importance in the Protocol is drinking water, which is used or intended for consumption to quench thirst, for processing and food preparation, or for personal hygiene. According to this Protocol, to achieve its goals, its parties must set national goals for the quality of plumbing water and the quality of evacuations, and for the implementation of water resources and wastewater treatment tasks. This is essential to reduce outbreaks and morbidity associated with water. The Protocol must introduce a social part to assist on administration of water resources. Water administration should link socio-economic development with the preservation of ecosystems. Meanwhile, the Protocol says that its provisions apply only to the concept of 'water (water bodies)', which include: (a) surface freshwater; (b) groundwater; (c) estuaries; (d) coastal waters that are consumed for recreational purposes or the fish production by aquaculture, the production or collection of shellfish and crustaceans; (e) surrounded waters generally accessible for swimming. While the Protocol identifies water in the process of abstraction, transportation, treatment, or supply, wastewater is referred to as the concept of 'water (water resource)' in the process of collection, transportation, treatment, and evacuation or reuse and. This concept has not only quantitative and qualitative but also economic and environmental significance. This is confirmed in the manual for politicians and decision-makers (International Water Law 1999).

The basic principles of the universal framework convention on the right to water were adopted in 2004. They determined that the accessible right to water is enshrined in the Universal Declaration of Human Rights (1948) (Universal Declaration of Human Rights (UDHR) 1948) and in the International Agreement on Economic, Social and Cultural Rights (1966) (International Covenant on Economic, Social and Cultural Rights 1966), which has been approved by 148 states. The Universal Declaration of Human Rights guarantees public the right to a standard of living that provides them with good health. The International Agreement on Economic, Social, and Cultural Rights recognises 'the right of everyone to good physical and mental health'. It received particular recognition from the UN Committee on Economic, Social and Cultural Rights (2002) and was mentioned generally in note No. 15 on the right to water. Also, this Agreement was recognised in the Mar del Plata Action Plan – International Water Law Project (1977) (E/CONF.70/29 1977), in the Dublin Statement on Water and Sustainable Development (1992) (The Dublin Statement on Water and Sustainable Development 1992), in a variety of international documents and statements, and national constitutions of a growing number of states that adhere to the principles and rules of the UN Convention on the Right to Use International Watercourses (1997). In the International Conference about Water and the Environment (ICWE) which was held in Dublin (January 26–31, 1992), a report was adopted setting forth suggestions for action at local, national and international levels. These suggestions are subject to four guiding principles (The Dublin Statement on Water and Sustainable Development 1992):

Principle Number 1 – Fresh-water is a limited and susceptible resource necessary to maintain life, development, and the environment.

Principle Number 2 – Development and administration of water resources should be based on a collaborative approach of users, planners and politicians at all levels.

Principle Number 3 – Women have a vital role in the preparation, administration, and conservation of water

Principle Number 4 – Water is economically valuable in all its competing applications and should be known as an economic good.

All this allows us to conclude that there are independent objects of legal regulation of international water relations as relations between states, often with the participation of international organizations, legal entities, and individuals regarding joint efforts in solving the common tasks of protecting and using water resources.

The model of wastewater administration, which is the most powerful source of pollution of natural waters, is variable in terms of different factors in various countries.

About 15 laws and directives in England and Wales aim to control and prevent water pollution and related offences. These include: Environmental Protection Act 1990 (EPA 1990), Pt IIA; Water Resources Act 1991 (WRA 1991), Water Industry Act 1991 (WIA 1991), Environmental Permitting (England and Wales) Regulations 2016 (SI 2016/1154) (EPR 2016), Salmon and Freshwater Fisheries Act 1975, as well as EU water directives, including the EU Water Framework Directive 2000/60/EC, the Council Directive 91/271/EEC (Urban Waste Water Treatment

Directive), the Council Directive 91/676/EEC (Nitrates Directive), the Directive 2006/113/EC of the European Parliament and the Council on the quality required of shellfish waters, the Directive 2006/44/EC of the European Parliament and the Council on the quality of freshwaters, etc.

Section 85 in the Water Resources Act 1991 (Water Resources Act 1991) governs the criminal liability of those who contaminate natural water supplies. The major offence is to allow toxic, harmful or polluting substances or any waste materials to enter any controlled water. For such violations rigid liability has been established, that is, bringing to legal liability on the basis of the fact of harm alone, without establishing the subjective guilt of the offender. In the case of strict liability, it is not required to prove the presence or absence of criminal intent or negligence on the part of the defendant who committed the crime. The second component of the offence, 'knowingly permit' is used less regularly in prosecutions, as it needs to prove the intention in order for the defendant to be liable (Parpworth 2009). The Agency of Environmental Protection, private individual, or association may institute criminal proceedings under Section 85. A distinctive feature is that the law does not describe the concepts of 'poisonous', 'noxious' or 'polluting' to flexibly define these concepts by the courts (Bell and McGillivray 2008). The term "controlled waters" includes territorial, coastal, inland, and groundwater and is described in accordance with Section 104 of the Water Resources Act 1991 (Water Resources Act 1991). Penalties for a violation under Section 85 are imprisonment of not more than three months or a fine of not more than GBP 20 000, or both. Many violations may result in imprisonment for a term not exceeding two years, an unlimited fine, or both (Water Resources Act 1991).

Most states in the United States make water pollution an offence. Some of them also allow for strict liability (Franz 2011), but at the same time state criminal laws are inefficient forms of water pollution prevention. The economic benefits of illegal disposal of toxic substances far outweigh the risk of being detained, prosecuted, fined, or imprisoned in an inadequate system of enforcement (Burns *et al.*, 2008; Humphreys 1990). Many rules imply that crimes must be deliberate or committed with the knowledge of the matter. If the intent is lessening, penalties are consistently weak. For corporate entities to no longer view contamination as a rational economic risk offset by a corporation, strict liability laws for criminal negligence or reckless violations are required to prevent such violations (Brickey 2008). Too often states were worried that these rules would allow overzealous prosecutors to pursue taxation and employment bases (Frank 1986), but at the same time state environmental laws are mostly vaguely worded or too generally worded; thus, law implement often faces conditional challenges (Burns *et al.*, 2008; Selmi and Monaster 1989). The court invalidated state laws for failing to provide sufficient notice of what is unallowed, or for failing to establish clear standards for persons seeking facts to determine guilt (Louisiana v. Union Tank Car Co.; State v. Cote). The preliminary reason for the criminal implementation of water crimes in many states is not punishment, deterrence or incapacitation, but to supplement agency clout. In these states, criminal implementation is avoided by conciliatory attorney practices (Environmental Law Institute 1987; Selmi and Monaster 1989). When enforcing agencies have found for victims, courts have been quick to see the agency role as civil, not criminal. Investigation of state action shows the adoption of strict liability water laws, where combined with delegation of authority to agencies (as was usually the case) resulted in zero net effect on the number or outcome of cases prosecuted. In fact, agency procedures removed rigid liability from enforcement options in an effective manner (Frank 1986; Franz 2011). From what has been said, it seems that state water crimes suffer from the inability of the states to either establish simple applicable laws or to comply with the laws announced by them.

In contrast to the above, the Hong Kong Government Environmental Protection Division has published a guide to abating water pollution (The Government of Hong Kong 2018). For offences against the waters, just like in the UK, fines and a liability in the form of imprisonment are set at the same time. An increasing fine is a distinctive feature of Hong Kong legislation and an effective tool to combat water pollution.

In Israel, there are some rules and dozens of regulations regarding water and wastewater created under these laws. The main legislative instruments in this area include Water Law 5719-1959 (Water Law 5719-1959 2000), Licensing of Businesses Law 5728-1968 (Licensing of Businesses Law 5728-1968 1994), Local Authorities (Sewage) Law (Local Authorities (Sewage) Law 1962), Streams and Springs Authorities Law 1965 (Khan *et al.*, 2020a). The rules for preventing water pollution that fall under the Law on Water, the Law on Combating Harmful Impacts and the Law on Licensing Business include: prohibiting anyone from directly or indirectly draining or increasing chemical and/or biological materials or their residues from atomisers, prefabricated reservoirs or any other installations in a water supply; bans and limitations on the building of new cesspools and septic tanks, as well as existing ones, including schedules for the gradual liquidation of cesspools in certain circumstances; special conditions for the creation and operation of gas stations, to prevent leaks and oil pollution; mandatory prevention of leakages from inappropriately closed evaporation ponds used for manufacturing wastes, as well as from prefabricated ponds serving for the storage of wastewater; the prohibition of the evacuation of wastewater from a vessel into a water supply. Some of the important provisions of the rules include:

(1) the ban on wastewater dilution to reduce pollutant concentration;
(2) the ban on the discharge of wastewater into the pre-treatment plant before taking all reasonable measures to decrease the amount of wastewater, and to prevent and decrease emission of pollutants from the manufacturing procedure into wastewater;

(3) the prohibition of the evacuation of wastewater that has not been pre-treated at the pre-treatment facility, and the density of pollutants in which exceed the concentrations specified in the appendix;
(4) in the galvanic industry, the prohibition of wastewater discharges in which the concentration of suspended solids exceeds the established limit.

It should be noted that Israeli law controls the pollution of water sources with pesticides, fertilisers, and pharmaceuticals, which is regulated by the Licensing of Businesses Law 5728-1968 in the article 'Regulations on the prevention of livestock diseases and water pollution (Amendment 1974, 1998) 11B' (Licensing of Businesses Law 5728-1968 1994). This is due to livestock farms, wherein the process of combating livestock diseases, pollution of water sources with pesticides, fertilisers, and pharmaceuticals should be prevented, and workers should be trained in the rules to achieve these goals.

It is known that modern wastewater contains pharmaceutical substances or their metabolites. Moreover, this is characteristic not only for sewage from hospitals and pharmaceutical enterprises, livestock farms, and fisheries but also for domestic wastewater. Due to the wide variety of antibiotics and other drugs available for free sale, it is currently difficult to control their use. Often, unused drugs enter household sewers (Bound and Voulvoulis 2006; Khan *et al.*, 2019a, 2020b; Kümmerer 2008; Oliver and Chapman 2003).

In some countries, standards for wastewater discharge are divided into domestic and industrial, while in others (for example, Ukraine) there is one standard for all wastewater, and the requirements of the standard are not strict, but formal (SanPiN 4630–88). An investigation of the present issue of hospital wastewater (HWW) demonstrates that in Europe there are not enough specific instructions for the administration of such wastewater as well, except for some directives (Commission Decision of 3 May 2000 (2000/532/EC); EU 2008). EU (2008), on the administration of toxic waste and the list of perilous waste in the European Decision n. 532 of May 3, 2000 (2000/532/CEE) (Commission Decision of 3 May 2000 (2000/532/EC)), states that certain hospital wastes in a liquid phase (pharmaceuticals, drugs, residues of substances used as solvents, soaps, non-halogenated organic substances, etc.) should not be evacuated into the sewer, but it should be considered as waste, collected and disposed of as such. Due to the lack of specific guidelines in the European Union, its members have established their own rules, assessment, and removal criteria for HWW to manage it (Khan *et al.*, 2019b).

In Poland, each of the nearly 2500 communities, which are the smallest managerial units, are solely responsible for the water resources and wastewater administration in their region (Mikosz 2013). The main legal actions that served as the basis for the applied enforcement of this responsibility are the Water Law (2001), the Law on Environmental Protection (2001), the Law on Urban Water Resources and Sewerage (2001), and orders of the Ministry of the Environment on the quality of wastewater evacuated into the environment (last edition of 2009) and many other more detailed rules (Maciejewski & Walczykiewicz 2006; Wałęga *et al.*, 2009). In accordance with Polish evacuation standards, which are dependent on Directive 91/271/EEC, as a rule, limit only the values BOD5, COD, and TSS in wastewater from small treatment plants. The directive also prohibits the removal of sediment in surface water. Also, it introduces detailed monitoring provisions and forces the EU Member States to offer reports on their sludge removal operations every two years. Given the directive offered, wastewater sludge should be recycled, which has led to the continuous expansion of novel methods for managing and reusing wastewater sludge. Also, the Water Framework Directive remedy sludge as a new production and not as waste. The nitrogen and phosphorus elimination in wastewater treatment plants is necessary when wastewater is evacuated into a lake or seaside. However, the allowable concentrations are so high that this situation can be easily attained with the help of relatively simple methods for treating. Because of the complication of the administration of waste, legal actions relating to the treatment of sewage is enormous. The main legal legislations in the EU regarding the evacuation of wastewater consist of Council Directive 86/278/EEC of 12 June 1986. The directive applies to the requirements that sewage must satisfy in terms of agricultural use. The EU Wastewater Sludge and its Management Directives set many restrictions that are stricter in general than, for example, those offered by the US EPA. The most significant of the above guidance is the Landfill Directive, which introduces limits on the amount of sewage and other organic waste in landfills. Due to the continuous expansion of sewer networks, there is an increase in the amount of sludge that must be disposed of. At the same time, along with the need to protect the environment, the need for the restoration of raw materials and energy is growing, which contributes to the creation of new legal acts. In these legal acts, the definition of maximum permissible concentrations of harmful substances and waste management methods differ for each EU country (Cieślik *et al.*, 2018).

In Germany, HWW is considered internal and previous authority is not required (Verlicchi 2018). In most cases, if HWWs meet specific features concerning wastewater, they are discharged into purification plants with further consideration (Carraro *et al.*, 2011). At the same time, the German Federal Law (Federal Ministry for the Environment) has requirements for wastewater before combining it with other wastewater, including wastewater from the production of AOX-relevant pharmaceutical components, with a content of 8 mg/L (AOX is adsorbable organic halogens). In Italy, if the medical organization has 50 beds or less,, the HWW is evacuated as wastewater. The HWW is treated together with household wastewater at ordinary wastewater treatment plants (Marchetti 2012).

If we consider the situation in addition to Europe, for example in Mauritius, any person who contravenes the Wastewater Management Authority Act 2000 (Act No. 39 of 2000) fails to comply with any requirements regarding restriction on free disposal of effluent and water, injurious effluent, contravention and wastewater works or refuses entry to an officer of the Authority, shall commit an offence and shall, on conviction, be liable to a forfeit and imprisonment not exceeding 12 months. In the rules of the Waste Water Regulations 2019 (Waste Water Regulations), a list of pollutants is provided for each industrial activity, and the pharmaceutical industry is highlighted among these industries. The controlled parameters of wastewater are temperature, pH, oil, electrical conductivity, COD, sulfide, phenols, and the total amount of suspended solids and phosphorus; however, the presence and concentration of active pharmaceutical ingredients are not controlled. other wastewater regulations also do not contain recommendations regarding the treatment of wastewater containing active pharmaceutical ingredients.

In China, HWWs are considered manufacturing waste, and they view F-coliform as an indicator of a 50-bed healthcare facility (Liu *et al.*, 2010). Investigating the problem of HWW in Vietnam, it can be seen that they have special regulation on the administration and remedy of HWW No. 55/2014/QH13 (2014). The rules states that the hospital must collect and treat wastewater under treatment standards. On the other hand, legislation was adopted that takes into account the issue of collecting solid waste in water bodies (Verlicchi 2018).

Such a significant difference in the standards and rules of water protection, the difference in the details of the water legislation of different countries, in the approaches to wastewater management (including wastewater containing active pharmaceutical ingredients), and in the civil and criminal liability of those who committed an offence against water resources, is justified by many differences between countries. These differences include the level of economic and technological development of a country, the level of education and culture of the country's population, climatic, and other features of the geographical location of the country. However, to attain a common purpose, namely the conservation of natural waters on the planet, each country should strive for the best practices in wastewater administration and legislative preservation of the waters depending on the present experience of different countries. Currently, there are new advances in the wastewater treatment field from pharmaceutical ingredients (Carraro *et al.*, 2016; Khan *et al.*, 2020a, b, 2021; Kovalova *et al.*, 2012) and it should be taken into account when revising modern wastewater quality standards and water protection legislation.

6.5 CONCLUSIONS

Along with the classical methods of protecting water bodies from pollution by toxic substances, such as environmental engineering, legal mechanisms should be applied to prevent offences concerning water resources. A special place is taken by means of criminal law influence on persons guilty of causing harm to these environmental objects.

An analysis of the rules for protecting surface waters from pollution by wastewater, wastewater quality standards, as well as the dynamics of crimes against water resources in Ukraine as an example, showed a negative trend in the protection of natural waters. The rules for the collection of wastewater into the sewer systems that operate in Ukraine are not strict enough. In addition, these rules do not set clear limits for the discharge of wastewater containing active pharmaceutical ingredients. Those limiting values that are established in the quality standards for wastewater are so high that it is enough to apply relatively simple methods of treatment to achieve the result. A similar situation exists in Poland.

The statistics of offences against water resources in Ukraine for the entire study period show that the level of criminal offences has increased yearly. The growth rate indicates a growing trend towards both the opening (accounting) of criminal offences and the closure of criminal proceedings for various reasons. While maintaining these trends, a negative outlook can be seen in the development of the situation related to the criminal legal protection of the environment in general and water protection in particular. According to the forecast, the number of recorded crimes against water resources in the next three years will increase by about a quarter a year (21.8%). However, criminological studies show that hidden crimes (hidden, unregistered) are usually 2–3 times more. This means that the forecast in Ukraine generally confirms the dangerous situation in the field of environmental protection, namely its most dangerous manifestations associated with criminal offences. At the same time, the influence of the law on environmental protection is becoming less and less, since the number of registered violations is steadily growing, and simultaneously, fewer people are held accountable for their violation. Consequently, a decrease in the number of persons who received notification of suspicion indicates a lack of work to search and identify persons who committed crimes of this category. A similar situation is observed in the Republic of Moldova. From the analysis of the facts about water pollution, it can be concluded that law enforcement agencies are in no hurry or do not want to initiate a criminal case on water pollution based on the current criminal law. They do not work well enough to prevent and detect irregularities in water resources. Exemption from criminal liability in many cases is a consequence of the fact that it is difficult to demonstrate a psychological attitude (subjective aspect) to a crime against the environment.

This situation arose due to the lack of strict guidelines, an insufficiently effective judicial and executive system, as well as a lack of rigor in the laws (there is a reason to believe that the law does not always indicate specifically the manifestations of criminal behaviour aimed at worsening the state of water bodies under criminal conditions). Therefore, strict standards and laws must be developed for the management of water resources in general and wastewater in particular. Standards and laws should necessarily include subsections requiring the most effective wastewater treatment methods containing pharmaceutical compounds. Great success in this has been achieved, for example, England, Hong Kong, Israel, Germany, and Mauritius.

REFERENCES

Act No. 39 of 2000. The Wastewater Management Authority Act. Proclamation No. 14 of 2001. Mauritius, 26. Available from: www.wmamauritius.mu/wp-content/uploads/2019/08/WMA-ACT-2000.pdf.

Barbăneagră A. (2003). Criminal Code of the Republic of Moldova. Comment. Drept penal. Partea generală. Chisinau, Arc, 859.

Belecciu Ş. (2017). The subjective aspect of water pollution offence art. 229 of the Criminal Code of the Republic of Moldova. Acta Universitatis George Bacovia. *Juridica*, **6**(2). Available from: http://www.ugb.ro/Juridica/Issue12ROEN/8._Latura_subiectiva_a_infractiunii_de_poluare_a_apei.Stefan_Belecciu.EN.pdf.

Bell S. and McGillivray D. (2008). Environmental Law, 7th edn. Oxford University Press, Oxford, 607.

Bound J. and Voulvoulis N. (2006). Household disposal of pharmaceuticals as a pathway for aquatic contamination in the United Kingdom. *Environmental Health Perspectives*, **113**(12), 1705–1711.

Brickey K. (2008). Environmental Crime: Law, Policy, Prosecution. Aspen, New York.

Burns R. G., Lynch M. J. and Stretesky P. (2008). Environmental Law, Crime, and Justice. LFB Scholarly, New York.

Carraro E., Bonetta S., Bertino C., Lorenzi E., Bonetta S. and Gilli G. (2011). Hospital effluents management: chemical, physical, microbiological risks and legislation in different countries. *Journal of Environmental Management*, **168**, 185–199.

Carraro E., Bonetta Si., Bertino C., Lorenzi E., Bonetta Sa. and Gilli G. (2016). Hospital effluents management: Chemical, physical, microbiological risks and legislation in different countries. *Journal of Environmental Management*, **168**, 185–199. DOI: 10.1016/j.jenvman.2015.11.021

Cieślik B. M., Świerczek L. and Konieczka P. (2018). Analytical and legislative challenges of sewage sludge processing and management. *Monatshefte für Chemie-Chemical Monthly*, **149**(9), 1635–1645.

Code of Ukraine on Administrative Offenses: Law of Ukraine of August 13, 2020 № 80731-X. (1984). Information of the Verkhovna Rada of Ukraine (VVR), **51**, 1122. Available from: https://zakon.rada.gov.ua/laws/show/80731-10#Text.

Commission Decision of 3 May 2000 (2000/532/EC). Official Journal of the European Communities, L 226/3-L 226/24. Available from: https://eur-lex.europa.eu/legal-content/EN/TXT/PDF/?uri=CELEX:32000D0532&from=en.

Convention on the Protection and Use of Transboundary Watercourses and International Lakes. (2013). New York and Geneva. Available from: www.unece.org/fileadmin/DAM/env/water/publications/WAT_Text/ECE_MP.WAT_41.pdf.

Council Directive 91/271/EEC of 21 May 1991 concerning urban waste-water treatment. *Official Journal of the European Communities*, **L135**, 41–52. Available from https://eur-lex.europa.eu/legal-content/EN/TXT/PDF/?uri=CELEX:31991L0271&from=EN.

Criminal Code of Ukraine of 13.08.2020. (2001). Information of the Verkhovna Rada of Ukraine (VVR), **25–26**, 131. Available from: https://zakon.rada.gov.ua/cgi-bin/laws/main.cgi?nreg=2341-14#Text.

Dhingra A., Khan N. A., Ahmed S., Gautam S., Vambol S., Vambol V. and Kovalenko S. (2020). Investigation of medical institutions in India as a source of surface water pollution. *World Review of Science, Technology and Sustainable Development*, (in press). [to be updated]

Directive 2000/60/EC of the European parliament and of the council of 23 October 2000. Official Journal of the European Communities. Available from: https://eur-lex.europa.eu/resource.html?uri=cellar:5c835afb-2ec6-4577-bdf8-756d694eeb.0004.02/DOC_1&format=PDF.

Donner P. and Johansson T. (2014). Oil pollution in the waters of the United Kingdom North Sea. In: Carpenter A. (eds). Oil Pollution in the North Sea. The Handbook of Environmental Chemistry, vol 41. Springer, Cham. https://doi.org/10.1007/698_2014_295.

E/CONF.70/29. (1977). Report of the United Nations. Water Conference. Mar del Plata, 14–25 March 1977. G 1014. United Nations, New York. Available from: www.internationalwaterlaw.org/bibliography/UN/Mar_del_Plata_Report.pdf.

Economic and Social Council. (1992). Protocol on Water and Health to the 1992 Convention on the Protection and Use of Transboundary Watercourses and International Lakes. United Nations, London. Available from: www.unece.org/fileadmin/DAM/env/documents/2000/wat/mp.wat.2000.1.e.pdf.

EU. (2008). European Union. Directive 2008/98/EC of the European Parliament and of the Council of 19 November 2008 on waste and repealing certain Directives (Text with EEA relevance).

Falås P., Andersen H. R., Ledin A. and la Cour Jansen J. (2012). Occurrence and reduction of pharmaceuticals in the water phase at Swedish wastewater treatment plants. *Water Science and Technology*, **66**, 783–791.

Frank N. (1986). From Criminal Law to Regulation: A Historical Analysis of Health and Safety Law. Garland, New York.

Franz A. (2011). Crimes against water: non-enforcement of state water pollution laws. Crime, Law and Social Change, **56**(1), 27–51.

Getirana A., Kumar S., Girotto M. and Rodell M. (2017). Rivers and floodplains as key components of global terrestrial water storage variability. *Geophysical Research Letters*, **44**(20), 10,359–10,368.

Greife M. B. and Stretesky P. B. (2013). Treadmill of Production and State Variations in Civil and Criminal Liability for oil Discharges in Navigable Waters. Routledge International Handbook of Green Criminology. Routledge, London, 150–166.

Humphreys S. L. (1990). An enemy of the people: prosecuting the corporate polluter as a common law criminal. *The American University Law Review*, **39**, 311–354.

International Covenant on Economic, Social and Cultural Rights. (1966). United Nations General Assembly (resolution 2200A). Available from: www.ohchr.org/Documents/ProfessionalInterest/cescr.pdf.

International Paper v. Ouellete, 479 U.S. 481 (Case Cited, 1987).

International Water Law. (1999). Signed at Bern on 12 April 1999, 132. Available at: www.cawater-info.net/library/rus/legal_7.pdf.

Khan N. A., Ahmed S., Vambol S., Vambol V. and Farooqi I. H. (2019a). Field hospital wastewater treatment scenario. *Ecological Questions*, **30**(3), 57–69.

Khan N. A., Khan S. U., Ahmed S., Farooqi I. H., Hussain A., Vambol S. and Vambol V. (2019b). Smart ways of hospital wastewater management, regulatory standards and conventional treatment techniques. *Smart and Sustainable Built Environment*. https://doi.org/10.1108/SASBE-06-2019-0079.

Khan A. H., Khan N. A., Khan S. A., Dhingra A., Khan S. U., Mohammadi A. A., Changani F., Yousefi M., Alam S., Vambol S., Vambol V., Khursheed A. and Ali I. (2020a). Application of advanced oxidation processes followed by different treatment technologies for hospital wastewater: a case study of multispecialty hospitals. *Journal of Cleaner Production*, **269**, 122411–122424.

Khan N. A., Ahmed S., Farooqi I. H., Ali I., Vambol V., Khorasgani F. C., Yousefi M., Vambol S., Khan S. U. and Khan A. H. (2020b). Occurrence, sources and conventional treatment techniques for various antibiotics present in hospital wastewaters: A critical review. *Trends in Analytical Chemistry*, **129**, 115921–115935.

Khan N. A., Ahmed S., Farooqi I. H., Vambol S., Vambol V. and Koloskov V. (2021). Operational parameters optimisation for simultaneous removal of two drugs in hospital wastewater – RSM approach. *International Journal of Environment and Waste Management*, (in press). [to be updated]

Kopela S. (2011). Civil and criminal liability as mechanisms for the prevention of oil marine pollution. *Review of European Community and International Environmental Law*, **20**, 313–324.

Kovalova L., Siegrist H., Singer H., Wittmer A. and McArdell C. S. (2012). Hospital wastewater treatment by membrane bioreactor: performance and efficiency for organic micropollutant elimination. *Environmental Science and Technology*, **46**, 1536–1545. DOI: 10.1021/es203495d

Kümmerer K. (2008). Pharmaceuticals in the Environment Sources, Fate, Effects and Risks. Springer, Berlin Heidelberg.

Licensing of Businesses Law 5728-1968. (1994). Available from: www.sviva.gov.il/English/Legislation/Documents/Licensing%20of%20Businesses%20Laws%20and%20Regulations/LicensingOfBusinessesLaw1968-Excerpts.pdf.

Liu Q., Zhou Y., Chen L. and Zheng X. (2010). Application of MBR for hospital wastewater treatment in China. *Desalination*, **250**(2), 605–608.

Local Authorities (Sewage) Law. (1962). Available from: www.sviva.gov.il/English/Legislation/Documents/Wastewater%20Laws%20and%20Regulations/LocalAuthoritiesSewerageLaw1962.pdf.

Louisiana v. Union Tank Car Co., 439 So.2d 377 (Case Cited, La. 1983).

Maciejewski M. and Walczykiewicz T. (2006). Current experiences in implementation of the water framework directive 2000/60/EC in Poland. *Infrastruktura i Ekologia Terenów Wiejskich*, **4**, 63–76.

Marchetti F. (2012). Regulation for the simplification of administrative obligations in environmental matters burdening companies. In news from 9 February 2012.

Mazhar M. A., Khan N. A., Ahmed S., Khan A. H., Hussain A., Changani F. and Vambol V. (2020). Chlorination disinfection by-products in Municipal drinking water – A review. *Journal of Cleaner Production*, **273**, 123159–123171.

Mel'nychok L. S. (2013). Bathymetric studies of Lake Svityaz. II International Scientific and Practical Conference, Brest, September 25–27, 2013. In two parts. Part 1, 124–126. Available from: http://www.brsu.by/sites/default/files/geophys/tom_1.pdf#page=124.

Mikosz J. (2013). Wastewater management in small communities in Poland. *Desalination and Water Treatment*, **51**(10–12), 2461–2466.

Milwaukee v. Illinois, 451 U.S. 304 (Case Cited, 1981).

On approval of the Rules of acceptance of wastewater into centralized drainage systems and the Procedure for determining the amount of payment for excessive discharges of wastewater into centralized drainage systems: Order of the Ministry of Regional Development, Construction and Housing of Ukraine from 01 December 2017 № 316. Annex 2 (paragraph 4 of section III), Kyiv. Available from: https://zakon.rada.gov.ua/laws/ show/z0056-18#n177.

On approval of the Rules for the provision of services for district heating, hot and cold water supply and sewerage and the standard contract for the provision of services for district heating, cold and hot water supply and drainage: resolutions of the Cabinet of Ministers of Ukraine of July 21, 2005 № 630, Kyiv. Available from: http://search.ligazakon.ua/l_doc2.nsf/link1/KP050630.html.

On approval of the Rules for the use of centralized municipal water supply and sewerage systems in the settlements of Ukraine: Order of the Ministry of Housing and Communal Services of Ukraine of June 27, 2008 № 190, Kyiv. Available from: https://zakon.rada.gov.ua/laws/ show/z0936-08#Text.

On environmental protection: Law of Ukraine of June 7, 2020 № 1264-XII. (1991). *Vidomosti Verkhovnoyi Rady Ukrayiny (VVR)*, **41**, 546.

On drinking water, drinking water supply and drainage: Law of Ukraine of May 1, 2019 № 2918-III. (2002). Information of the Verkhovna Rada of Ukraine (VVR), **16**, 112. Available from: https://zakon.rada.gov.ua/laws/show/2918-14#Text.

On housing and communal services: Law of Ukraine of April 2, 2020 № 2189-VIII. (2018). *Information of the Verkhovna Rada of Ukraine (VVR)*, **1**, 331. Available from: https://zakon.rada.gov.ua/laws/show/2189-19#n436.

Oliver D. and Chapman A. I. (2003). Pharmaceutical waste survey, final report. Local Hazardous Waste Management Program in King County, Seattle, WA.

Ordinance on Requirements for the Discharge of Waste Water into Waters (Waste Water Ordinance – AbwV). Federal Ministry for the Environment, Nature Conservation and Nuclear Safety, Germany, 17 June 2004. Federal Law Gazette (BGBl. I), 1108.

Parpworth N. (2009). Who may be liable for an offence contrary to s.85 of the Water Resources Act 1991. *Journal of Planning and Environment Law*, **3**, 294–309.

Rules for the protection of surface waters: approved by the USSR State Committee for Nature Protection on 02.21.91. Available from: https://zakon.rada.gov.ua/laws/show/n0002400-91#Text.

Rules for the reclamation of industrial waste waters from communal and household canalization systems of settlements in Ukraine: approved by the Order of the State Budget of Ukraine from 19 February 2002 N 37. Available from: https://zakon.rada.gov.ua/laws/showTz02.

SanPiN 4630–88. Sanitary rules and norms for the protection of surface waters from pollution: approved by the Chief State Sanitary Doctor of the USSR from 04.07.1988. Available from: https://zakon.rada.gov.ua/laws/show/v4630400-88#Text.

Selmi D. and Monaster K. (1989). State Environmental Law (The Clark Boardman Environmental Law Series). Clark Boardman Callaghan, a division of Thomson Legal Publishing, New York, USA.

Shapovalova I. O. (2007). Availability and use of water consumer resources. *Culture of the peoples of the Black Sea region*, **116**, 107–111. Available: http://dspace.nbuv.gov.ua/bitstream/handle/123456789/107829/28-Shapovalova.pdf?sequence=1.

State building norms of Ukraine B.2.5-75:2013. Sewerage. External networks and structures. Basic design provisions. Ministry of Regional Development, Construction and Housing of Ukraine, Kyiv, **2012**, 207.

State v. Cote, 945 A.2d 412 (Case Cited, Conn. 2008).

State hazardous waste enforcement study 61 (Final Draft, October). (1987). Environmental Law Institute, Washington, DC.

The Dublin Statement on Water and Sustainable Development. (1992). International Conference on Water and the Environment (ICWE), Dublin, Ireland. Available from: www.wmo.int/pages/prog/hwrp/documents/english/icwedece.html.

The Government of Hong Kong. (2018). Special Administrative Region. Environmental Protection Department. Available from: https://www.epd.gov.hk/epd/english/environmentinhk/water/guide_ref/guide_wpc_wpco_14.html.

Treaty establishing the European Community (Treaty establishing the European Economic Community) of 1 January 2005 № 994_017. (2005). Consolidated version as of January 1, 2005 Available from: https://zakon.rada.gov.ua/laws/show/994_017#Text.

Universal Declaration of Human Rights (UDHR). (1948). United Nations General Assembly (resolution 217 A). Paris, Available from: www.ohchr.org/EN/UDHR/Documents/UDHR_Translations/eng.pdf.

Vadez V., Kholova J., Zaman-Allah M. and Belko N. (2013). Water: the most important 'molecular'component of water stress tolerance research. *Functional Plant Biology*, **40**(12), 1310–1322.

Vambol S., Vambol V., Sobyna V., Koloskov V. and Poberezhna L. (2018). Investigation of the energy efficiency of waste utilization technology, with considering the use of low-temperature separation of the resulting gas mixtures. *Energetika*, **64**(4), 186–195.

Vambol S., Vambol V., Sundararajan M. and Ansari I. (2019). The nature and detection of unauthorized waste dump sites using remote sensing. *Ecological Questions*, **30**(3), 43–55. DOI: 10.12775/EQ.2019.018.

Verlicchi P. (2018). Hospital Wastewaters – Characteristics, Management, Treatment and Environmental Risks. International Publishing Springer, Cham. DOI: 10.1007/978-3-319-62178-4.

Wałęga A., Chmielowski K. and Satora S. (2009). Water and wastewater management condition in Poland regarding water framework directive implementation. *Infrastructure and Ecology of Rural Areas*, **4**, 57–72.

Water Code of Ukraine: Law of Ukraine of June 7, 2020 № 213/95-VR. (1995). Information of the Verkhovna Rada of Ukraine (VVR), **24**, 189. Available from: https://zakon.rada.gov.ua/laws/show/213/95-%D0%B2%D1%80#Text.

Water Law 5719-1959, Israel. (2000). Available from: www.sviva.gov.il/English/Legislation/Documents/Water%20Laws%20and%20Regulations/Wat erLaw1959-Excerpts.pdf.

Water Resources Act. (1991). UK Public General Acts, **199**, 57. Available from: www.legislation.gov.uk/ukpga/1991/57/contents.

Waste Water (Licence for Discharge of Industrial Effluent into a Waste Water System) Regulations 2019. Mauritius, 24. Available from: www.wmamauritius.mu/wp-content/uploads/2020/01/Waste-Water-Licence-for-Discharge-of-Industrial-Effluent-into-a-Waste-Water-System-Regulations-2019.pdf.

Yanko N. V., Stankevich V. V. and Koval N. M. (2014). Anthropogenic eutrophication of the Shatsky lakes. *Environment and Health*, **3**, 51–55. Available from: https://cyberleninka.ru/article/n/antropogenna-evtrofikatsiya-shatskih-ozer/viewer.

Ziarati P., Vambol V. and Vambol S. (2020). Use of inductively coupled plasma optical emission spectrometry detection in determination of arsenic bioaccumulation in *Trifolium pratense* L. from contaminated soil. *Ecological Questions*, **31**(1), 15–22. DOI: 10.12775/EQ.2020.003.

Chapter 7

Eco-toxicity and health issues associated with the pharmaceuticals in aqueous environments: a global scenario

Saif Ullah Khan, Hasan Rameez, Farrukh Basheer and Izharul Haq Farooqi

Department of Civil Engineering, Zakir Husain College of Engineering & Technology, Aligarh Muslim University, Aligarh, U.P., India

ABSTRACT

Availability of pharmaceuticals has undoubtedly been a blessing to mankind, but its excessive use has also compromised the quality of the environment and ultimately human health. Excessive and unregulated use of pharmaceutical compounds has resulted in discharge of these from various sources, thus contaminating our water reserves. These pharmaceutical compounds are composed of ingredients that interact with human or veterinary biological systems in one way or the other. A few of these may potentially cause adverse effects on non-target organisms, being more vulnerable for pregnant women, children and people with certain pre-medical conditions. Apart from human consumption, a major portion of pharmaceutical consumption comes under animal husbandry with several hormones and estrogens widely used for managing animal milk production, growth and their reproduction. Consequently, the pharmaceuticals getting passed into the environment through animal waste go unnoticed, thereby posing another big challenge. Moreover, risk assessment of chronic exposure to these pharmaceutical contaminants is itself a very challenging task. The most widely adopted methodology for such quantification is evaluation of Risk Quotient values for the presence of pharmaceutical compounds in water. These approaches of selecting high priority pharmaceuticals provide a practical base to manage the

 Pharmaceutical Wastewater Treatment Technologies: Concepts and Implementation Strategies
Editor(s): Nadeem A Khan, Sirajuddin Ahmed, Viola Vambol and Sergij Vambol
doi: 10.2166/9781789061338_0145

risks caused. Here we provide a critical view of the state of human health risk assessment and effect of excessive use of these pharmaceuticals on our ecosystem due to which arises a need to periodically monitor the pharmaceutical residue concentration in surface and other water streams by regulating agencies for the presence of those beyond permissible limits.

Keywords: Eco-toxicity, Hospital wastewater, Pharmaceutically active compounds, Risk quotient, Genotoxicity, Suppressed immune system.

Contents

7.1 Introduction 146
7.2 Pharmaceutical Usage and Consumption Patterns 148
  7.2.1 Persistence, pathways and transformations in environment 150
7.3 Health Assessment and Risk Analysis 153
  7.3.1 Analyzing risk quotient 155
7.4 Hospital Wastewater Containing Toxic Chemicals and PHaCS 156
7.5 Potential Implications of Discharging Untreated/Partially Treated Wastewater in the Aquatic Environment 161
7.6 Impact of Pharmaceutical Residues on Humans and the Ecosystem 161
  7.6.1 Disease outbreaks and ecological hazards 163
  7.6.2 DNA damage and genotoxicity 169
7.7 Conclusion and Recommendations 170
References 171

7.1 INTRODUCTION

The global population has undergone multi-fold growth over the last few decades. This, along with the increasing demand of resources, has put an enormous burden on the available water resources for municipal, industrial and agricultural purposes. Providing clean and affordable drinking water is the utmost need of this century for which we mainly depend on surface and groundwater reserves. Shortage of water and its contamination pose a serious global challenge nowadays. Moreover, various human activities and reckless use of available sources intervenes with quality of water as it is an excellent solvent and keeps on moving according to laws of hydrodynamics. Rapid urbanization and industrialization have led to increased direct and indirect discharge of wastewater into water bodies and reserves (Khan *et al.*, 2020a). To meet this ever-increasing demand of water, most countries have been putting more emphasis on wastewater reuse as a part of an efficient and sustainable water management programme. While such a water resource management policy could help tackle the problem of water scarcity to some extent, the presence of one or more contaminants above a certain limit in reused or recycled water may pose serious health issues to both humans and the aquatic ecosystem.

Pharmaceuticals, undoubtedly a boon to mankind, have extended life expectancies, restored the health of millions and improved their overall personal satisfaction. However, this achievement has now pushed their extensive use towards emerging natural contaminants. A variety of pharmaceuticals and personal care products, along with artificial sweeteners, are collectively categorized as CECs (Contaminants of Emerging Concern) as they are found in micro quantities in environmental matrices (Brack *et al.*, 2012; Petrie *et al.*, 2015). Along with the conventional contaminants, CECs may also find their way into wastewater streams, contaminating nearby water reserves. Pharmaceutically Active Compounds (PhACs) are known to be present in surface water bodies, groundwater, other major drinking sources and wastewater streams (de Jongh *et al.*, 2012; Godfrey *et al.*, 2007; Halling-Sørensen *et al.*, 1998; Kolpin *et al.*, 2002). These PhACs find their way into ground and surface waters due to the discharge of untreated or partially treated wastewater streams from various sources, including hospital effluents. Pharmaceuticals themselves are among a huge group of compounds with around 3000 different type of pharmaceuticals enlisted just in the European associated market (Taylor & Senac, 2014). Studies show that pharmaceutical wastewater carries different types of PhACs such as antibiotics, anticoagulants antiepileptics, analgesics, anti-inflammatories, lipid regulators, steroidal compounds, beta-blockers, diuretics, contrast media, cosmetics, psycho-stimulants and anti-depressants (Deblonde *et al.*, 2011; Luo *et al.*, 2014; Nassiri-Koopaei & Faramarzi, 2015; Shaik *et al.*, 2016). Few pharmaceuticals, including non-inflammatory drugs, estrogens, beta blockers, lipid regulators, etc., have been found to exist in ranges of μg/L to ng/L in surface waters across the world. The problem has grown intense due to the enormous use of pharmaceuticals at different medical facilities and high over-the-counter sales.

With unregulated consumption and discharge of so many pharmaceuticals, their harmful and toxic effects have been studied and recently established (Grenni *et al.*, 2018; Kolpin *et al.*, 2002; Rivera-Utrilla *et al.*, 2013). Wastewater containing pharmaceuticals could potentially cause acute and chronic toxicity, carcinogenicity, reproductive and developmental toxicity and neurotoxicity (Nassiri Koopaei & Abdollahi, 2017). These PhACs are composed of ingredients that are intended to interact with human or veterinary biological systems or the metabolism in one way or another. The presence and disposal of pharmaceutical residues in the environment is of utmost concern due to the development of drug resistant genes and the decreasing effect of antibiotics on diseases. Though the concentration of these PhACs found in the environment are much lower than their prescribed doses (Webb *et al.*, 2003), still some of them could potentially cause adverse effects to non-target organisms, especially more vulnerable groups such as pregnant women, children and people with certain medical conditions. As some of the PhACs can disrupt the biological functioning of humans even at very low concentrations (Brack *et al.*, 2012), the presence of such compounds in drinking water sources could prove hazardous to human health. The problem gets worse due to the fact that monitoring of such a huge number of pharmaceuticals in water is very difficult. Although quantitative analysis using recent advanced technologies such as GC-MS/MS, LC-MS/MS, UPLC/MS have made it possible to detect these compounds with concentrations in ranges of μg/L and ng/L (Lindsey *et al.*, 2001; Hernández *et al.*, 2007; Petrovic & Barceló, 2006), it is not yet feasible to deploy such techniques for monitoring so many compounds on a large scale. Despite their presence in the environmental matrices being well known for the past two decades, there is an absence of specific guidelines worldwide regarding this issue (Daughton, 2004; Webb *et al.*, 2003). With an in-depth regulatory framework, only Australia seems to have defined rules for pharmaceuticals present in drinking water supplies (Pomati, 2007).

7.2 PHARMACEUTICAL USAGE AND CONSUMPTION PATTERNS

Data regarding pharmaceutical consumption, although available in most the countries, is not very specific regarding usage patterns worldwide. (aus-der-Beek *et al.*, 2016). In many countries pharmaceuticals are sold totally unregulated, resulting in unaccounted utilization and inaccurate assessment of consumption (Kümmerer, 2009). In 2001, an enormous amount of around 38,000 tons of common pharmaceuticals per annum were consumed in Germany, however this quantity was around 22,700 tons in the US. In 2002 it was estimated that only antibiotics consumption worldwide was in the range of 100,000–200,000 tons per annum (Castiglioni *et al.*, 2006). It is quite evident that the consumption of antibiotics only has increased multi-fold over the years in most of the countries which is a major proportion of total pharmaceutical compounds. For instance, Beek *et al.* (2016) in his study revealed this increased pattern of consumption by correlating with examples such as; paracetamol quantification only in USA was around 5790 tons in 2002, with around 3303 tons of production in France in 2005 and 2763 tons of chlortetracycline in South Korea for veterinary purposes. Considering the global consumption of drugs, around 30% comprises of active pharmaceuticals, 20% metaformin, 12% ibuprofen, 8% metamizole, 7% acetylsalicyclic acid, 6% acetaminophen, 3% iomeprolol, 2% each of metoprolol, allopurinol, amoxicillin (Van Boeckel *et al.*, 2014).

Many researches have investigated the presence of traces of pharmaceuticals in aquatic environments and their toxicities which pose serious threats to receptors. Considering the gravity of the problem that can be caused by the discharge of untreated effluent, it is essential to conduct analysis of hospital effluent, with respect to physio-chemical, biological parameters, selected chemicals and high priority pharmaceutical residues. The contamination due to this is not just limited to water, it affects the whole ecosystem severely. The different pathways through which PhACs enter the environment are shown in Figure 7.1. Contaminated water streams can get mixed with other water reserves that are used for various agricultural and domestic purposes. Thus, extensive use of wastewater for

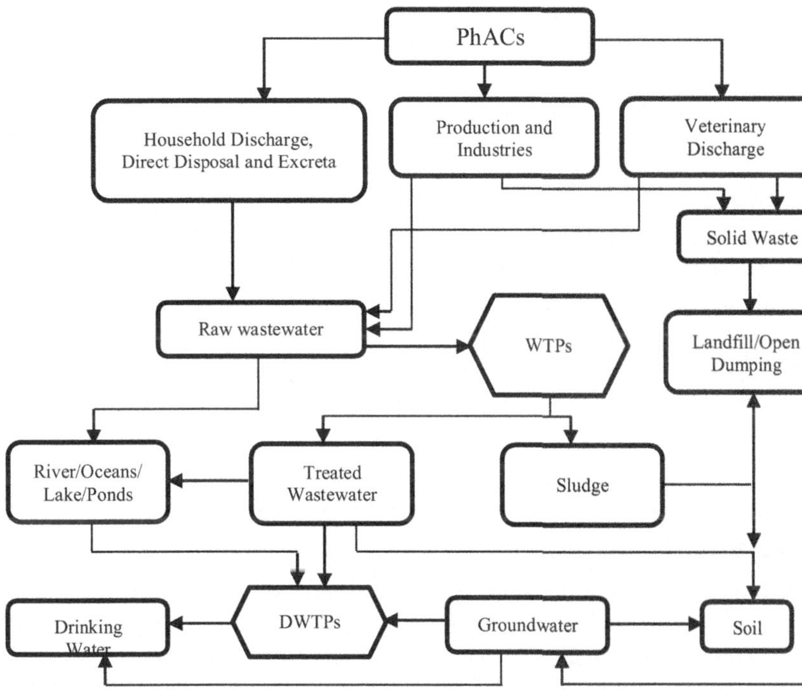

Figure 7.1 Flowchart depicting PhACs pathways in the environment (figure prepared using Balakrishna *et al.*, 2017).

agricultural and other applications without adequate treatment is a serious health hazard to humans. In countries with water resource scarcity, inefficient water management systems, and high demand of irrigation water, untreated or partially treated wastewater may be used to meet irrigation needs resulting in transportation of contaminant loads to the farm. This not only exposes the farm workers to pharmaceutical waste hazards, but the crops could also intake the compounds from where they could enter the food chain and accumulate in the bodies of consumers. The concentration of these pharmaceuticals depends upon factors such as the characteristics of water and soil, source of water supply, cropping farms and seasonal variation in consumption of medicines.

Several studies show that the conventional wastewater treatment techniques are only partially capable of removing pharmaceutical compounds. On account of their relatively longer half-lives and hydrophilic nature, most of the pharmaceutical compounds cannot be removed by treatment plants as efficiently as other contaminants (Batt *et al.*, 2006). While advanced techniques are available to remove most of the contaminants from wastewater, such technologies are not economically viable for most low to middle income countries (Khan *et al.*, 2019a; Schacht *et al.*, 2016; Wichelns *et al.*, 2015). Common treatment techniques, such as primary settling (removal efficiency of 3–45%), activated sludge process (7–100%), sand filtration (0–99%), biological filtration (6–71%), and advanced techniques such as ozonation (1–100%), microfiltration and reverse osmosis (91–100%), reverse osmosis (62–97%), photolysis (UV/hydrogen peroxide) (52–100%), UV/TiO_2 (more than 95%), dark and light Fenton (80–100%), bio-membrane (23–99%) and ultrasound (24–100%), show huge variations in their removal efficiencies on account of the type of pharmaceutical compounds present and local operational conditions (World Health Organization, 2012). Considering the inconsistencies of the present treatment systems to efficiently deal with pharmaceutical wastes, it is quite imperative and desirable to modify current treatment techniques in order to efficiently deal with huge volumes of the wastewater generated (Khan *et al.*, 2019b). Along with retrofitting of existing treatment facilities, their removal efficiencies should also be monitored on a continuous basis.

7.2.1 Persistence, pathways and transformations in environment

Pharmaceutical industries use water for various purposes at different stages in the drug manufacturing and development process. Based on the usage, this water consumption can be categorized into general use water that is used for normal day to day purposes and can be considered as municipal wastewater; manufacturing water containing known compounds in known quantities. Their presence in all forms of surface water and groundwater have been confirmed in the last two decades (World Health Organization, 2012). The discharge of wastewater effluent from Sewage Treatment Plants (STPs) into surface water bodies is considered to be the major source of pharmaceutical contamination in the aquatic environment. In rural areas without an operational sewage system in place, septic tanks are a major source of exposure (Du *et al.*, 2014; Stanford & Weinberg, 2010). Several studies in the past have been carried out for the assessment of contamination due to PhACs in both developed and developing nations. The absence of advanced wastewater treatment facilities has resulted in greater risk and vulnerability in developing countries, posing health challenges for human beings. For instance, Larsson *et al.* (2007) indicated in his research about heavy concentrations of pharmaceutical compounds (for example ciprofloxacin concentrations as high as 31 mg/L) in the influent of a treatment plant in Patancheru in Hyderabad, India.

Once consumed, PhACs do not remain in their parent form, rather they are transformed into secondary compounds. Studies have reported that 10–90% of therapeutic doses of drugs are excreted out in the parent form while the rest is excreted out in metabolized forms. Gut microbes and host enzymes cause the metabolic transformation of these compounds before being excreted out (Patel *et al.*, 2019). Along with urine and fecal matter, the un-metabolized and partially metabolized parent compounds are directly excreted out into our environment, some of which are shown in Table 7.1. The overall process of transformation may occur in different ways. In the first phase, hydrolysis, reduction and oxidation reactions occur resulting in modification of parent compounds while in the second phase, conjugate reactions occur. First, this transformation takes place during a metabolic process within the body of the consumer (Testa & Krämer, 2008). Second, transformation also occurs during various chemical or biological processes taking place at the STP during wastewater treatment (Kern *et al.*, 2010). Besides these, transformation of the pharmaceutical products may also occur naturally in the environment under certain conditions (Escher & Fenner, 2011). Lastly, such transformations may also occur when surface or ground water is given tertiary treatment to make them potable. Processes such as ozonation, UV/hydrogen peroxide treatment and chlorination can oxidize the parent pharmaceutical compounds to form new products (Richardson *et al.*, 2007). Generally, the transformation products are equally or less toxic than the primary contaminant (Escher & Fenner, 2011). The occurrence of pharmaceuticals in water bodies points towards the need for further research in the field of metabolism of pharmaceuticals and their fate in the aquatic environment and their toxicity assessment in regard to the predicted environment concentration. The occurrence of pharmaceuticals and their metabolites in the water bodies calls for the adequate treatment of effluent containing pharmaceutical residues at source. Before discharging hospital effluents and other sources of pharmaceutical wastewater into municipal drains without a certain minimum treatment, the above-mentioned implications must be kept in mind for the well-being of aquatic and human life.

With the evolution of nanotechnology in recent years, nanomaterials have found their use in the pharmaceutical industry besides others (Khan *et al.*, 2017). Estimations show that in 2010 the global production of nano-particles was over 260,000–309,000 metric tons, of which 63–91% was released onto the land that could find its way to surface water bodies (Keller *et al.*, 2013; Wiesner *et al.*, 2009). Out of this huge volume, a small yet significant amount could be attributed to the pharmaceutical industry. This calls for a need to analyze the fate and potential health impacts of nano-particles on the health of humans. These nano-particles could find their way into the surface water bodies, affecting the aquatic ecosystem and leading to bioaccumulation of these particles in aquatic flora and fauna thus ultimately finding their way into humans through the food chain. The fate of the nano-particles is governed by their physiochemical characteristics and the conditions of the aquatic ecosystem they enter (Klaine *et al.*, 2008; Ray *et al.*, 2009). Sufficient toxicity data is not available to quantify the health risks of nano-particles in aquatic media, the limited available studies find the human health impacts of nano-particles to be low (Oberdörster, 2010; Tsuji *et al.*, 2006).

Table 7.1 List of some common drugs along with their fractions unmetabolized by human metabolism.

Pharmaceutical Class	Drugs used	Excretion (%)	Reference
Antibiotics	Amoxicillin	$\geq$70	Jjemba (2006)
	Erythromycin	25	Huschek *et al.* (2004)
	Ciprofloxacin	50–83.7	Mompelat *et al.* (2009)
Beta blocker	Acebutolol	39	Mompelat *et al.* (2009)
	Atenolol	90	Bound and Voulvoulis (2005)
NSAID	Diclofenac	100	Carlsson *et al.* (2006)
	Ibuprofin	10	Bound and Voulvoulis (2005)
Anticonvulsant	Felbamate	40–50	Mompelat *et al.* (2009)
Antibacterial	Ceftriaxone	70	Mompelat *et al.* (2009)
Antihistamine	Cimetedine	48–75	Mompelat *et al.* (2009)
Pain killer	Tramadol	15–35	Mompelat *et al.* (2009)
Antihypertensive	Verapamil	$\leq$5	Jjemba (2006)
Analgesic	Paracetamol	$\leq$5	Jjemba (2006)
Cardiovascular	Digoxin	40–69	Jjemba (2006)
Lipid lowering agent	Atorvastatin	$\leq$5	Jjemba (2006)
	Bezafibrate	50	Bound and Voulvoulis (2005)
Anti-cancerous	Cyclophosphamide	5–20	Jjemba (2006)
	Ifosfamide	Up to 15	Jjemba (2006)
Blood pressure regulator	Valsartan	80	Mompelat *et al.* (2009)
	Flurosemide	40–69	Jjemba (2006)

7.3 HEALTH ASSESSMENT AND RISK ANALYSIS

Despite many studies being carried out across the globe, not much knowledge of the potential human health impact of PhACs exists. This knowledge gap is due to the vast number of PhaCs that are being used globally. According to the Orange Book, published by the US Food and Drug Administration, more than 10,000 therapeutic drugs are being manufactured and used worldwide. The concentration of PhACs is generally reported in terms of ng/L, thus their direct health effects on human beings is not well understood until ingested for long durations along with water as well as dietary needs (Kimura *et al.*, 2004). However, some literature dealing with ingesting small quantities of pharmaceuticals through the intake of vegetables, meat, fruits and some other food items are also available (Christou *et al.*, 2017; Malchi *et al.*, 2014). However, relevant data such as on concentration in the environment, dose-response relationships and fate and transport for these PhACs are not available. Thus, their cumulative effect of ingesting these pharmaceutical compounds through these sources raise doubts over any direct health effects on human body. This is also due to the fact that analyzing methods for the detection and measurement of most of these PhACs is either not developed yet, or is too expensive to be deployed on a large scale (Dong *et al.*, 2013; Patel *et al.*, 2019).

As stated already, several studies have reported the presence of pharmaceuticals in drinking water. The risk assessment results suggest that small concentrations of pharmaceuticals in drinking water are not very likely to affect human health. However, such studies were carried out with certain limitations. The risk assessment of chronic exposure to pharmaceutical contaminants is a challenging task since very little long-term toxicological data is available for these compounds (Sharma & Sanghi, 2013). First, due to technical, economical and time constraints, the monitoring data set of concentration of pharmaceuticals under consideration was often very limited (Benotti *et al.*, 2009; Jones *et al.*, 2005; Villanueva *et al.*, 2014; Williams & Brooks, 2012). Second, few studies assumed that drinking water treatment techniques do not remove any pharmaceutical contaminants from the water, and monitored the concentration data from the surface water sources instead of monitoring the actual drinking water (Cunningham *et al.*, 2010; Johnson *et al.*, 2008; Vulliet & Cren-Olivé, 2011). Third, in the case of the simultaneous presence of multiple pharmaceutical contaminants, most of the studies did not account for the interactive effect between the PhACs present in the water as toxicological data related to health impacts associated with long-term exposure to such mixtures does not exist (Boxall *et al.*, 2012; Villanueva *et al.*, 2014; World Health Organization, 2012). Fourthly, out of all the risk assessment studies carried out to estimate the health risks of pharmaceuticals in drinking water, only very few considered the presence and effects of the transformation products of the pharmaceuticals present in the water (Cunningham *et al.*, 2010; de Jongh *et al.*, 2012; Schwab *et al.*, 2005).

Since knowledge regarding the formation of these transformation products and their persistence under different environmental conditions is very limited and analytical methods for measuring these transformation products are not well developed, this is a huge drawback in carrying out risk assessment studies. Few researches have made attempts to address these drawbacks and proposed methodologies to overcome some of these constraints.

Dong *et al.* (2013) proposed a new approach to screen the most significant pharmaceutical compounds and rank them based on their potential ecological and human health risks. The prioritization/ranking of the pharmaceutical compounds under consideration was based upon the number of the prescriptions rather than their occurrence in the environment. Also, toxicology data and their percentage removal through metabolism and wastewater treatment also accounted for the rankings. This methodology was applied to rank the 200 most prescribed drugs in the US in 2009. The results of the study suggested that along with the loading data of the pharmaceuticals, their fate in considering the removal mechanisms is a very vital consideration towards risk assessment and hence should be incorporated in further studies.

Houtman *et al.* (2014) carried out a risk assessment of the pharmaceuticals present in Dutch drinking water. The objective of the study was to quantify the lifelong exposure to pharmaceuticals present in the drinking water by the consumers. The study also analysed the health risks associated with the exposure to individual compounds as well as their combined effect. The analysis was performed on using a relatively large data set: 42 pharmaceutical compounds were monitored on a 4-weekly basis for two years. The selection of pharmaceuticals was made on the basis of their intake volume, ease of detection and their toxicities. The results showed that lifetime exposure quantities of the pharmaceuticals were very low, i.e., less than 10% of dose administered in a day. The health risks associated with that exposure to individual compounds, as well as their mixtures, were also found to be negligibly low. The study also concluded that taking maximum concentrations instead of their average value tends to overestimate the health risks.

de Jongh *et al.* (2012) carried out the human health risk assessment of pharmaceuticals present in surface water and drinking water sources in the Netherlands. Multiple samples of untreated surface water, pre-treated surface water and final drinking water were monitored for 17 common pharmaceuticals and nine transformation products. Twelve pharmaceuticals and seven transformation products were found to be present in the samples. As expected, the concentrations were found to be highest in untreated surface water, lower in pre-treated surface water, and negligible in drinking water. However, the analysis showed that the ratios of concentrations of parent compounds to their transformation product were fairly constant across the samples. This suggests that the concentrations of the transformation products could be predicted from the concentrations of their parent compound. Also, the presence of transformation products with similar effects and concentrations as their parent compounds highlight the need to monitor transformation products as well while carrying out such risk assessments.

7.3.1 Analyzing risk quotient

Researchers across the globe have opted for different methodologies for the quantification of human health and ecological risks posed due to the presence to pharmaceuticals in waste and drinking water sources. One of the most common methodologies for such a quantification is the evaluation of Risk Quotient (RQ) values for different pharmaceutical compounds present in the water. RQ represents the relative value of concentration of a compound present in the water to its permissible limit that would not cause any effect. Based on the RQ values, the severity of the potential risks can be categorized into three categories: $RQ < 0.1$ implies low risk; values between 0.1 and 1 implies moderate risk; $RQ > 1$ implies high risk.

For the estimation of human health risks, the exposed population can be divided into different age groups. Age-dependent RQ is calculated by dividing the measured concentrations of each PPCP in ground water by the corresponding age-dependent drinking water equivalent level (DWEL). Similarly, for the estimation of ecological risks, RQ is calculated by dividing the concentration of each PPCP measured in the water by their respective predicted no effect concentration (PNEC) values for the receptor aquatic species under consideration.

Sharma *et al.* (2019) carried out analysis to quantify the potential risks to human health and aquatic ecosystem due to the presence of pharmaceutical and personal care products (PPCPs) and artificial sweeteners (ASWs) in surface and ground water in the Ganges River Basin, India. River water and ground water (untreated) at several sites along the Ganges was monitored for 15 PPCPs and 5 ASWs. The estimation of human health risks from exposure to PPCPs was done considering the ground water concentrations and risks posed to aquatic organisms from exposure to PPCPs and ASWs was done using the concentrations in river water.

Risk assessment was carried out for the worst-case scenario, i.e. considering the maximum measured concentrations of each contaminant. For estimating human health risks, age-dependent Risk Quotient (RQ) was calculated for each contaminant. For the estimation of ecological risks, RQ was calculated for each PPCP and ASW present in the river water for three aquatic species, i.e. algae, daphnia magna and fish.

The results for human health risk assessment showed that among all the contaminants under consideration, carbamazepine, ciprofloxacin, ketoprofen, caffeine, ibuprofen, and triclosan had significantly higher RQ values than the rest. Also, children (1–11 years) had higher RQ values associated with them as compared to adults (>21 years). Since all the values of RQ were well below 1, the analysis suggests that the concentrations of the PPCPs present in the ground water were unlikely to pose any health risks to humans through drinking. Results of the ecological risk assessment showed that only caffeine and triclosan had RQ values above 1 for algae, indicating potential risk to algae. In most of the cases, the RQ values for daphnia magna and fish were lower than 0.1, indicating almost no potential risks to those species.

Wen *et al.* (2014) carried out a similar study to estimate the potential human health risks for different age groups, attributed to the presence of five pharmaceutical compounds in a drinking water source in Shanghai, China. The Huangpu River water was monitored at multiple sites for a two year duration for the presence of five pharmaceuticals, namely ibuprofen (IBP), ketoprofen (KEP), naproxen (NPX), diclofenac (DFC), and clofibric acid (CA). Results of the human health risk assessment showed that DFX had much higher RQ values relative to the other four compounds. It was observed that infants (1–12 months) were relatively much more susceptible to the health risks as compared to the other age groups, though none of the RQ values were above 1, implying a negligible risk to human health via consumption of the drinking water.

Semerjian *et al.* (2018) estimated the human health risks of pharmaceuticals present in treated wastewater, primarily used for landscape irrigation. The effluent from a wastewater treatment plant in Sharjah, UAE, was monitored for the presence and quantification of 10 pharmaceuticals, namely sulfapyridine, sulfamethazine, sulfadiazine, sulfamethoxazole, ciprofloxacin, ofloxacin, erythromycin, acetaminophen, metoprolol and risperidone.

Based on the exposure conditions and receptor characteristics, the exposed population was categorized into three groups: (1) people working in commercial, industrial and street landscaping; (2) children playing in parks, playgrounds, etc.; (3) adults using facilities such as athletic and golf courses areas; all of which are assumed to be irrigated by the treated wastewater from the plant. Dermal contact and accidental ingestion pathways were considered for the assessment of health risks. Exposure through inhalation was not considered as most of the pharmaceuticals are non-volatile and hence unlikely to get inhaled (Crook, 2005). The results of the analysis are shown in Table 7.2. It was observed that dermal exposure had considerably higher risks compared to ingestion. Also, the RQ values were found to be highest for a field worker, followed by a child playing in a park, and lowest for an adult using an athletic field or golf course. For each receptor category, the RQ values were found to be highest for acetaminophen and lowest for sulfamethazine.

7.4 HOSPITAL WASTEWATER CONTAINING TOXIC CHEMICALS AND PHaCS

Hospitals play a vital role in society's health and care process. In the course of healthcare procedures, various types of undesirable and harmful products are

Table 7.2 Risk quotient values corresponding to dermal and ingestion pathways for the average concentration of pharmaceuticals in the effluent of wastewater treatment plant in Sharjah (Source: Semerjian *et al.*, 2018).

Pharmaceuticals	Landscape Worker			Child Playing in a Green Park			Adult Using Athletic Field/Gold Course		
	Dermal	Ingestion	Total	Dermal	Ingestion	Total	Dermal	Ingestion	Total
Sulfapyridine	1.02×10^{-5}	6.52×10^{-6}	1.67×10^{-5}	2.63×10^{-6}	3.96×10^{-7}	3.03×10^{-6}	1.43×10^{-6}	1.70×10^{-7}	1.60×10^{-6}
Sulfamethazine	6.90×10^{-7}	3.63×10^{-7}	1.05×10^{-6}	1.78×10^{-7}	2.20×10^{-8}	2.00×10^{-7}	9.67×10^{-8}	9.43×10^{-9}	1.06×10^{-7}
Sulfadiazine	1.66×10^{-5}	1.43×10^{-5}	3.09×10^{-5}	4.28×10^{-6}	8.66×10^{-7}	5.14×10^{-6}	2.33×10^{-6}	3.71×10^{-7}	2.70×10^{-6}
Sulfamethoxazole	1.35×10^{-5}	6.18×10^{-6}	1.97×10^{-5}	3.48×10^{-6}	3.75×10^{-7}	3.85×10^{-6}	1.89×10^{-6}	1.61×10^{-7}	2.05×10^{-6}
Ciprofloxacin	3.34×10^{-4}	3.54×10^{-4}	6.88×10^{-4}	8.60×10^{-5}	2.15×10^{-5}	1.08×10^{-4}	4.68×10^{-5}	9.21×10^{-6}	5.61×10^{-5}
Ofloxacin	4.32×10^{-5}	8.42×10^{-5}	1.27×10^{-4}	1.11×10^{-5}	5.11×10^{-5}	1.62×10^{-5}	6.05×10^{-6}	2.19×10^{-6}	8.24×10^{-6}
Erythromycin	8.76×10^{-5}	1.41×10^{-4}	2.29×10^{-4}	2.25×10^{-5}	8.56×10^{-6}	3.11×10^{-5}	1.23×10^{-5}	3.67×10^{-6}	1.59×10^{-5}
Acetaminophen	1.54×10^{-3}	5.29×10^{-4}	2.07×10^{-3}	3.97×10^{-4}	3.21×10^{-5}	4.29×10^{-4}	2.16×10^{-4}	1.37×10^{-5}	2.30×10^{-4}
Metoprolol	6.38×10^{-4}	1.64×10^{-4}	8.03×10^{-4}	1.64×10^{-4}	9.97×10^{-6}	1.74×10^{-4}	8.94×10^{-5}	4.27×10^{-6}	9.37×10^{-5}
Risperidone	5.38×10^{-6}	1.23×10^{-6}	6.61×10^{-6}	1.38×10^{-6}	7.48×10^{-8}	1.46×10^{-6}	7.54×10^{-7}	3.21×10^{-8}	7.86×10^{-7}

generated. The management of biomedical waste is a crucial issue in health and environmental management. The rules make it mandatory for the healthcare establishments to segregate, disinfect and dispose their waste in an eco-friendly manner, however not many hospitals are complying with such stipulations. In many cases effluents are left untreated and are discharged into the drainage system for treatment in the sewage treatment plant (STP). Hospital effluent contains various hazardous contaminants, which include pharmaceutical residues and hazardous substances such as hypochlorite, gluteraldehyde and other chemicals discharged from laboratories, dialysis units, etc. (Khan *et al.*, 2019c). These also have traces of radioactive isotopes, cytotoxics and pathogens. The occurrence of drug residues and other hazardous chemical residues in hospital wastewater is a serious threat for the aquatic environment and it should be reduced to the absolute minimum.

Due to so many emerging contaminants in hospital wastewater, it poses physical, chemical and biological risks to the aquatic environment and public health. Hospital effluents having high concentrations of pharmaceutical compounds are the second largest discharge source, besides industrial and urban domestic outlets. However, there are, as such, no legal obligations before discharging hospital effluents into drains or surface water. Since there is so much criticality about the composition of effluent from hospitals and the risk associated with it is huge, there is a dire need to impart proper treatment before discharging it into the drains. However, there are no specific effluent discharge standards in India and many other countries. Pharmaceutical residues have not been given a regulatory status in most industrialized countries. Thus, there is a dire need to identify the contaminants present in hospital effluents and evaluate the performance efficiency of the existing treatment plants so that the necessary augmentations can be made in order to safeguard the environment and public health against the contaminants present in inadequately treated hospital effluents.

In general, hospital effluents are composed of wastewater from different medical activities such as operations, diagnosis, and non-medical activities such as toilets, laundry, etc., including special discharges from laboratories and research facilities (Carraro *et al.*, 2016). Compared to sewage, it contains higher concentrations including a variety of micropollutants 4–50 times higher than sewage such as drugs and their metabolites, heavy metals and contrasting agents (Verlicchi *et al.*, 2012). These may be harmful to humans as well as animals (Chonova *et al.*, 2016). Some of the substances are controlled by local laws but for most of them there are no proper disposal laws and mechanisms in most of the countries in the world (Verlicchi *et al.*, 2012). It is also important to note that while antibiotics are harmful, their residues, including metabolites, are also dangerous to human health. Many studies have reported the presence of such contaminants in higher concentrations and their impact on the environment. Kümmerer (2009) reported that more than one-quarter of antibiotics are discharged from hospitals in

Germany, as compared with Switzerland where the figure is only about 18%. If these are medically prescribed, the figure may be up to 20% in Europe (Schröder *et al.*, 2004), 30% in the UK and about 25% in the USA (Wise, 2002). Its contribution depends upon the study area and targeted compounds, as per the report of Thomas *et al.* (2007), which is only about 2% or even less in treatment plants. In some studies it was found that the contribution of hospitals is about one-third of the total input and targeting specific molecules shows much higher concentrations as compared to others causing environmental risk (Herrmann *et al.*, 2015). Beier *et al.* (2011) reported about 34% of prescribed drugs originating from hospitals and 94% of clarithromycin as targeted molecules. Escher *et al.* (2011) concluded that hospitals contribute about 38% as compared to 62% by sewage. Daouk *et al.* (2015) reported that it is about 29% for Switzerland and up to 77% as targeted molecules such as piperacillin and cisplatin from the hospitals.

Several studies reported that the contribution of the targeted compounds from hospitals was well below 15% (Verlicchi *et al.*, 2012). Ort *et al.* (2010) found that the hospital contribution is much lesser, about 5% for 17 compounds and up to 15% for another 11 compounds under consideration (trimethoprim = 18% and roxithromycin = 56%). Verlicchi *et al.* (2012) concluded in their study that it is about 5% for 32 targeted molecules, up to 15% for 18 molecules and slightly higher than 15% for 12 molecules, including seven antibiotics and one analgesic. The higher concentration of ofloxacin, azithromycin, clarithromycin and metronidazole was also reported. According to Herrmann *et al.* (2015), the consumption of drugs in general hospitals for alimentary tract and the cardiovascular system is lower by nearly 500 times than that of other nearby hospitals, including the same consumption pattern as compared to municipal sewage for some antibiotics such as like cefuroxime and clomethiazole. The contribution by hospitals towards micropollutants in the water can be determined by using different sample strategies and methods. Hospitals are relatively important sources for the entry of antibiotics, analgesics and other therapeutic class drugs. It must also be noted that their contribution also depends on numerous factors such as bed density, location, season and type of healthcare facility, among others (Ort *et al.*, 2010; Verlicchi *et al.*, 2012).

There are very few, or no, regulations relating to hospital wastewater in the world. Even if there are regulations, it is not known if these have been published in the national/regional language. There are no specific directives or guidelines with respect to hospital effluents in Europe. However, the European Directive No. 91 of 21 May 1991 (91/271/CEE, amended by Directive 27 of February 1998 No. 98/15/CE), is related to protection of the aquatic system from the adverse effects of discharge of effluent (both domestic and industrial) from urban areas (Stănescu *et al.*, 2017). Although certain regulations relating to wastewater and hospital effluent have been considered under this pursuit, the characterization of hospital effluents differs from urban wastewater. Environmental drug contamination with high concentrations of these pharmaceuticals, however, can be derived from other sources, such as livestock, slaughterhouses, aquaculture and agriculture and, in some cases, with a greater total concentration of pharmaceutical compounds (Carraro *et al.*, 2016; Emmanuel *et al.*, 2005). There are very few elaborate guidelines in the world regarding the management of PhAC's and the disposal of hospital wastewater. The line between the management of PhACs and industrial effluent has always been a complex issue and a matter of debate. There are many guidelines on the disposal of solid waste, but hospital effluents and PhACs has not been a major focus for the authorities in the past. The Environment Protection Law quotes a special section regarding protection of the environment with respect to hospitals. Article 72 of this law makes it obligatory for hospitals to collect and treat wastewater, in accordance with the prescribed standards, depending upon its end use.

Specific micropollutants are also quantified in the case of industrial or hospital effluents, such as adsorbable organic halogens (AOX), total and free chlorine, detergents, disinfectants, tensio-actives, oil and grease, sulphates, cyanides, organophosphates, total nitrogen, heavy metals and microbiological indicators (total coliform, faecal coliform or *Escherichia coli*) and toxicity. The guidelines and legislations suggest various parameters to be checked before disposal (Khan *et al.*, 2016). The standards for parameters have been reviewed at the level of on-site treated hospital effluent, hospital effluent leading to municipal sewage treatment plants and direct disposal onto the surface aquatic environment (Amouei *et al.*, 2015; Carraro *et al.*, 2016)

The Clean Water Act (CWA) is an environment protection law enacted in the United States of America. As per CWA, the publication of specific regulations and permits are made regarding the discharge by water pollution sources. It is mandatory to comply with the effluent guidelines regarding technology-based parameters and other site-specific effluent limitations before discharging the matter into common wastewater treatment plants, or the aquatic environment. The EPA also highlights that inadequately treated hospital effluents result in water-borne problems by degrading the drinking water quality and thus pose more environmental problems. For an efficient effluent management system a special budget should be allocated for a proper maintenance and management system for hazardous effluents. With proper planning and implementation, the risk of water-borne epidemics can be reduced with the use of even a basic treatment system. By using advanced treatment systems, the risk can be further reduced. The EPA has made it mandatory for the hospitals to obtain authorization from the State Environmental Agency or the Environmental Protection Agency and comply with the national discharge standards laid out by them to restrict the concentrations of hazardous pollutants in the hospital effluents (Frédéric & Yves, 2014).

In Germany, hospital effluents are considered as domestic effluents since there are no legal obligations to treat the wastewater before it is discharged into surface water. However, in other countries there are certain prescribed limits with respect to physio-chemical and biological parameters. If these are met, the effluent is considered as equivalent to domestic wastewater and thus can be discharged without any authorization (Ternes, 1998) In the case of Italy, the healthcare facilities having a bed strength of less than 50 with pathological labs can discharge the effluent as domestic wastewater without any permission. Hospitals with more than 50 beds having lab and other facilities have to seek authorization in accordance with the Italian Legislative Decree No. 152/2006 (Castiglioni *et al.*, 2006).

7.5 POTENTIAL IMPLICATIONS OF DISCHARGING UNTREATED/PARTIALLY TREATED WASTEWATER IN THE AQUATIC ENVIRONMENT

Effluent from healthcare facilities is generally loaded with a cocktail of chemicals, formalin, gluteraldehyde, radioactive compounds, detergents, pharmaceutical residues, disinfectants, pathogens, tissue cultures, infected blood and urine, pharmaceutical residues, etc. When a pharmaceutical is administered to a patient, it is metabolized in the body and it is either excreted as unchanged pharmaceutical residue or its metabolites. These pharmaceutical residues and chemical traces pose a risk to the ecological balance and public health. Hospital wastewater, if left untreated, could lead to serious outbreaks of communicable diseases, diarrhea epidemics, cholera, water and soil contamination and radioactive pollution. Hospital wastewater contains many toxic chemicals and biological components that are needed to be treated onsite before they reach water bodies. Inadequately treated hospital effluent becomes a major source of intrusion of hazardous waste into public sewers. These toxic chemicals discharged by hospitals hamper the quality of water being supplied for consumptive purposes and causes skin diseases or enteric illness (Gautam *et al.*, 2007). Disinfectants and detergents are used in huge quantities in healthcare facilities and are a mixture of highly complex active substances. Their traces are also present in hospital effluent. The sludge generated by hospitals also contains toxicants having the capability of percolating down and contaminating the ground water reserves.

7.6 IMPACT OF PHARMACEUTICAL RESIDUES ON HUMANS AND THE ECOSYSTEM

The presence and persistence of PhACs in the aquatic ecosystem poses serious concerns for various living species and affects humans directly or indirectly. The exposure cycle starts from aquatic species before entering the food chain, resulting in biomagnification. In some cases, it is also observed that these compounds enter the human body directly by consumption of water from contaminated reserves. Apart from human consumption, a major proportion of pharmaceutical consumption is attributed to animal husbandry and agricultural practices which is far greater than human consumption (Sarmah *et al.*, 2006).

Several hormones and estrogens, such as steroids, oxytocin, progesterone, HCG, are widely used for managing animal milk production, growth and reproduction. Consequently, the pharmaceuticals getting passed into the environment through animal waste go unnoticed and pose a much broader problem as it is scarcely treated compared to human waste.

Although the presence of pharmaceuticals is reported widely in drinking water supplies posing health concerns for humans, their concentrations are generally found to be very low compared to therapeutic doses. When a human is administered a pharmaceutical, 10–90% of it is excreted unchanged; and the remainder is excreted in the form of metabolites. It is regarded that any pharmaceutical having an excretion factor of more than 10% should be considered important from the point of view of the environment (De Voogt *et al.*, 2009).

Chiral pharmaceuticals result in enantiospecific toxicity on different species. However, in the case of human beings their effect and potency of enantiomers differ. Thus, their effect and aquatic species also varies. For instance, ibuprofen S-enantiomer is far more active than its R-enantiomer. A small planktonic crustacean, Daphnia magna growth indicates higher inhibition due to S-atenolol as compared to the case of R-atenolol (De Andrés *et al.*, 2009). However, the same species is not affected by stereospecific toxicity (Stanley *et al.*, 2007).

Recent studies point out the presence of low concentrations of antibiotics in municipal effluents and surface water (Golet *et al.*, 2002; Heberer, 2002). India is one of the largest consumers of antibiotics in the world. Their presence is also ubiquitous in water bodies, groundwater etc. There are ample research papers pointing towards their occurrence in surface water. The persistence of antibiotics in surface water is an environmental concern since antibiotic contaminants pose a threat to microbial ecology, amplify the proliferation of antibiotic resistant microbes, and could pose a danger to public health. The occurrence of antibiotic residues in effluent hampers conventional biological treatment systems. Thus, they show resistance to treatment and become persistent in nature. This also poses a challenge for wastewater reutilization and stops the hospitals from achieving zero discharge concepts. The prevalence of the antibiotics erythromycin, amikacin, amoxicillin and vancomycin isolates have been found in hospital wastewater (Kümmerer, 2009).

Estrogens have also been found in aquatic systems in the ng/L range. It is evident from past research studies conducted on pharmaceuticals in aquatic environments that sex hormones have a potent implication on wildlife at concentrations far below 1 mg/L. For example, estradiol, the female sex hormone, is commonly present in water bodies and has the potential to alter the sex characteristics of certain fish at concentrations of ng/L (Isidori *et al.*, 2009). Sex hormones are known to affect the behavior of aquatic organisms and it has also affected egg laying and spawning capacities.

Cytotoxic substances pose a risk to humans and the environment in terms of their potential to have carcinogenicity, mutagenic effects and fetotoxic properties and are commonly used as a pharmaceutical in hospitals for treatment. Expected concentrations in hospital effluents are 5–50 mg/L for single compounds (Fent *et al.*, 2006). These substances have lower biodegradability. Synergistic and additive toxic effects of 5-fluorouracil with β-lactam antibiotics, cephalosporines, norfloxacin and other antibiotic agents against bacteria, as described in the medical literature, also occurred in the presence of wastewater from hospitals.

Since pharmaceutical residues have been found in surface water, ground water and even in drinking water, the study of their impact has been a topic of research. The impact of pharmaceutical residues belonging to different therapeutic classes on various aquatic organisms is elaborated in Table 7.3.

7.6.1 Disease outbreaks and ecological hazards

Antibiotic traces in the aquatic environment induces resistance in bacterial strains, hence reducing their effectiveness in the treatment of humans (Khan *et al.*, 2002). Epidemic diseases in hospitals are often caused by resistant Klebsiellae strains. The Klebsiellae from a treatment plant, in which 90% exhibited insensitivity to ampicillin and 6% showed multiple resistances, have been isolated. Important groups of antibiotics such as quinolones, nitromidazoles and sulfonamides are of low biodegradability (Emmanuel *et al.*, 2002). The indicators of viral pollution like enterovirus and adenovirus have been found pertinent in the wastewater generated by hospitals. Moreover, HIV, the causing agent of AIDS, has also been isolated from biological liquids and the feces of infected patients. The presence of infectious particles (IPs) of HIV per liter in the range of 1.4×10^{-2} and 8.6×10^{-1} (lower than 1 IP of HIV per liter of effluent) in the city of Chicago has been reported(Escher *et al.*, 2011).

Effluent streams from healthcare facilities also contain chlorinated molecules which are a component of detergents and disinfectants and also other heavy metals such as mercury and silver used in dentistry, thermometers etc. The presence of AOX is probably due to iodized X-ray contrast media, surfactants, disinfectants, cleaners and pharmaceuticals containing chlorine (Khan *et al.*, 2020b). Studies have been carried out in the hospitals of Germany wherein the AOX concentration at the points of discharge into the public drainage system was found to be in the range of 0.13–0.94 mg/L, and from hospitals alone, the range is significantly high (Emmanuel *et al.*, 2005). It has also been found that the AOX concentration in the urine of patients not treated with pharmaceuticals is very low. The lesser the dilution, the more are its effects. In this case the dilution is more, which substantially reduces its effect (Dhingra *et al.*, 2017). Besides, genotoxic substances have also been found which are suspected to be a possible reason for the cancers observed in the last decades (Chen *et al.*, 2015).

Certain pharmaceuticals having higher bioconcentration factors are supposed to bioaccumulate and also have a tendency to show biomagnification. The mutagenic

Table 7.3 Toxicological impact of pharmaceuticals on aquatic and other non-target organisms.

Name	Species	Concentration (μg/L)	Effects	Reference
Ethinyl estradiol	Pimephales promelas	0.004	Secondary sexual characteristics/Altered sex ratio/feminization/VTG	Corcoran *et al.* (2010)
Estradiol	Pimephales promelas	0.781	Reduction in reproductive output	Corcoran *et al.* (2010)
Tamoxifen	Pimephales promelas	5.6	Inhibition of Reproductive output/VTG/Gonadal histology	Corcoran *et al.* (2010)
Letrozole	Oryziaslatipes	5	Fecundity/fertility/VTG	Corcoran *et al.* (2010)
5α-Dihydrotestosterone	Pimephales promelas	6	Masculinization of females/Vtg induction in females	Corcoran *et al.* (2010)
Cyproterone	Fundulus heteroclitus	0.1	PlasmaT and 11-KT reduction in males	Corcoran *et al.* (2010)
Deltamethrin	Chironomids		Killing of larvae in sediments	Åkerblom *et al.* (2008)
Ibuprofen	Medaka	1.00–100.00	Change of reproduction pattern	Lúcia *et al.* (2010)
	Rainbow trout	1000	Impairment of ion regulation	
Indomethacin	Zebrafish	100,000	Disruption of oocyte maturation/ovulation	Lúcia *et al.* (2010)
Naproxen	Carp	230,260	Inhibition of CYP2M	Lúcia *et al.* (2010)
Ketoprofen	Carp	254,280	Inhibition of CYP2M	Quinn *et al.* (2008)
Fluoxetine	Bluehead wrasse	6	Decreased territorial aggression	Quinn *et al.* (2008)

	Striped bass	23.20–100.00	Decreased ability to catch prey	
	Fathead minnow	51.00–170.00	Decreased feeding rate in Fathead minnow	
	Medaka	0.10–0.50	Decreased growth	
	Fathead minnow	51.00–53.00	Developmental abnormalities	
	Medaka	0.10–5.00	Developmental abnormalities	
Fenofibrate	Rainbow trout	19,124.52	Cytotoxic	Bound and Voulvoulis (2004)
	Goldfish	1500	Reduced testosterone levels	
Gemfibrozil		250,340	Inhibition of CYP2M	Bound and Voulvoulis (2004)
Propranolol	Medaka	500	Reduced growth	Bound and Voulvoulis (2004)
	Rainbow trout	10,000	Impaired growth	
	Zebrafish			
	Fraysee	7002.18	Altered heart rate	
	Rainbow trout	2,593,400	Altered blood flow in gills	
	Medaka	1296.70	Drop in glucose production	
		5	Total number of eggs	
Ivermectin	Beetles and dung fly		Kiling eggs and larvae in dung	Brinke *et al.* (2010)
Atenolol	Fathead minnow	10,000.00	Impaired growth	Emmanuel *et al.* (2005)
	Fathead minnow	>10,000.00	Number of viable eggs hatched	

(Continued)

Table 7.3 Toxicological impact of pharmaceuticals on aquatic and other non-target organisms (*Continued*).

Name	Species	Concentration (μg/L)	Effects	Reference
Sulphonamides	Medaka	>100,000.00	Acute toxicity	Emmanuel *et al.* (2005)
(Oxy) tetracyclines	Rainbow trout	0.10–50.0	Suppressed immune system	Emmanuel *et al.* (2005)
Metoprolol	*D. subspicatus*	8.0	Growth inhibition	Godoy *et al.* (2015)
	Lemna minor		NOEC (growth inhibition, frond area)	
	D. magna	0.16	Immobilization	
Verapamil	*R. subcapitata*		Growth inhibition	Godoy *et al.* (2015)
	Spirostomum ambiguum		Morphological changes	
	D. magna		Reproduction	
	O. mykiss		Mortality	
	Cyprinus carpio		Malformations and edema with hematocele	
Diclofenac	Fish	1 5	Proteomic analysis showed effects in chloroplasts	Jones *et al.* (2001)
Diclofenac	Vultures		Population decline	Oaks *et al.* (2004)
Carbamazepine	Algae	150,000	Growth inhibition	Jones *et al.* (2001)
Carbamazepine	Rainbow trout		Oxidative stress	Li *et al.* (2010)
Sulfamethoxazole	Algae	0–2500		Jones *et al.* (2001)

			Chronic toxicity effects for the photosynthetic apparatus	
Neomycin	Algae	1000	Mortality	Mutiyar and Mittal (2014)
Sulfamethoxazole and levofloxacin	*L. gibba*	80	Phytotoxic effects	Mutiyar and Mittal (2014)
Fluoroquinolones	Other plants	60	Affecting chloroplast replication	Mutiyar and Mittal (2014)
Ciprofloxacin	*M. aeruginosa*	5.0	Growth inhibition	Mutiyar and Mittal (2014)
Sulfadiazine	Soil bacteria		Community alterations	Hammesfahr *et al.* (2008)
Sulfamethoxazole	*P. subcapitata*	90	Growth inhibition	Mutiyar and Mittal (2014)
Sulfamethoxazole	*S. leopoliensis*	30	Growth inhibition	Mutiyar and Mittal (2014)
Streptomycin	*C. reinhartii*	660	Growth inhibition	Mutiyar and Mittal (2014)
Erythromycin	*S. capricorntum*	20	Mortality	Mutiyar and Mittal (2014)
Sulfamethoxazole	*C. dubia*	210	Growth inhibition	Mutiyar and Mittal (2014)
Neomycin	*M. macrocopa*	740	Mortality	Mutiyar and Mittal (2014)
Ampicillin	*Vibrio fischeri*	163,000	Mortality	Mutiyar and Mittal (2014)
Sulfamethaxazole	Rice and oat plant		Inhibits germination	Liu *et al.* (2009)
Cyclophosphamide	*P. subcapitata*	<30–64.8	Growth inhibition	Lúcia *et al.* (2010)
Cyclophosphamide	*D. magna*	<6–143	Reproduction inhibition	Lúcia *et al.* (2010)
Ifosfamide	*V. fischeri*	<0.3–5	histopathological alterations	Lúcia *et al.* (2010)

(*Continued*)

Table 7.3 Toxicological impact of pharmaceuticals on aquatic and other non-target organisms (*Continued*).

Name	Species	Concentration (µg/L)	Effects	Reference
Tamoxifen	*B. calyciflorus*	143–215	histopathological alterations	Lúcia *et al.* (2010)
Tamoxifen	*T. platyurus*	0.2–15	Mortality	Lúcia *et al.* (2010)
Ibuprofen	Algae	92 and 920	Amplified fragment length polymorphism assessment of DNA damage and 2-DE proteomic analysis	Minguez *et al.* (2016)
Fluoxetine	Mollusks	300 and 3000	Reproductive alterations	Minguez *et al.* (2016)
Sertraline	Fish	3, 10, and 30	Binding activity of serotonin reuptake transporter	Minguez *et al.* (2016)
Fluoxetine, carbamazepine, and diazepam	Insect and crustacean	0.1	Mortality	Minguez *et al.* (2016)
Enrofloxacin	Duckweed and green algae			Ebert *et al.* (2011)

effect of hospital effluents has been revealed using Ames and Hamster Cell tests on hospital effluent (Emmanuel *et al.*, 2002). The hospital effluent depicted a high toxicity as determined using the Daphnia and Luminescent bacteria tests. Certain pharmaceuticals have experimentally found to be ecotoxic in lower organisms (Emmanuel *et al.*, 2005). Pharmaceuticals such as simvastatin, diclofenac, ibuprofen, sorafenib and gemfibrozil have high bio concentration values which depict their potential for bioaccumulation. The ecotoxicity of an individual compound is less than that of a mixture of compounds as they may show additive and synergistic effects.

7.6.2 DNA damage and genotoxicity

The genotoxic potential of wastewater generated by Rouen University Hospital was approximated by SOS chromo and Ames fluctuation tests. The samples taken from the hospitals were found to be genotoxic. The residues of fluoroquinolones such as ciprofloxacin were experimentally found to be genotoxic and that has been confirmed by UMUC test. The teratogenic effects of certain pharmaceutical residues are also known. Many antibiotics and chemotherapy pharmaceuticals exhibit the destruction of genetic material (DNA) of both prokaryotes and eukaryotes.

Many drugs used in hospitals (e.g. antibiotics, cytostatic drugs) exhibit DNA damage toward bacteria or eukaryotic cells and these raise concerns about the human and ecological hazard of hospital wastewater (Zenker *et al.*, 2014). Hartmann *et al.* (1999) confirmed the genotoxic potential of hospital wastewater by conducting a chromosome aberration test on hamster cells. UMUC assay was also used to confirm the genotoxic nature of wastewater from an oncology hospital. This suggests that hospital effluent could be a source of genotoxicity in lower organisms and the addition of mutagens into the aquatic ecosystem (Huschek *et al.*, 2004). After entering the body of a human being, the antibiotic residue interacts with microbiota containing various microorganisms. This has resulted in alteration in the intestinal microbial population as it greatly affects the host microbiome (Blaser, 2016). This disturbance in the intestine causes various detrimental species like bacteria and pathogens to cause intestinal ailments, colorectal cancer and pseudo membranous colitis. Sometimes, if these bacteria acquire resistance to antibiotics, they grow unprecedently, becoming fatal (Ben *et al.*, 2019).

Persistence can be correlated with accumulation of pharmaceutical residues in seafood and vegetables. A higher levels of contaminants has been detected in the fishes as compared to the surrounding water. This implies that the exposure of these contaminants is more likely through seafood rather than drinking water. To date, data on the extent to which these residues can be transferred through the food chain and their biomagnification level is missing. The bioaccumulation potential of pharmaceuticals, including carbamazepine and a few antibiotics, has also been studied in vegetables such as cabbage and wiscosin fast plants. These residues have been found in leaves and seedpods (Kümmerer, 2009). The metabolites of triclosan have also been found in carrot plants grown in contaminated soil (Deblonde & Hartemann, 2013). The changes in the microbial population which can cause hazard to humans because of the traces of sulphonamides in agricultural fields has been reported (Jones *et al.*, 2001). Bean plants have also been found to contain traces of 17a-ethinyl estradiol and triclosan in their roots and also a comparatively lower concentration in their leaves. Metformin traces have also been found in parts of potato, carrot, wheat, barley and oat grains (Deblonde & Hartemann, 2013).

The pollution loads from hospitals has grown multifold with the microbiological load having bactericides which may hamper the biological processes involved in wastewater treatment. Though the hospital wastewater is diluted to a higher degree, there is still some possibility that certain substances can cause a biological imbalance in the aquatic environment. To protect the aquatic system from excess loads of pollutants released from hospitals, it is important to provide treatment to hospital effluent before it is discharged in the municipal sewage system (Patel *et al.*, 2019).

7.7 CONCLUSION AND RECOMMENDATIONS

Pharmaceutical compounds in one form or another have been in continuous use since long back but concerns regarding their health effects on our ecosystem and different species, including humans, has engrossed attention majorly in the last two decades. Despite several studies available on pharmaceutical compounds, their health effects on humans, flora and fauna lacks sufficient information. However, most of the relevant studies carried out in the past have indicated that these pharmaceutical compounds, along with other chemicals emerging from various other sources, combine together to form new compounds posing higher toxicity and a greater threat to the ecosystem. Guidelines for the discharge of toxic compounds like PhACs need to be proposed and implemented globally to restrict the unregulated production and consumption of these which results in reckless discharge into streams. Generally, hospital wastewater contains several micropollutants which get discharged from various medical and non-medical treatment activities having sources from OPD waste, laboratory waste, operation theatre waste, radiology rooms and kitchens. It mainly contains antibiotics, disinfectants, heavy metals and are either mixed with domestic waste or discharged to the environment. This has resulted in bacteria becoming antibiotic resistant, ineffectiveness of antibiotics and development of antibiotic resistant genes which makes the problem graver. Effluent streams from the pharmaceutical industry need to be monitored and strict regulatory measures for discharge needs to be well implemented. A large number of pharmaceuticals being used for healthcare and treatment strengthen the need to prioritize pharmaceutical residues. The prioritization will enable regulating authorities to monitor the pharmaceutical residue concentration in surface, ground and potable water. The approach of selecting high priority pharmaceuticals provides an efficient and practical base to manage the risks related to discharge of pharmaceuticals into water bodies. Health risk mitigation necessitates cost utility assessment aimed at better infrastructure facilities for wastewater treatment. Country specific guidelines and localized pathogen eradication methods should be executed involving industrial cooperation and government bodies, private stake holders.

REFERENCES

Åkerblom N., Arbjörk C., Hedlund M. and Goedkoop W. (2008). Deltamethrin toxicity to the midge Chironomus riparius Meigen – effects of exposure scenario and sediment quality. *Ecotoxicology and Environmental Safety*, **70**(1), 53–60.

Amouei A., Asgharnia H., Fallah H., Faraji H., Barari R. and Naghipour D. (2015). Characteristics of effluent wastewater in hospitals of Babol University of Medical Sciences, Babol, Iran. *Health Scope*, **4**(2), e23222.

aus der Beek T., Weber F. A., Bergmann A., Hickmann S., Ebert I., Hein A. and Küster A. (2016). Pharmaceuticals in the environment – global occurrences and perspectives. *Environmental Toxicology and Chemistry*, **35**(4), 823–835.

Balakrishna K., Rath A., Praveenkumarreddy Y., Guruge K. S. and Subedi B. (2017), A review of the occurrence of pharmaceuticals and personal care products in Indian water bodies. *Ecotoxicology and Environmental Safety*, **137**, 113–120.

Batt A. L., Bruce I. B. and Aga D. S. (2006). Evaluating the vulnerability of surface waters to antibiotic contamination from varying wastewater treatment plant discharges. *Environmental Pollution*, **142**(2), 295–302.

Beier S., Cramer C., Köster S., Mauer C., Palmowski L., Schröder H. F. and Pinnekamp J. (2011). Full scale membrane bioreactor treatment of hospital wastewater as forerunner for hot-spot wastewater treatment solutions in high density urban areas. *Water Science and Technology*, **63**(1), 66–71.

Ben Y., Fu C., Hu M., Liu L., Wong M. H. and Zheng C. (2019). Human health risk assessment of antibiotic resistance associated with antibiotic residues in the environment: a review. *Environmental Research*, **169**, 483–493.

Benotti M. J., Trenholm R. A., Vanderford B. J., Holady J. C., Stanford B. D. and Snyder S. A. (2009). Pharmaceuticals and endocrine disrupting compounds in U.S. drinking water. *Environmental Science and Technology*, **43**(3), 597–603.

Blaser, M. J. (2016). Antibiotic use and its consequences for the normal microbiome. *Science*, **352**(6285), 544–545.

Bound J. P. and Voulvoulis N. (2004). Pharmaceuticals in the aquatic environment – a comparison of risk assessment strategies. *Chemosphere*, **56**(11), 1143–1155.

Bound J. P. and Voulvoulis N. (2005). Household disposal of pharmaceuticals as a pathway for aquatic contamination in the United Kingdom. *Environmental Health Perspectives*, **113**(12), 1705–1711.

Boxall A. B. A., Rudd M. A., Brooks B. W., Caldwell D. J., Choi K., Hickmann S., Innes E., Ostapyk K., Staveley J. P., Verslycke T. and Ankley G. T. (2012). Pharmaceuticals and personal care products in the environment: What are the big questions? *Environmental Health Perspectives*, **120**(9), 1221–1229.

Brack W., Dulio V. and Slobodnik J. (2012). The NORMAN Network and its activities on emerging environmental substances with a focus on effect-directed analysis of complex environmental contamination. *Environmental Sciences Europe*, **24**(1), 1–5.

Brinke M., Höss S., Fink G., Ternes T. A., Heininger P. and Traunspurger W. (2010). Assessing effects of the pharmaceutical ivermectin on meiobenthic communities using freshwater microcosms. *Aquatic Toxicology*, **99**(2), 126–137.

Carlsson C., Johansson A. K., Alvan G., Bergman K. and Kühler T. (2006). Are pharmaceuticals potent environmental pollutants? Part I: Environmental risk assessments of selected active pharmaceutical ingredients. *Science of the Total Environment*, **364**(1–3), 67–87.

Carraro E., Bonetta S., Bertino C., Lorenzi E., Bonetta S. and Gilli G. (2016). Hospital effluents management: chemical, physical, microbiological risks and legislation in different countries. *Journal of Environmental Management*, **168**, 185–199.

Castiglioni S., Bagnati R., Fanelli R., Pomati F., Calamari D. and Zuccato E. (2006). Removal of pharmaceuticals in sewage treatment plants in Italy. *Environmental Science and Technology*, **40**(1), 357–363.

Chen Y., Xi X., Yu G., Cao Q., Wang B., Vince F. and Hong Y. (2015). Pharmaceutical compounds in aquatic environment in China: locally screening and environmental risk assessment. *Frontiers of Environmental Science and Engineering*, **9**(3), 394–401.

Chonova T., Keck F., Labanowski J., Montuelle B., Rimet F. and Bouchez A. (2016). Separate treatment of hospital and urban wastewaters: a real scale comparison of effluents and their effect on microbial communities. *Science of the Total Environment*, **542**, 965–975.

Christou A., Karaolia P., Hapeshi E., Michael C. and Fatta-Kassinos D. (2017). Long-term wastewater irrigation of vegetables in real agricultural systems: concentration of pharmaceuticals in soil, uptake and bioaccumulation in tomato fruits and human health risk assessment. *Water Research*, **109**, 24–34.

Corcoran J., Winter M. J. and Tyler C. R. (2010). Pharmaceuticals in the aquatic environment: a critical review of the evidence for health effects in fish. *Critical Reviews in Toxicology*, **40**(4), 287–304.

Crook J. (2005). Irrigation of Parks, Playgrounds, and Schoolyards with Reclaimed Water: Extent and Safety. WateReuse Foundation, Alexandria.

Cunningham V. L., Perino C., D'Aco V. J., Hartmann A. and Bechter R. (2010). Human health risk assessment of carbamazepine in surface waters of North America and Europe. *Regulatory Toxicology and Pharmacology*, **56**(3), 343–351.

Daouk S., Chevre N., Vernaz N., Bonnabry P., Dayer P., Daali Y. and Fleury-Souverain S. (2015). Prioritization methodology for the monitoring of active pharmaceutical ingredients in hospital effluents. *Journal of Environmental Management*, **160**, 324–332.

Daughton C. G. (2004). Non-regulated water contaminants: emerging research. *Environmental Impact Assessment Review*, **24**(7–8), 711–732.

De Andrés F., Castañeda G. and Ríos Á. (2009). Use of toxicity assays for enantiomeric discrimination of pharmaceutical substances. Chirality: The Pharmacological, Biological, and Chemical Consequences of Molecular Asymmetry, **21**(8), 751–759.

Deblonde, T. and Hartemann, P. (2013). Environmental impact of medical prescriptions: assessing the risks and hazards of persistence, bioaccumulation and toxicity of pharmaceuticals. *Public Health*, **127**(4), 312–317.

Deblonde T., Cossu-Leguille C. and Hartemann P., (2011), Emerging pollutants in wastewater: A review of the literature. *International Journal of Hygiene and Environmental Health*, **214**(6), 442–448.

de Jongh C. M., Kooij P. J. F., de Voogt P. and ter Laak T. L. (2012). Screening and human health risk assessment of pharmaceuticals and their transformation products in Dutch surface waters and drinking water. *Science of the Total Environment*, **427–428**, 70–77.

De Voogt P., Janex-Habibi M. L., Sacher F., Puijker L. and Mons M. (2009). Development of a common priority list of pharmaceuticals relevant for the water cycle. *Water Science and Technology*, **59**(1), 39–46.

Dhingra A., Ahmed S. W. A., Gautam S. and Khan A. A. (2018). Biomedical waste generation and its management in hospitals of NCR-Delhi, India. *Journal Research in Environmental and Life Sciences*, **11**(3), 78–83.

Dong Z., Senn D. B., Moran R. E. and Shine J. P. (2013). Prioritizing environmental risk of prescription pharmaceuticals. *Regulatory Toxicology and Pharmacology*, **65**(1), 60–67.

Du B., Price A. E., Scott W. C., Kristofco L. A., Ramirez A. J. and Chambliss C. K., *et al.*, (2014). Comparison of contaminants of emerging concern removal, discharge, and water quality hazards among centralized and on-site wastewater treatment system effluents receiving common wastewater influent. *Science of the Total Environment*, **466–467**, 976–984.

Ebert I., Bachmann J., Kühnen U., Küster A., Kussatz C., Maletzki D. and Schlüter C. (2011). Toxicity of the fluoroquinolone antibiotics enrofloxacin and ciprofloxacin to photoautotrophic aquatic organisms. *Environmental Toxicology and Chemistry*, **30** (12), 2786–2792.

Emmanuel E., Perrodin Y., Keck G., Blanchard J. M. and Vermande P. (2002). Effects of hospital wastewater on aquatic ecosystem. In Proceedings of the XXVIII Congreso Interamericano de Ingenieria Sanitaria y Ambiental, Cancun, México, pp. 27–31.

Emmanuel E., Perrodin Y., Keck G., Blanchard J. M. and Vermande P. (2005). Ecotoxicological risk assessment of hospital wastewater: a proposed framework for raw effluents discharging into urban sewer network. *Journal of Hazardous Materials*, **117**(1), 1–11.

Escher B. I. and Fenner K. (2011). Recent advances in environmental risk assessment of transformation products. *Environmental Science and Technology*, **45**(9), 3835–3847.

Escher B. I., Baumgartner R., Koller M., Treyer K., Lienert J. and McArdell C. S. (2011). Environmental toxicology and risk assessment of pharmaceuticals from hospital wastewater. *Water Research*, **45**(1), 75–92.

Fent K., Weston A. A. and Caminada D. (2006). Ecotoxicoloy of human pharmaceuticals. *Aquatic Toxicology*, **76**, 122–159.

Frédéric O. and Yves P. (2014). Pharmaceuticals in hospital wastewater: their ecotoxicity and contribution to the environmental hazard of the effluent. *Chemosphere*, **115**, 31–39.

Gautam A. K., Kumar S. and Sabumon P. C. (2007). Preliminary study of physico-chemical treatment options for hospital wastewater. *Journal of Environmental Management*, **83**(3), 298–306.

Godfrey E., Woessner W. W. and Benotti M. J. (2007). Pharmaceuticals in on-site sewage effluent and ground water, Western Montana. *Ground Water*, **45**(3), 263–271.

Godoy A. A., Kummrow F. and Pamplin P. A. Z. (2015). Occurrence, ecotoxicological effects and risk assessment of antihypertensive pharmaceutical residues in the aquatic environment – A review. *Chemosphere*, **138**, 281–291.

Golet E. M., Alder A. C. and Giger W. (2002). Environmental exposure and risk assessment of fluoroquinolone antibacterial agents in wastewater and river water of the Glatt Valley Watershed, Switzerland. *Environmental Science and Technology*, **36**(17), 3645–3651.

Grenni P., Ancona V. and Caracciolo A. B. (2018). Ecological effects of antibiotics on natural ecosystems – A review. *Microchemical Journal*, **136**, 25–39.

Halling-Sørensen B., Nors Nielsen S., Lanzky P. F., Ingerslev F., Holten Lützhøft H. C. and Jørgensen S. E. (1998). Occurrence, fate and effects of pharmaceutical substances in the environment – A review. *Chemosphere*, **36**(2), 357–393.

Hammesfahr U., Heuer H., Manzke B., Smalla K. and Thiele-Bruhn S. (2008). Impact of the antibiotic sulfadiazine and pig manure on the microbial community structure in agricultural soils. *Soil Biology and Biochemistry*, **40**(7), 1583–1591.

Hartmann A., Golet E. M., Gartiser S., Alder A. C., Koller T. and Widmer R. M. (1999). Primary DNA damage but not mutagenicity correlates with ciprofloxacin concentrations in German hospital wastewaters. *Archives of Environmental Contamination and Toxicology*, **36**(2), 115–119.

Heberer T. (2002). Occurrence, fate, and removal of pharmaceutical residues in the aquatic environment: a review of recent research data. *Toxicology letters*, **131**(1–2), 5–17.

Hernández F., Sancho J. V., Ibáñez M. and Guerrero C. (2007). Antibiotic residue determination in environmental waters by LC-MS. *TrAC Trends in Analytical Chemistry*, **26**(6), 466–485.

Herrmann M., Olsson O., Fiehn R., Herrel M. and Kümmerer K. (2015). The significance of different health institutions and their respective contributions of active pharmaceutical ingredients to wastewater. *Environment International*, **85**, 61–76.

Houtman C. J., Kroesbergen J., Lekkerkerker-Teunissen K. and van der Hoek J. P. (2014). Human health risk assessment of the mixture of pharmaceuticals in Dutch drinking water and its sources based on frequent monitoring data. *Science of the Total Environment*. **496**, 54–62.

Huschek G., Hansen P. D., Maurer H. H., Krengel D. and Kayser A. (2004). Environmental risk assessment of medicinal products for human use according to European Commission recommendations. *Environmental Toxicology: An International Journal*, **19**(3), 226–240.

Isidori M., Bellotta M., Cangiano M. and Parrella, A. (2009). Estrogenic activity of pharmaceuticals in the aquatic environment. *Environment International*, **35**(5), 826–829.

Jjemba, P. K. (2006). Excretion and ecotoxicity of pharmaceutical and personal care products in the environment. *Ecotoxicology and Environmental Safety*, **63**(1), 113–130.

Johnson A. C., Jürgens M. D., Williams R. J., Kümmerer K., Kortenkamp A. and Sumpter J. P. (2008). Do cytotoxic chemotherapy drugs discharged into rivers pose a risk to the environment and human health? An overview and UK case study. *Journal of Hydrology*, **348**(1–2), 167–175.

Jones O. A. H., Voulvoulis N. and Lester J. N. (2001). Human pharmaceuticals in the aquatic environment a review. *Environmental Technology*, **22**(12), 1383–1394.

Jones O. A., Lester J. N. and Voulvoulis N. (2005). Pharmaceuticals: A threat to drinking water? *Trends in Biotechnology*, **23**(4), 163–167.

Keller A. A., McFerran S., Lazareva A. and Suh S. (2013). Global life cycle releases of engineered nanomaterials. *Journal of Nanoparticle Research*, **15**(6), 1–17.

Kern S., Baumgartner R., Helbling D. E., Hollender J., Singer H. and Loos M. J., *et al.* (2010). A tiered procedure for assessing the formation of biotransformation products of pharmaceuticals and biocides during activated sludge treatment. *Journal of Environmental Monitoring*, **12**(11), 2100–2111.

Khan A., Das S. C., Ramamurthy T., Sikdar A., Khanam J., Yamasaki S., Takeda Y. and Nair G. B. (2002). Antibiotic resistance, virulence gene, and molecular profiles of Shiga toxin-producing Escherichia coli isolates from diverse sources in Calcutta, India. *Journal of Clinical Microbiology*, **40**(6), 2009–2015.

Khan S. U., Zaidi R., Hassan S. Z., Farooqi I. H. and Azam A. (2016). Application of Fe-Cu binary oxide nanoparticles for the removal of hexavalent chromium from aqueous solution. *Water Science and Technology*, **74**(1), 165–175.

Khan S. U., Farooqi I. H. and Ayub S. (2017). Studies on application of Fe based binary oxide nanoparticles for treatment of lead (Pb2+) contaminated water-A batch study. *Materials Today: Proceedings*, **4**(9), 9650–9655.

Khan N. A., Khan S. U., Ahmed S., Farooqi I. H., Dhingra A., Hussain A. and Changani F. (2019a). Applications of nanotechnology in water and wastewater treatment: A review. *Asian Journal of Water, Environment and Pollution*, **16**(4), 81–86.

Khan S. U., Islam D. T., Farooqi I. H., Ayub S. and Basheer F. (2019b). Hexavalent chromium removal in an electrocoagulation column reactor: Process optimization using CCD, adsorption kinetics and pH modulated sludge formation. *Process Safety and Environmental Protection*, **122**, 118–130.

Khan N. A., Khan S. U., Ahmed S., Farooqi I. H., Hussain A., Vambol S. and Vambol V. (2019c). Smart ways of hospital wastewater management, regulatory standards and conventional treatment techniques. *Smart and Sustainable Built Environment*, **6**, 9–16.

Khan N. A., Ahmed S., Farooqi I. H., Ali I., Vambol V., Changani F., Yousefi M., Vambol S., Khan S. U. and Khan A. H. (2020a). Occurrence, sources and conventional treatment techniques for various antibiotics present in hospital wastewaters: A critical review. *TrAC Trends in Analytical Chemistry*, **129**, 115921.

Khan A. H., Khan N. A., Ahmed S., Dhingra A., Singh C. P., Khan S. U., Mohammadi A. A., Changani F., Yousefi M., Alam S. and Vambol S. (2020b). Application of advanced oxidation processes followed by different treatment technologies for hospital wastewater treatment. *Journal of Cleaner Production*, **269**, 122411.

Kiimmerer K. and Henninger A. (2003). Promoting resistance by the emission of antibiotics from hospitals and households into effluents. *Clinical Microbiology and Infection*, **9**, 1203–1214.

Kimura K., Toshima S., Amy G. and Watanabe Y. (2004). Rejection of neutral endocrine disrupting compounds (EDCs) and pharmaceutical active compounds (PhACs) by RO membranes. *Journal of Membrane Science*, **245**(1–2), 71–78.

Klaine S. J., Alvarez P. J. J., Batley G. E., Fernandes T. F., Handy R. D., Lyon D. Y., Mahendra S., McLaughlin M. J. and Lead J. R. (2008), Nanomaterials in the environment: Behavior, fate, bioavailability and effects. *Environmental Toxicology and Chemistry*, **27**(9), 1825–1851.

Kolpin D. W., Furlong E. T., Meyer M. T., Thurman E. M., Zaugg S. D., Barber L. B. and Buxton H. T. (2002). Pharmaceuticals, hormones, and other organic wastewater contaminants in U.S. streams, 1999–2000: A national reconnaissance. *Environmental Science and Technology*, **36**(6), 1202–1211.

Kümmerer K. (2009). The presence of pharmaceuticals in the environment due to human use – present knowledge and future challenges. *Journal of Environmental Management*, **90**(8), 2354–2366.

Larsson D. J., de Pedro C. and Paxeus N. (2007). Effluent from drug manufactures contains extremely high levels of pharmaceuticals. *Journal of Hazardous Materials*, **148**(3), 751–755.

Li D., Yu T., Zhang Y., Yang M., Li Z., Liu M. and Qi R. (2010). Antibiotic resistance characteristics of environmental bacteria from an oxytetracycline production wastewater treatment plant and the receiving river. *Applied and Environmental Microbiology*, **76**(11), 3444–3451.

Lindsey M. E., Meyer M. and Thurman E. M. (2001). Analysis of trace levels of sulfonamide and tetracycline antimicrobials in groundwater and surface water using solid-phase extraction and liquid chromatography/mass spectrometry. *Analytical Chemistry*, **73**(19), 4640–4646.

Liu F., Ying G. G., Tao R., Zhao J. L., Yang J. F. and Zhao L. F. (2009). Effects of six selected antibiotics on plant growth and soil microbial and enzymatic activities. *Environmental Pollution*, **157**(5), 1636–1642.

Lúcia H. M. L. M., Santos Araújo A. N., Adriano F., Pena A., Delerue-Matos C. and Montenegro M. C. B. S. M. (2010). Ecotoxicological aspects related to the presence of pharmaceuticals in the aquatic environment. *Journal of Hazardous Materials*, **175**(1–3), 45–95.

Luo Y., Guo W., Ngo H. H., Nghiem L. D., Hai F. I., Zhang J., Liang S. and Wang X. C. (2014). A review on the occurrence of micropollutants in the aquatic environment and their fate and removal during wastewater treatment. *Science of the Total Environment*, **473–474**, 619–641.

Malchi T., Maor Y., Tadmor G., Shenker M. and Chefetz B. (2014). Irrigation of root vegetables with treated wastewater: evaluating uptake of pharmaceuticals and the associated human health risks. *Environmental Science and Technology*, **48**(16), 9325–9333.

Minguez L., Pedelucq J., Farcy E., Ballandonne C., Budzinski H. and Halm-Lemeille M. P. (2016). Toxicities of 48 pharmaceuticals and their freshwater and marine environmental assessment in northwestern France. *Environmental Science and Pollution Research*, **23**(6), 4992–5001.

Mompelat S., Le Bot B. and Thomas O. (2009). Occurrence and fate of pharmaceutical products and by-products, from resource to drinking water. *Environment International*, **35**(5), 803–814.

Mutiyar P. K. and Mittal A. K. (2014). Occurrences and fate of selected human antibiotics in influents and effluents of sewage treatment plant and effluent-receiving river Yamuna in Delhi (India). *Environmental Monitoring and Assessment*, **186**(1), 541–557.

Nassiri Koopaei N. and Abdollahi M. (2017). Health risks associated with the pharmaceuticals in wastewater. *Journal of Pharmaceutical Science*, **25**(1), 9.

Nassiri-Koopaei N. and Faramarzi M. A. (2015). Recent developments in the fungal transformation of steroids. *Biocatal. Biotransformation*, **33**(1), 1–28.

Oaks J. L., Gilbert M., Virani M. Z., Watson R. T., Meteyer C. U., Rideout B. A. and Mahmood S. (2004). Diclofenac residues as the cause of vulture population decline in Pakistan. *Nature*, **427**(6975), 630–633.

Oberdörster G. (2010). Safety assessment for nanotechnology and nanomedicine: Concepts of nanotoxicology. *Journal of Internal Medicine*, **267**(1), 89–105.

Ort C., Lawrence M. G., Reungoat J., Eaglesham G. and Carter S. K. J. (2010). Determining the fraction of pharmaceutical residues in wastewater originating from a hospital. *Water Research*, **44**, 605–615.

Patel M., Kumar R., Kishor K., Mlsna T., Pittman C. U., Jr and Mohan D. (2019). Pharmaceuticals of emerging concern in aquatic systems: chemistry, occurrence, effects, and removal methods. *Chemical Reviews*, **119**(6), 3510–3673.

Petrie B., Barden R. and Kasprzyk-Hordern B. (2015). A review on emerging contaminants in wastewaters and the environment: Current knowledge, understudied areas and recommendations for future monitoring. *Water Research*, **72**, 3–27.

Petrovic M. and Barceló D. (2006). Application of liquid chromatography/quadrupole time-of-flight mass spectrometry (LC-QqTOF-MS) in the environmental analysis. *Journal of Mass Spectrometry*, **41**(10), 1259–1267.

Pomati F. (2007). Pharmaceuticals in drinking water: is the cure worse than the disease? *Environmental Science and Technology*, **41**(24), 8204–8204.

Quinn B., Gagné F. and Blaise C. (2008). An investigation into the acute and chronic toxicity of eleven pharmaceuticals (and their solvents) found in wastewater effluent on the cnidarian, Hydra attenuata. *Science of the Total Environment*, **389**(2–3), 306–314.

Ray P. C., Yu H. and Fu P. P. (2009). Toxicity and environmental risks of nanomaterials: Challenges and future needs. *Journal of Environmental Science and Health. Part C, Environmental Carcinogenesis and Ecotoxicology Reviews*, **27**(1), 1–35.

Richardson S. D., Plewa M. J., Wagner E. D., Schoeny R. and DeMarini D. M. (2007). Occurrence, genotoxicity, and carcinogenicity of regulated and emerging disinfection by-products in drinking water: A review and roadmap for research. *Mutation Research. Reviews in Mutation Research*, **636**(1–3), 178–242.

Rivera-Utrilla J., Sánchez-Polo M., Ferro-García M. Á., Prados-Joya G. and Ocampo-Pérez R. (2013). Pharmaceuticals as emerging contaminants and their removal from water. A review. *Chemosphere*, **93**(7), 1268–1287.

Sarmah A. K., Meyer M. T. and Boxall A. B. (2006). A global perspective on the use, sales, exposure pathways, occurrence, fate and effects of veterinary antibiotics (VAs) in the environment. *Chemosphere*, **65**(5), 725–759.

Schacht K., Chen Y., Tarchitzky J. and Marschner B. (2016). The use of treated wastewater for irrigation as a component of integrated water resources management: Reducing environmental implications on soil and groundwater by evaluating site-specific soil sensitivities. In: Integrated Water Resources Management: Concept, Research and Implementation. Springer International Publishing, Geneva, pp. 459–470.

Schröder H., Meyer E., Nink K., Hoffmann S., Steib-Bauert M., Kämmerer R. and Kern W. V. (2004). Antibiotikaanwendung in Deutschland im europäischen Vergleich (Antibiotic use in Germany in a European comparison). *DMW-Deutsche Medizinische Wochenschrift*, **129**(38), 1987–1992.

Schwab B. W., Hayes E. P., Fiori J. M., Mastrocco F. J., Roden N. M., Cragin D., Meyerhoff R. D., Vincent J. D. and Anderson P. D. (2005). Human pharmaceuticals in US surface waters: A human health risk assessment. *Regulatory Toxicology and Pharmacology*, **42**(3), 296–312.

Semerjian L., Shanableh A., Semreen M. H. and Samarai M. (2018). Human health risk assessment of pharmaceuticals in treated wastewater reused for non-potable applications in Sharjah, United Arab Emirates. *Environ. Int.*, **121**(Pt 1), 325–331.

Shaik A. N., Grater R., Lulla M., Williams D. A., Gan L. L., Bohnert T. and LeDuc B. W. (2016). Comparison of enzyme kinetics of warfarin analyzed by LC-MS/MS QTrap and differential mobility spectrometry. *Journal of Chromatography. B, Analytical Technologies in the Biomedical and Life Sciences*, **1008**, 164–173.

Sharma S. K. and Sanghi R. (2013). Wastewater Reuse and Management. Springer, Netherlands.

Sharma B. M., Bečanová J., Scheringer M., Sharma A., Bharat G. K., Whitehead P. G., Klánová J. and Nizzetto L.. (2019). Health and ecological risk assessment of emerging contaminants (pharmaceuticals, personal care products, and artificial sweeteners) in surface and groundwater (drinking water) in the Ganges River Basin, India. *Science of the Total Environment*, **646**, 1459–1467.

Stănescu M., Buta C. and Mihai G. (2017). Reabilitation of a wastewater treatment plant–possible environment impact. *Ovidius University Annals of Constanta-Series Civil Engineering*, **19**(1), 65–78.

Stanford B. D. and Weinberg H. S. (2010). Evaluation of on-site wastewater treatment technology to remove estrogens, nonylphenols, and estrogenic activity from wastewater. *Environmental Science and Technology*, **44**(8), 2994–3001.

Stanley J. K., Ramirez A. J., Chambliss C. K. and Brooks B. W. (2007). Enantiospecific sublethal effects of the antidepressant fluoxetine to a model aquatic vertebrate and invertebrate. *Chemosphere*, **69**(1), 9–16.

Taylor D. and Senac T. (2014). Human pharmaceutical products in the environment – the “problem” in perspective. *Chemosphere*, **115**, 95–99.

Ternes T. A. (1998). Occurrence of drugs in German sewage treatment plants and rivers. *Water Research*, **32**(11), 3245–3260.

Testa B. and Krämer S. D. (2008). The biochemistry of drug metabolism – An introduction: Part 4. Reactions of conjugation and their enzymes. *Chemistry and Biodiversity*, **5**(11), 2171–2336.

Thomas K. V., Dye C., Schlabach M. and Langford K. H. (2007). Source to sink tracking of selected human pharmaceuticals from two Oslo city hospitals and a wastewater treatment works. *Journal of Environmental Monitoring*, **9**(12), 1410–1418.

Tsuji J. S., Maynard A. D., Howard P. C., James J. T., Lam C. W., Warheit D. B. and Santamaria A. B. (2006). Research strategies for safety evaluation of nanomaterials, part IV: risk assessment of nanoparticles. *Toxicological Sciences*, **89**(1), 42–50.

Van Boeckel T. P., Gandra S., Ashok A., Caudron Q., Grenfell B. T., Levin S. A. and Laxminarayan R. (2014). Global antibiotic consumption 2000 to 2010: an analysis of national pharmaceutical sales data. *The Lancet Infectious Diseases*, **14**(8), 742–750.

Verlicchi P., Al Aukidy M., Galletti A., Petrovic M. and Barceló D. (2012). Hospital effluent: investigation of the concentrations and distribution of pharmaceuticals and environmental risk assessment. *Science of the Total Environment*, **430**, 109–118.

Villanueva C. M., Kogevinas M., Cordier S., Templeton M. R., Vermeulen R., Nuckols J. R., Nieuwenhuijsen M. J. and Levallois P. (2014). Assessing exposure and health consequences of chemicals in drinking water: current state of knowledge and research needs. *Environmental Health Perspectives*, **122**(3), 213–221.

Vulliet E. and Cren-Olivé C. (2011). Screening of pharmaceuticals and hormones at the regional scale, in surface and groundwaters intended to human consumption. *Environmental Pollution*, **159**(10), 2929–2934.

Webb S., Ternes T., Gibert M. and Olejniczak K. (2003). Indirect human exposure to pharmaceuticals via drinking water. *Toxicology Letters*, **142**(3), 157–167.

Wen Z. H., Chen L., Meng X. Z., Duan Y. P., Zhang Z. S. and Zeng E. Y. (2014). Occurrence and human health risk of wastewater-derived pharmaceuticals in a drinking water source for Shanghai, East China. *Science of the Total Environment*, **490**, 987–993.

Wichelns D., Drechsel P. and Qadir M. (2015). Wastewater: Economic Asset in an Urbanizing World. Springer, Netherlands, pp. 3–14.

Wiesner M. R., Lowry G. V., Jones K. L., Hochella M. F., Digiulio R. T., Casman E. and Bernhardt E. S. (2009). Decreasing uncertainties in assessing environmental exposure, risk, and ecological implications of nanomaterials. *Environmental Science and Technology*, **43**(17), 6458–6462.

Williams E. S. and Brooks B. W. (2012). Human Health Risk Assessment for Pharmaceuticals in the Environment: Existing Practice, Uncertainty and Future Directions. Springer, New York, pp. 167–224.

Wise R. (2002). Antimicrobial resistance: priorities for action. *Journal of Antimicrobial Chemotherapy*, **49**(4), 585–586.

World Health Organization. (2012). Pharmaceuticals in Drinking-Water. WHO, Geneva.

Zenker A., Cicero M. R., Prestinaci F., Bottoni P. and Carere M. (2014). Bioaccumulation and biomagnification potential of pharmaceuticals with a focus to the aquatic environment. *Journal of Environmental Management*, **133**, 378–387.

Chapter 8

Treatment schemes – conventional and dedicated for PhACs treatment

Pavlo Kozub[1], Svetlana Kozub[2], Niloofar Mozaffari[3], Parveen Sihag[4], Nastaran Mozaffari[5], Sergij Vambol[6], Viola Vambol[7], Nishana Ramsawak[8] and Nadeem A Khan[9]

[1]*Department of Natural Science, Kharkiv National University of Radio Electronics, Kharkiv, Ukraine*
[2]*Department of Medical and Bioorganic Chemistry, Kharkiv National Medical University, Kharkiv, Ukraine*
[3]*Department of Physics, Faculty of Sciences, Science and Research Branch, Islamic Azad University, Tehran, Iran*
[4]*Department of Civil Engineering, Shoolini University, Solan, Himachal Pradesh, India*
[5]*Department of Environmental Engineering, Faculty of Natural Resources and Environment, Science and Research Branch, Islamic Azad University, Tehran, Iran*
[6]*Life Safety and Law Department, Kharkiv Petro Vasylenko National Technical University of Agriculture, Ukraine*
[7]*Educational and Scientific Department of Occupational Safety and Health, National Scientific and Research Institute of Industrial Safety and Occupational Safety and Health, Ukraine*
[8]*Department of Civil Engineering, Queens University, Kingston, Canada*
[9]*Department of Civil Engineering, Jamia Millia Islamia, New Delhi*

Editor(s): Nadeem A Khan, Sirajuddin Ahmed, Viola Vambol and Sergij Vambol
doi: 10.2166/9781789061338_0181

ABSTRACT

The process of destruction of pharmaceutical compounds depends on their structure, physico-chemical characteristics, and decomposition mechanisms. Because of the broad range of drugs available in wastewater, finding one standard method for removing them is complicated. This chapter focuses on the study of treatment techniques, both traditional and advanced methods, for wastewater contaminated with pharmaceutical compounds. The experimental investigation of photocatalytic degradation of PhACs and the combined effect of UV radiation and hydrogen peroxide on organic compounds is shown in detail. This method seems to be the most promising due to its high efficiency; however, modern achievements through this method are not effective for practical implementation. This is due not to the lack of scientific data, but to the lack of technological solutions in this area. An experimental study revealed that the photocatalytic material on a PVC film had the best characteristics for the effective removal of pharmaceutical substances from contaminated waters. However, the destruction of stable pharmaceutical compounds in contaminated water by the combined effect of UV radiation and hydrogen peroxide is of great interest, as has been demonstrated by many scientific studies.

Keywords: Pharmaceutical compounds, Photocatalytic degradation, Polluted water, Treatment techniques

Contents

8.1 Introduction 183
8.2 Traditional Treatment Methods of Contaminated Water 183
8.3 Carbonaceous Nanomaterials 187
8.4 Graphene 188
8.5 Other Nano-Adsorbents 189
8.5.1 MoS_2 nanosheets 189
8.5.2 Hexagonal boron nitride 189
8.5.3 Nano-hydroxyapatite 190
8.5.4 Chitosan 190
8.6 Metal Oxides 190
8.7 Nanocomposites 191
8.8 Advanced Oxidation Processes 193
8.8.1 Electrochemical advanced oxidation processes (EAOP) 193
8.8.2 Fenton process 194
8.8.3 Photocatalytic degradation of PhACs and the combined effect of UV radiation and hydrogen peroxide on organic compounds 195
8.8.3.1 Photocatalytic material fabrication 197
8.8.3.2 Experimental conditions 197
8.8.3.3 Experimental works plan 199
8.8.3.4 Properties of photocatalytic samples produced 200
8.8.3.5 Research results 202

8.9 Water Treatment Methods for Field Hospitals 205
8.10 Conclusions 207
References 208

8.1 INTRODUCTION

The focus on the removal of active ingredients of pharmaceutical compounds (PhACs) became significant with the first observation of the feminization of fishes downstream of wastewater treatment plants (WWTPs) in 1999 (Khan *et al.*, 2017; Larsson *et al.*, 1999). The destruction process of pharmaceutical compounds depends on their structure, physico-chemical characteristics, and decomposition mechanisms. Due to the wide array of manufactured drugs available, it is difficult to find one common method to remove them all from effluents (Khan *et al.*, 2021; Khan *et al.*, 2020; Mazhar *et al.*, 2020; Verlicchi *et al.*, 2010). Biological, physico-chemical, and chemical methods for purifying contaminated waters are distinguished, using the available methods based on:

- the use of one substance (oxidizing agent) or one exposure factor (traditional methods);
- the combined effects of several substances or factors (advanced methods).

The use of these methods provides purification by extraction of pharmaceutical compounds from solution (based on physical transfer) or its decomposition into fragments (partially or fully mineralized). The technique most employed in the removal of contaminants in wastewaters includes photocatalytic degradation (Khan *et al.*, 2017; Chamjangali *et al.*, 2015), biodegradation (Khan *et al.*, 2017; Khan *et al.*, 2020; Mazhar *et al.*, 2020; Chamjangali *et al.*, 2015; Islam *et al.*, 2015; Schiavon *et al.*, 2015; Silva *et al.*, 2014), membranes (Khan *et al.*, 2017; Khan *et al.*, 2021; Arami *et al.*, 2006), ozonation (Khan *et al.*, 2017; Carbajo *et al.*, 2016; Lin & Liaw 2015), Fenton process (Khan *et al.*, 2017; Khataee *et al.*, 2016; Yuan *et al.*, 2016), electrochemical techniques (Khan *et al.*, 2017; Ni *et al.*, 2015; Särkkä *et al.*, 2015), reverse osmosis (Grobert 2007) and chemical precipitation (Khan *et al.*, 2017; Bruno *et al.*, 1995; Tan *et al.*, 2014; Tang and Reeder 2009; Adriano *et al.*, 2005; Kim *et al.*, 2007) among others (Khan *et al.*, 2017).

8.2 TRADITIONAL TREATMENT METHODS OF CONTAMINATED WATER

Biological treatment is achieved using activated sludge (biofilm) for the removal of dissolved contaminants in contaminated water. For certain pharmaceutically active compounds (for example antibiotics), this method is not applicable due to the toxicity of the bacterial component of activated sludge (Kümmerer 2009). Many pharmaceutical compounds are highly resistant to the biodegradation process (Prieto-Rodrıguez *et al.*, 2013; Oller *et al.*, 2011), and can be adsorbed (for example, tetracycline) on the solid phase of activated sludge, unchanged (Belkheiri *et al.*, 2011). For purification of contaminated water with pharmaceutical compounds, biological methods can be of interest only at the post-treatment stage after the mandatory step in which the pharmaceutical compounds partially or completely lose their biological activity.

Chemical methods are the largest group in wastewater treatment based on the mineralization of organic pollutants under the influence of an oxidizing agent. Most pharmaceutical compounds are biodegradable and difficult to oxidize organic substances (Prieto-Rodrıguez *et al.*, 2013; Collado *et al.*, 2013). Methods based on the use of only one oxidizing agent are ineffective for such substances. Hence, the most applied approach using the oxidative method is the use of oxygen as an oxidizing agent; however, this requires high-temperature treatment (900–1100°C) of contaminated water, which is preceded by additional stages of sorption or membrane separation for the concentration of impurities. The process is complex environmentally, and economically inexpedient, except for cases with a high content of dissolved organic substances (more than 10% of the mass) (Thomason and Modell 1984).

Flotation, coagulation, membrane cleaning methods, sorption, and UV treatment are examples of physico-chemical methods. The inefficiency of flotation and coagulation to remove pharmaceutical compounds are due to low intensity of the molecular interaction of pharmaceutical compounds with water (Kamrukov *et al.*, 2013; Strikalenko 2005). Hence, flotation and coagulation can only provide preliminary treatment of contaminated water (Adams *et al.*, 2002; Ternes *et al.*, 2002). Membrane processes (micro- and ultrafiltration) for water purification from pharmaceutical compounds were previously employed (Khan *et al.*, 2017; Arami *et al.*, 2006; Liu *et al.*, 2005). However, with the size of molecules of a pharmaceutical compound being hundreds of times smaller than the pore size of the membrane, purification is less effective (Urase *et al.*, 2005). The reverse osmosis process is quite effective in the removal of pharmaceuticals from contaminated waters, but its extensive use is still an issue (Grobert 2007; Bellona *et al.*, 2008).

Since these techniques are not 100% effective in removing recalcitrant and personal care products, there is an accumulation of residual concentrations in soils and water reservoirs entering the drinking water system, hence an alternative step such as sorption is utilized (Cerro-Lopez and Méndez-Rojas 2017).

Researchers have revealed high sorption efficiency of activated carbons, independently of the raw materials from which they were obtained (Önal *et al.*, 2007; Mestre *et al.*, 2007; Jodeh *et al.*, 2016; Deegan *et al.*, 2011). They also note the difficulty in extracting activated carbon from purified water (Deegan *et al.*, 2011; Fogarty 2013), with the additional separation processes, sedimentation, and filtration complicating and increasing the cost of the process.

The possibility of using other sorbents, such as the natural mineral palygorskite for the effective sorption of tetracycline (Liu *et al.*, 2010), natural and organo-modified bentonite for the effective sorption of ampicillin, exists (Rahardjo *et al.*, 2011). It has been reported that the modification of natural bentonite leads to an increase in sorption efficiency from 60 to 90%, which indicates the suitability of using organo-modified bentonite as a sorbent for removing antibiotics from contaminated water.

The major disadvantage of the sorption purification method is the inability to provide a partial or complete loss of the biological activity of pharmaceutical compounds and, therefore, the used sorbents turn into hazardous waste (which contains active pharmaceutical compounds) and require the regeneration or inactivation of APS (Zhou *et al.*, 2012). If regeneration of used sorbents is not utilized, the sorbents should be subjected to high-temperature incineration technologies that ensure complete degradation of substances (Vambol *et al.*, 2016; Park *et al.*, 2005; Bernadiner 2010) and it should also energetically affordable (Vambol *et al.*, 2018; Vambol *et al.*, 2018).

The use of the adsorption method is constantly increasing due to its simplicity, efficiency, and economical value, because the main process is optimization of the adsorption capacity of an adsorbent, achieved by varying the physico-chemical conditions such as pH, temperature, contact time, etc. (Khan *et al.*, 2017; Liu *et al.*, 2011; Liu *et al.*, 2013; Sun *et al.*, 2013). In comparison to other methods, adsorption does not generate unwanted by-products and can be modified for industrial and municipal uses (Cerro-Lopez and Méndez-Rojas 2017). In this process, a substance attaches itself to the solid surface from the gas or the liquid. This adsorption process by solid adsorbents is thought to be a highly efficient and effective method for water treatment due to its cost-effectiveness, simplicity to use and its ability to be readily modified (Cincinelli *et al.*, 2015; Wang and Peng 2010), especially due to its ability to remove numerous chemical pollutants (Cincinelli *et al.*, 2015). It is the most common environmentally friendly remediation process due to its modification capabilities and practicality for retrofitting current water treatments as well as its inexpensive cost (Cai *et al.*, 2013; Bui and Choi 2009).

Furthermore, the percentage removal of PhACs is dependent on the physiochemical structure and the adsorbent (Khan *et al.*, 2017; Buser *et al.*, 1998; Buser *et al.*, 1998; Buser *et al.*, 1999; Kümmerer *et al.*, 2000; Michael *et al.*, 2013; Mohle *et al.*, 1999; Richardson and Bowron 1985; Stamatelatou *et al.*, 2003; Stumpf *et al.*, 1999; Ternes 1998; Zwiener and Frimmel 2003); as such, the structure and surface charge density must be taken into consideration when choosing an adsorbent to remove a particular pollutant (Khan *et al.*, 2017). Adsorbents are classified into natural (charcoal, clays and its minerals, zeolites) and synthetic (industrial waste, sewage sludge, polymeric adsorbents), based on their properties such as porosity, pore structure and the nature of the surface being adsorbed (Cincinelli *et al.*, 2015). Traditional adsorbents such as zeolites, clays, polymeric adsorbents, and bentonite have been widely studied in the removal of water-soluble PhACs including carbamazepine, diclofenac, ibuprofen, ketoprofen, and such like (Rahardjo *et al.*, 2011; Cai *et al.*, 2013; Bui and Choi 2009). However, there are limitations to using them, mainly low adsorption rate and capacity, which limits the efficiency and selectivity (Cincinelli *et al.*, 2015; Qu *et al.*, 2013).

The growing advancement in nanotechnology provides more efficient and effective degradation of PhACs in the treatment of water and wastewater (Cai *et al.*, 2013). Nanotechnology refers to the use of materials and structures with a dimension of 1–100 nm (Larsson *et al.*, 1999; Masciangioli and Zhang 2003; Qu *et al.*, 2013). This structure includes nanoparticles or nanocrystals, nanolayers, and nanotubes (Cai *et al.*, 2013; Qu *et al.*, 2013). The astonishing assets of nanomaterials include large surface areas (Suhikova *et al.*, 2019; Vambol *et al.*, 2017), small size effect, quantum effect, photosensitivity, catalytic activity, tunable pore size (Vambol *et al.*, 2017), surface chemistry (Suchikova *et al.*, 2018), and electrochemical and magnetic properties (Vambol *et al.*, 2017; Suchikova *et al.*, 2018), which are positive characteristics for the use of treating PhACs in wastewater; which makes them highly suitable for use in adsorption (Cai *et al.*, 2013; Bandala *et al.*, 2014). The significant size and surface ratio of nanomaterials makes them favorable for uses where this feature can improve the catalytic activity and the ability to remove pollutants through adsorption/absorption. With its increased stability and solubility specifically in water, many nanomaterials can be utilized in environmental processes for the selective and efficient removal or degradation of pharmaceutical compounds (Cerro-Lopez and Méndez-Rojas 2017; Ramírez-Sánchez *et al.*, 2016; Prachi *et al.*, 2013). In comparison to bulk particles, nanoparticles were found to have a greater surface area and can be synergized to many chemical groups. The exceptional properties of nanoparticles have been used in the development of selective sorbents with high capacities for the removal of pharmaceutical compounds (Zhang *et al.*, 2010). Nano-adsorbents provide several advantages which contribute to their characterization including large surface size, selective and abundant adsorption sites, short intraparticle diffusion distance, and tunable pore size with fast regeneration and reusability (Cai *et al.*, 2013; Bandala *et al.*, 2014). They can also be designed to have multiple reactive agents synergistically (Cincinelli *et al.*, 2015; Zhang *et al.*, 2010). However, when used in water treatment the nano-adsorbents are not released into the environment and are encapsulated in larger particles during filtration or when using a support material for the sorbent (Cincinelli *et al.*, 2015; Zhang *et al.*, 2010). Nanomaterials such as carbon nanotubes, polymeric adsorbents from both natural and synthetic sources, nano clays and activated charcoal (Cerro-Lopez and Méndez-Rojas 2017), MoS_2 nanosheets, hexagonal boron nitride, and nanocomposites have been widely researched and utilized in the removal of PPCPs from water and wastewater treatments for the past ten years (Cincinelli *et al.*, 2015).

8.3 CARBONACEOUS NANOMATERIALS

The hydrophobic effect, π—π stacking, hydrogen bonding, electrostatic and covalent interaction have contributed to the adsorption of PhACs on carbonaceous nanomaterials (Cai *et al.*, 2013; Liu *et al.*, 2016). These bonding mechanisms influence the types, strength and general function of the adsorbents of both the carbonaceous nanomaterial and the PhACs and as such, its surface area, functional groups, and pores determines the adsorptive removal of PhACs efficiency in the aqueous phase (Cai *et al.*, 2013). Several allotropes are formed through the hybridization of carbon, including CNTs (Cai *et al.*, 2013; Collins *et al.*, 2000), carbon nanofibers (CNFs) (Cai *et al.*, 2013; Geim and Novoselov 2007), graphene (Cai *et al.*, 2013; Zhao *et al.*, 2011; Khajeh *et al.*, 2013), carbon nanoparticles and carbon nanobeads (CNBs) (Cai *et al.*, 2013; Yang *et al.*, 2012); with CNTs and graphene having shown the best adsorbents properties due to its extraordinary surface properties from nanostructures and its in depth research for removal of PhACs (Cai *et al.*, 2013; Lens *et al.*, 2013; Kim *et al.*, 2014).

Since their invention in 1991, carbon nanotubes (CNTs) have been utilized in numerous environmental and medical remediation processes (Zhang *et al.*, 2010; Kim *et al.*, 2014; Singh *et al.*, 2014; Schwarzenbach *et al.*, 2003). They have cylindrical graphite sheets with very high Van der Waal indexes (Zhang *et al.*, 2010; Kim *et al.*, 2014; Carabineiro *et al.*, 2012) with different structures varying in length, thickness, chirality, and number of layers (Zhang *et al.*, 2010; Camilli *et al.*, 2014). CNTs are an exceptional material for wastewater treatment due to their large specific area and high mechanical and chemical properties (Zhang *et al.*, 2010; Upadhyayula *et al.*, 2011; Madrakian *et al.*, 2011; Hayasi and Saadatjoo 2017) and can be grouped based on their structures as (Zhang *et al.*, 2010; Upadhyayula *et al.*, 2011): single-walled nanotubes (SWNT); multi-walled nanotubes (MWNT); and, double-walled nanotubes (DWNT) (Zhang *et al.*, 2010).

CNT nano-structured sponges or nano-sponge, containing iron and sulfur, is also quite effective for absorbing water contaminants such as oil, pesticides and fertilizers, and pharmaceuticals, with its magnetic properties making them easier to clean up (Zhang *et al.*, 2010; Upadhyayula *et al.*, 2011).

The hydrophobicity of graphene sheets is a crucial attribute that enables CNTs to interact with any solute that has π-electrons in its composition (Cerro-Lopez and Méndez-Rojas 2017). As a result, CNTs are very effective adsorbents for aromatic compounds (Cerro-Lopez and Méndez-Rojas 2017; Machulek *et al.*, 2013; Cerro-Lopez *et al.*, 2014). Research carried out by Gotovc *et al.* (Upadhyayula *et al.*, 2011) suggested a π—π coupling between the π electrons of the aromatic rings of phenanthrene and the surface of CNTs. Pan *et al.* (Da Silva and Dos Santos 2017) researched the adsorption of 17 α-ethinylestradiol (EE2) and bisphenol A (BPA) on three different CN materials (SWCNT, MWCNTs and fullerene) showed π—π electron donor-acceptor interactions of EE2 and BPA with the CNTs surface. These forces of interactions are the reasons for the strong adsorption of tetracyclines in a wide spectrum of antibiotics on CNTs and graphene surfaces (Cerro-Lopez and Méndez-Rojas 2017; Novoselov *et al.*, 2004).

The major disadvantage of using CNT powders in water and wastewater treatment comes from handling and retrieval through filtration after the contamination adsorption (Zhang *et al.*, 2010). However, they are using millimeter size freestanding bulk CNT frameworks to aid in this, which makes them easier to handle (Zhang *et al.*, 2010; Upadhyayula *et al.*, 2011).

8.4 GRAPHENE

Graphene, a single atomic layer with a 2D structure (Cai *et al.*, 2013; Zhang *et al.*, 2016), provides many advantages over CNTs (Cerro-Lopez and Méndez-Rojas 2017). Graphene has two basal planes that can be used for adsorbate interactions with the inside walls of the CNTs being unreachable by the absorbates. Graphene oxide can be readily achieved by numerous exfoliation and reduction methods (Cerro-Lopez and Méndez-Rojas 2017). Typically, there are five adsorption interactions on graphene-based materials: hydrophobic effect, π—π bonds, hydrogen bonds, and covalent and electrostatic interactions (Cerro-Lopez and Méndez-Rojas 2017; Santhosh *et al.*, 2016). The adsorption of natural and semisynthetic tetracyclines on graphite oxides has shown astonishing adsorption capacities (≥200 mg/g) (Cerro-Lopez and Méndez-Rojas 2017; Santhosh *et al.*, 2016; Gao *et al.*, 2012; Cai and Larese-Casanova 2014). However, with higher pH or sodium concentration, the adsorption capacity for tetracyclines has decreased. The adsorption mechanism was due to π—π interaction and cation-π bonding (Cerro-Lopez and Méndez-Rojas 2017). Ciprofloxacin has also been investigated for adsorption on graphene and its oxide with similar adsorption capacities (Cerro-Lopez and Méndez-Rojas 2017). Although graphene-based materials are a viable option in the removal of pollutants in water, the limiting factors are cost, reusability, and health and environmental issues which need to be overcome for it to be more readily utilized (Cerro-Lopez and Méndez-Rojas 2017).

Graphene oxide has been widely researched in the removal of frequently used PhACs, specifically those with aromatic centers such as tetracycline (TCs), CIP, SAs, etc. (Khan *et al.*, 2017). Both graphene oxide and several PhACs have common features in terms of aromatic rings and functional groups, which contributes to its effective use as an adsorbent for them (Khan *et al.*, 2017). PhACs can be found in a mixture of ground and surface waters (Khan *et al.*, 2017) which has led to several investigations being focused on the co-adsorption of PhACs (Nam *et al.*, 2015; Wang *et al.*, 2013; Arvidsson *et al.*, 2014). Three types of graphene oxides, namely GO-C (xGnp-C-750, XG Sciences, Inc., Lansing, MI, USA), GO-M (xGnp-M-25, XG Sciences, Inc., Lansing, MI, USA) and GO-A (N006-010-P, Angstron Materials, Inc., Dayton, OH, USA), were used in the co-adsorption of CBZ and EDCs (Nam *et al.*, 2015). GO-C (SSA = 720

m^2/g) showed high adsorption capacity for CBZ, as compared to GO-M (SSA = 120 m^2/g) and GO-A (SSA = 100 m^2/g). This was due to its large surface area and the significantly strong and moderate electron activating phenol and ether groups, which makes it suitable for employment in the industry (Khan *et al.*, 2017). Although GO has proven to be highly efficient in the removal of PhACs, the energy associated with its production was approximately 1100 MJ/Kg; 50 times higher than activated carbon (Khan *et al.*, 2017; Chao *et al.*, 2014). The limitations of using GO are its high cost, lengthy production time, exhausting exfoliation, and separation of secondary pollution (Khan *et al.*, 2017). To use GO more often, a cheaper high surface area substitute must be formulated commercially (Khan *et al.*, 2017).

8.5 OTHER NANO-ADSORBENTS

Additional to graphene, 2D nanomaterials with a high specific surface area, such as molybdenum disulfide (MoS_2) nanosheets (Cai *et al.*, 2013; Yu *et al.*, 2017), double-layered hydroxides (Cai *et al.*, 2013; Liu *et al.*, 2016) and hexagonal boron nitride nanosheets (Cai *et al.*, 2013; Attia *et al.*, 2013), has recently shown high adsorption execution of organic compounds and can be effectively utilized for the removal of pharmaceutically active compounds (PhACs) in water and wastewater treatment (Cai *et al.*, 2013). Recovery efficiency is facilitated due to the use of magnetic nano-adsorbents or nanocomposite merged with magnetite (Cai *et al.*, 2013; Golberg *et al.*, 2010). Additionally, nano-hydroxyapatite and chitosan have been found to be good adsorbents for removing PhACs (Cai *et al.*, 2013).

8.5.1 MoS_2 nanosheets

MoS_2 nanosheets are vertically stacked 2D like graphene-like materials with Van der Waals bonds (Cai *et al.*, 2013). This allows the adsorption of PhACs through hydrophobic effect, π—π stacking and electrostatic interactions (Cai *et al.*, 2013; Yu *et al.*, 2017). MoS_2 possesses a large adsorption capacity with fast kinetics to remove PhACs effectively due to its large surface area and has been found to remove up to 83% of doxycycline antibiotic, with adsorption capacity around 310 mg/g (Cai *et al.*, 2013; Yu *et al.*, 2017).

8.5.2 Hexagonal boron nitride

Hexagonal boron nitride, an isoelectronic with graphene, has been found to have a large surface area, high mechanical strength, and high stability, making it suitable for adsorption (Cai *et al.*, 2013; Yu *et al.*, 2017). A highly porous hexagonal boron nitride nanosheet was synthesized by Lie *et al.* (Yu *et al.*, 2017) and showed the following adsorption capacities: 1170 mg/g for chlortetracycline, 284 mg/g tetracycline, 206 mg/g ciprofloxacin and 174 mg/g norfloxacin for removal of PhACs and was found to be reusable after basic regeneration. There are limitations for using boron nitride for PhACs removals, such as its preparation, high toxicity, and extremely high cost (Cai *et al.*, 2013; Chen *et al.*, 2015). Although it has positive characteristics as a good adsorbent material, there are still uncertainties of using it as there is limited information on adsorption performance, mechanism, effects of environmental factors and the regeneration of spent materials (Cai *et al.*, 2013).

8.5.3 Nano-hydroxyapatite

Nano-hydroxyapatite was recently considered to be a highly favorable adsorbent used in the biological and medical field for the removal of PhACs in wastewater based on its high stability, low solubility and large surface area (Cai *et al.*, 2013; Zhi *et al.*, 2005). However, it was found that only 54.3% norfloxacin and 47.3% ciprofloxacin were effectively removed using high concentrations of nanohydroxyapatite (20 g/L) (Cai *et al.*, 2013; Zhi *et al.*, 2005). This comparatively low affinity to PhACs makes it unfavorable as opposed to other nano adsorbents or conventional AC (Cai *et al.*, 2013).

8.5.4 Chitosan

Chitosan, a product from the naturally occurring abundant chitin, used in the presence of hydroxyl and amine groups, makes it suitable to use to remove organic pollutants from the pharmaceuticals. Chitosan nanoparticles have recently been found to have high adsorption kinetic and capabilities for removing diuretic furosemide from water, with the adsorption capacity peaking at 270 mg/g within 10 min (Cai *et al.*, 2013; Zhuo *et al.*, 2017). Chitosan shows fantastic capabilities for adsorbents due to its high abundance, low cost, and renewable and biodegradable properties (Cai *et al.*, 2013; Fang *et al.*, 2011).

8.6 METAL OXIDES

Nanosized metals and metal oxides are being utilized as nano-adsorbents for the removal of heavy metals, inorganic anions, and organics as they are low in cost and have high selectivity and capacity (Cerro-Lopez and Méndez-Rojas 2017). Nano zero-valent has been able to remove antibiotics via adsorption, redox and co-precipitation (Cerro-Lopez and Méndez-Rojas 2017; Lu *et al.*, 2016). One limitation of this method is the separation of the material used in adsorption from treated water and its regeneration. Surface modified magnetic iron oxide nanoparticles (MIONs) are, however, being researched for use in adsorption of PhACs in wastewater (Cerro-Lopez and Méndez-Rojas 2017) as they can be readily removed using an external magnetic field (Cerro-Lopez and Méndez-Rojas 2017; Taghipour *et al.*, 2019) in the elimination of diclofenac, ibuprofen, sulfadiazine, norfloxacin and other pharmaceutical compounds (Cerro-Lopez and Méndez-Rojas 2017; Machulek *et al.*, 2013; Fang *et al.*, 2011;

Khin *et al.*, 2012). Research has been carried out to show how pH, temperature and adsorbent concentration influences the selective adsorption of the different molecules (Cerro-Lopez and Méndez-Rojas 2017).

8.7 NANOCOMPOSITES

The solid structure of nanocomposites can exist in two main forms: (1) inorganic solid covered with an organic shell or vice versa, or (2) a combination of two or more inorganic/organic phases, one of which must be nano-dimensional (Wang 2014; Ajayan *et al.*, 2006; Bavastrello *et al.*, 2013; Zhang *et al.*, 2012; Hua *et al.*, 2012). Composite adsorbents were made by impregnating NPs into porous supports such as polymers and C_3N_4 to help improve the use of metal NPs in wastewater treatment (Wang 2014; O'Brien 2012). In recent times, synergizing several nanoparticles into a nanocomposite to explore positive changes in their functionality has received favorable findings (Wang 2014; Singh *et al.*, 2012). Nanocomposites are formed by merging metals, metal oxides and oxy-acids in combinations of metal-metal, metal-metal oxides, metal oxides-metal oxides, metal oxides-oxyacids and oxyacids-oxyacids (Wang 2014). These, along with nano-scale bases, form nanocomposites (Wang 2014). The nanocomposites findings are as follows.

Magnetic nanoparticle (γ-Fe_2O_3) coated zeolite achieved greater than 95% efficiency removal for diclofenac, naproxen, gemfibrozil, and ibuprofen within 10 min at an initial dilution of 100 μg/L and a sorbent dosage of 1 g/L (Cai *et al.*, 2013; Yu *et al.*, 2017). A synthesis of carbon-iron magnetic nanocomposite by Singh *et al.* (2012) found increased ibuprofen removal efficiency as compared to its precursor carbon. In comparison to other immobilization methods, the synthesis of magnetic nano-adsorbents allowed for quick and easy separation of the nanoparticles while maintaining its surface area which makes it possible for environmental remediation by nanomaterials, similar to engineering applications (Cai *et al.*, 2013).

Researchers have produced a photo regenerable multiwalled carbon nanotube membrane using different water permeabilities by implanting hierarchical titanium dioxide onto a pre-deposited bed of MWCNTs employing a modified soft gel method. The adsorption of the synergized MWCNT/titanium dioxide membrane used in three pharmaceuticals removal (acetaminophen, carbamazepine, and ibuprofen) resulted in an 80, 45 and 24% initial percentage removal, respectively (Zaib *et al.*, 2013). The membranes were regenerated using UV light at 254 nm and were re-employed, showing adsorption percentages of 55, 32 and 19% for carbamazepine, ibuprofen, and acetaminophen, respectively (Zaib *et al.*, 2013), which indicated a loss in adsorption capacity after regeneration (Zaib *et al.*, 2013; Wei *et al.*, 2013).

According to Table 8.1, the mentioned adsorbents have merit and demerits. Without a doubt, nanotechnology has had great advancement in the removal of PhACs. On the other hand, the properties of adsorbents should be improved to

Table 8.1 Merits and demerits of various adsorbents.

Adsorbent	Merits	Demerits	References
Carbon nanotubes	Ability to remove organic pollutants, high sorption kinetic, high mechanical property	Difficult to recovery, Expensive, environmental risks	Cai *et al.* (2013); Wei *et al.* (2013); Perreault *et al.* (2015)
Graphene	High mechanical property, quick sorption, high sorption kinetic	Environmental risks, high cost	Cai *et al.* (2013); Khajeh *et al.* (2013); Zhi *et al.* (2013)
MoS_2 nanosheets	High adsorption kinetic, high adsorption capacity	Expensive, low stability, environmental risks	Cai *et al.* (2013); Yu *et al.* (2017)
Hexagonal boron nitride	Stable, high adsorption capacity, high mechanical property	Expensive, toxic	Cai *et al.* (2013); Attia *et al.* (2013); Chen *et al.* (2015)
Nano-hydroxyapatite	High stability, low solubility, large surface area	Need to use high concentration of it for removing only a small concentration of a pollutant	Cai *et al.* (2013); Zhi *et al.* (2005)
Chitosan	Ability to remove organic pharmaceutical pollutants, high adsorption kinetic, low cost	Not appropriate to remove certain pollutants by itself and need to be doped	Cai *et al.* (2013); Zhuo *et al.* (2017); Fang *et al.* (2011); Amouzgar and Salamatinia (2015)
Metal oxide	Large surface area, low toxicity, able to have magnetic property	Not good in sorption capacity	Cai *et al.* (2013); Golberg *et al.* (2010)
Nanocomposites	Porous structure, having two special structures	Each composite is able to remove some pollutants	Wang (2014); Ajayan *et al.* (2006); Bavastrello *et al.* (2013); Zhang *et al.* (2012); Hua *et al.* (2012)

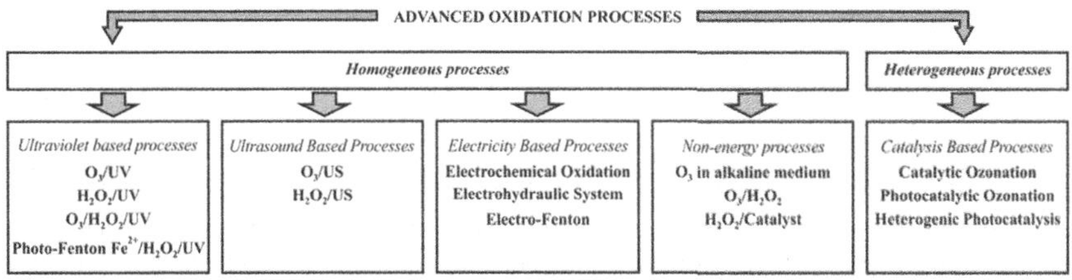

Figure 8.1 The advanced oxidation processes (AOPs) types.

have higher selectivity and adsorption of PhACs, and the cost of their production needs to be decreased. As a result, examination of the new-generation nano-adsorbents into the pilot scale is necessary for the evaluation of the adsorbents' efficiency in real water treatment conditions.

8.8 ADVANCED OXIDATION PROCESSES

Traditional methods (using one factor or one oxidizing agent) are not adequately effective. Many recent studies focused on improving oxidation processes (Advanced Oxidation Processes (AOPs)) for the polluted water treatment from pharmaceutical compounds (Prieto-Rodrıguez *et al.*, 2013; Oller *et al.*, 2011; Vogna *et al.*, 2004; Khan *et al.*, 2020). This is carried out by the combined action of several substances or factors such as electrochemical processes, the combination of ultrasound and/or UV radiation with hydrogen peroxide and ozone, catalytic/photocatalytic ozonation, and so on, as shown in Figure 8.1 (Tijani *et al.*, 2014).

8.8.1 Electrochemical advanced oxidation processes (EAOP)

For oxidative degradation of pharmaceutical compounds, anode oxidation methods, electro-Fenton, and photoelectro-Fenton can be used (Sirés and Brillas 2012; Panizza and Cerisola 2009; Brillas and Martínez-Huitle 2015). If the oxidation method is applied at the anode, the following scenarios can be executed (Brillas and Martínez-Huitle 2015; Marselli *et al.*, 2003; El-Ghenymy *et al.*, 2014):

- Organic pollutants oxidized by physically sorbed hydroxyl radicals, formed at the active anode with a high oxygen overvoltage (SnO_2, PbO_2 diamond electrodes doped with bromine);
- Organic pollutants oxidized using superoxide radicals, formed by the reaction of hydroxyl radicals with an active anode (Pt, RuO_2, and IrO_2).

In the electro-Fenton process, ferrous ions are additionally introduced into the reaction sphere to obtain Fenton's reagent with electrically generated hydrogen peroxide (H_2O_2) (Sun and Pignatello 1993; Nidheesh and Gandhimathi 2012).

In photoelectric Fenton, the reaction sphere is irradiated with UV to intensify the process due to the photolysis of Fe $(OH)_2$ (Flox *et al.*, (2007) and photodecomposition of the complexes formed by ferric and organic molecules (Garza-Campos *et al.*, 2016).

The efficiency of a single electric cell in EAOP processes is determined by the oxidizing ability of anode sites with adsorbed hydroxyl radicals and the ability of the cathode to generate hydrogen peroxide, and depends on the correct selection of electrode material (Garza-Campos *et al.*, 2016). Depending on various features of the technological process EAOP, the reduction in the content of micro-contaminants can range from 86 to 99%. Researchers have provided vast information on the use of various electrochemical methods for the oxidative destruction of active pharmaceutical substances (Sirés and Brillas 2012; Panizza and Cerisola 2009; Garza-Campos *et al.*, 2016; Yahiaoui *et al.*, 2013; Dirany *et al.*, 2012; Brinzila *et al.*, 2012; Dirany *et al.*, 2010).

For example, the 86.7% removal efficiency of tetracycline was achieved by anodic oxidation in an electrical system based on the Pb/PbO_2 anode (Yahiaoui *et al.*, 2013). By mathematical modeling, it was found that such efficiency is achieved at a temperature of 26°C, a current density of 25 mA/cm^2, and a stirring speed of 720 rpm if the initial concentration of tetracycline in the solution is 100 mg/L. The efficiency of the electro-Fenton process for the purification of contaminated water by sulfochloropyradisine and sulfomethaxazole was 90% (Dirany *et al.*, 2012; Dirany *et al.*, 2010). The initial concentration of sulfochloropyradizine and sulfomethaxazole in the solution was 0.021 and 0.028 mmol/L, respectively, and the electro oxidative system included a diamond anode, doped bromine, and a carbon-containing cathode. Based on these findings, researchers concluded that at low currents, oxidation is more active on the anode (anodic oxidation process), and at high currents the reaction is shifted directly into solution due to the electrically generated H_2O_2 (electro-Fenton process). Intriguingly, researchers stated the efficacy of destruction of tetracycline 99% (determined by reducing the COD of the solution) (Brinzila *et al.*, 2012). The anodic oxidation process was carried out in a reactor with a diamond anode (bromine doped anode) in a circulating mode, the initial concentration of tetracycline in the solution was 150 mg/L, and the exposure time was 4 hours.

8.8.2 Fenton process

One of the most effective advanced oxidizing agents used in the degradation of stable organic compounds is Fenton's reagent, consisting of H_2O_2 and Fe^{+2} salts. When the Fe^{+2} ions are oxidized by hydrogen peroxide to the Fe^{+3} ion, a hydroxyl radical and a hydroxide ion are formed. The Fe^{+3} is then reduced by another hydrogen peroxide molecule to Fe^{+2} and a hydroperoxyl radical and a hydrogen proton is formed. The total effect is the disproportionation of hydrogen peroxide, resulting in the formation of two different oxidizing radicals

(· H + · OH) (Zazo *et al.*, 2005). The disadvantage of this process is the low rate of water purification due to its accumulation of iron in the reaction medium. Researchers have utilized several substances that can rapidly reduce Fe^{+3} ions to intensify the process: catechol (Zanta *et al.*, 2010), dihydroxybenzene (Rodriguez *et al.*, 2001) and other aromatic compounds (Strlic *et al.*, 2002). Hence, the rate of chemical interactions has been improved. Despite the low speed, simplicity and cost, the photo-Fenton process is a favorable prospect to treat waters contaminated with various pharmaceutical compounds, especially for pre-treatment (Khan *et al.*, 2019; Dhingra *et al.*, 2020).

8.8.3 Photocatalytic degradation of PhACs and the combined effect of UV radiation and hydrogen peroxide on organic compounds

Photocatalytic oxidation of organic compounds is effective for treating some PhACs and is not new in science and technology (Khan *et al.*, 2017; Chamjangali *et al.*, 2015; Linsebigler *et al.*, 1995; Fujishima *et al.*, 1999; Duczmal and Sobczynski 2007; Yin *et al.*, 2019). However, the practical implementation of this technology for wastewater treatment has not gained wide spreading. Many studies in the field of photocatalysis have been investigated in real conditions, however, they do not allow for creating suitable technical solutions with practical implementation. Literature analysis shows that photocatalysis technologies lack the fundamental differences in the oxidation of various organic substances. The mineralization rate of various types of organic substances usually differs no more than 10 times. Thus, the efficiency of purification of toxic waters of pharmacological enterprises and medical institutions will not be lower than the efficiency of purification of domestic wastewater.

Photocatalytic processes are a sequence of chemical interactions consisting of the formation of holes in the valence band on the surface of the catalyst under the influence of radiation from the visible or ultraviolet part of the spectrum and subsequently, direct or indirect interaction of these electron holes and organic pollutants.

Direct interaction implies direct oxidation of organic substrates through holes in the valence band whereas indirect mechanism implies the participation of holes in the valence band in the formation of hydroxyl radicals, which interacts with the organic matter, oxidizing it (Bickley *et al.*, 2005).

The composition and structure of photocatalytic materials can widely vary. From changes in the basic composition (oxides, sulfides, mixtures thereof), additions to changes in the crystal structure, or particle size. Most studies are dedicated to photocatalytic oxidation using titanium dioxide and systems based on it as catalysts for the photo destruction process (Fujishima *et al.*, 1999; Duczmal and Sobczynski 2007; Cui *et al.*, 2016). This is explained by low cost, high physicochemical stability, and significant photocatalytic activity of titanium dioxide (Cui *et al.*, 2016). To date, most organic pollutants can be completely mineralized on the surface of titanium dioxide (Fujishima *et al.*, 1999). Some research articles (Khan *et al.*, 2017; Chamjangali *et al.*, 2015; Calza *et al.*, 2006) focused on the use of photocatalytic oxidation for the destruction of pharmaceuticals shows high efficiency of photocatalytic and photo electrocatalytic systems based on catalysts containing titanium dioxide (PtNCs/TiO_2 and Pd/CNS-TiO_2 Calza *et al.*, 2006) regarding the destruction of acetylsalicylic acid. For example, the destruction of diclofenac in a photocatalytic system based on suspended titanium dioxide is shown in a study (Calza *et al.*, 2006). The time from the beginning of the process to complete mineralization (that is, the total organic carbon absence) was 1.5–2 hours, depending on the initial concentration of diclofenac and titanium dioxide in the solution.

For a practical technological solution it is essential to use a catalytically active material with a highly developed surface, low hydraulic resistance, not requiring constant regeneration, operating both in open and closed mode. Therefore, the best technological solution is a catalyst supported on a carrier, which:

- has a highly developed surface (mesh or sponge);
- is resistant to the reaction medium (acidity, the presence of organic substances, synthetic surfactants, oxidizing agents, reducing agents);
- can provide easy access to reagents (liquids, air) in the reaction surface;
- will not lead to the removal of particles of the photocatalyst with a fluid stream;
- will be resistant and preferably transparent to UV radiation.

Given these requirements for the catalyst, a TiO_2-based system is recommended – a polymer carrier in which adhesion to the surface of the carrier occurs by burning pressure grains to it. An advantage of this method is that it uses only the part of the surface that was turned towards the carrier to fix the granules, leaving the remaining parts of the surface (not attached to the carrier) of the photocatalyst available for a chemical reaction. Additional advantages of such photocatalyst are:

- Its flexibility; allowing any configuration of the photocatalyst (film, mesh, sponge). Hence, the specific surface area of the photocatalyst will primarily depend on the structure of the carrier material.
- Using a transparent carrier material to UV radiation makes it possible to carry out a chemical process with irradiation from both the liquid side and direct UV radiation into the reaction zone from the side of the carrier material.

It is difficult to ensure the stability of a carrier material to oxidation at the photocatalyst-carrier interface during photocatalysis, so the carrier material must have a minimum volume of oxygen in its composition. As such, in the presence of UV radiation, a chemical oxidation reaction at the photocatalyst-carrier interface will not occur, and the carrier will not be destroyed. Additionally, the carrier must have acceptable technological and economic properties. Such an analysis shows that in order to create a technology for treating waters contaminated by pharmaceutical compounds, scientific research should be carried out not only to study the kinetics of the reducing concentration of organic substances, but also to study the conditions of operating the photocatalyst.

The most suitable carrier materials are polyethylene, polypropylene, polystyrene, PVC, silicone, and in this perspective, fluoroplastic can be considered. However, titanium dioxide (TiO2), as the main substance with photocatalytic properties, and a modification of rutile with a particle size of 20–100 nm, is the best. This is due to its cost and availability, and, where necessary, its experimental data can be recalculated for any other photocatalyst. The possibility of operating the photocatalyst described above was studied experimentally.

8.8.3.1 Photocatalytic material fabrication

(1) Carrier materials were used:
  - Transparent cover for binding (polypropylene) – sample 1
  - Medium-pressure polyethylene (approximately 100 microns) – sample 2
  - PVC film 550 microns thick – sample 3
  - Scotch tape (polypropylene tape with a thickness of the adhesive layer 10–30 microns) – sample 4
  - Low-pressure polyethylene (80 microns) – sample 5
  - Silicone – sample 6

(2) TiO_2 was used as the main material of the photocatalyst.

A mixture of TiO_2 and water was uniformly applied to a sheet of paper with a thick layer and coated with a carrier material. A similar layer of TiO_2 with water was also deposited on the carrier material and covered with another sheet of paper. The influence of high temperature and pressure ensured the fixation of the active substance on the media (except silicone) (see Table 8.2). Table 8.2 shows how strongly the surface structure depends on the type of plastic mass that was used as the carrier material.

8.8.3.2 Experimental conditions

(1) An organic pollutant and its concentration. Diamond green (*Viride nitens*) was used as a pollutant due to its complex structure (Figure 8.2), similar to natural pollutants, as well as a clear and intense color – green, which is very convenient for determining the number of organic substances in a solution.

   For the materials which were installed in the test with *Viride nitens*, experiments were conducted to purify water from humic acids, as more complex substances in their structure.

Table 8.2 Photos of composite materials for photocatalysis.

No. Sample	General Form	Structure	Sectional View
Sample 1	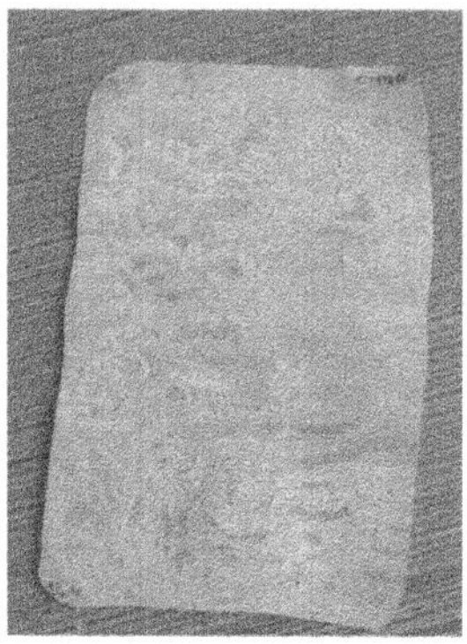		
Sample 2	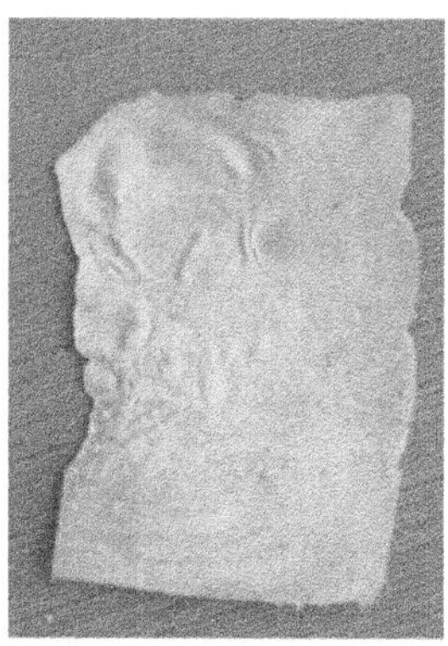		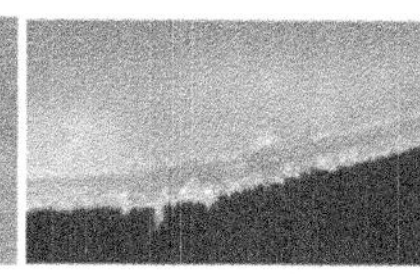
Sample 3	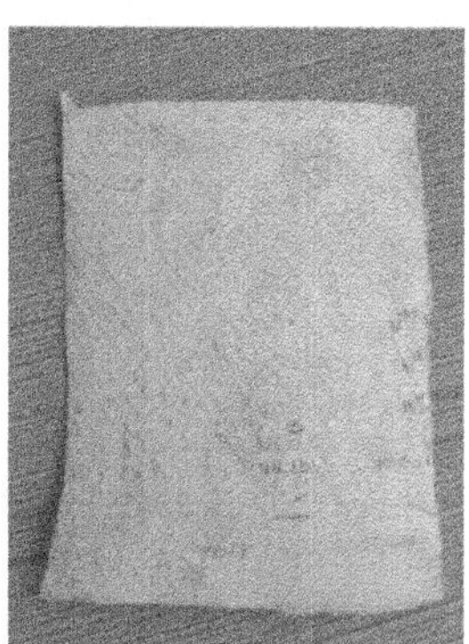		

(*Continued*)

Table 8.2 Photos of composite materials for photocatalysis (*Continued*).

No. Sample	General Form	Structure	Sectional View
Sample 4		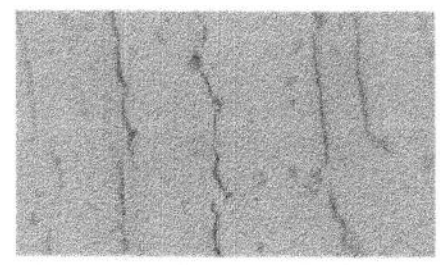	
Sample 5			
Sample 6		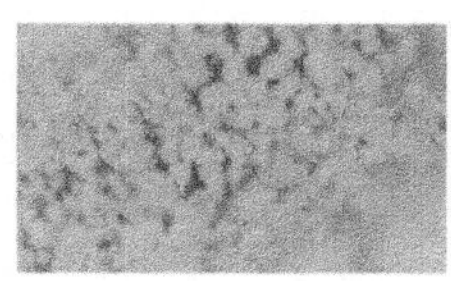	

(2) The effect of the oxidizing agent:

The study was conducted by free access to air, in a confined space and with the addition of an oxidizing agent – hydrogen peroxide (H_2O_2).

(3) The intensity of the radiation.

Different types of coating of the container with the solution and various thicknesses of the solution layer were used. A 15 V UV lamp was employed as a radiation source. For efficiency comparison, a study was also conducted on the use of sunlight with an intensity of 500 thousand lux.

8.8.3.3 Experimental works plan

The experimental study was divided into several stages:

(1) Investigation of the effect of the thickness of the solution layer on the rate of photocatalysis. The solution (900 mL of water and 0.3 mL of *Viride nitens*)

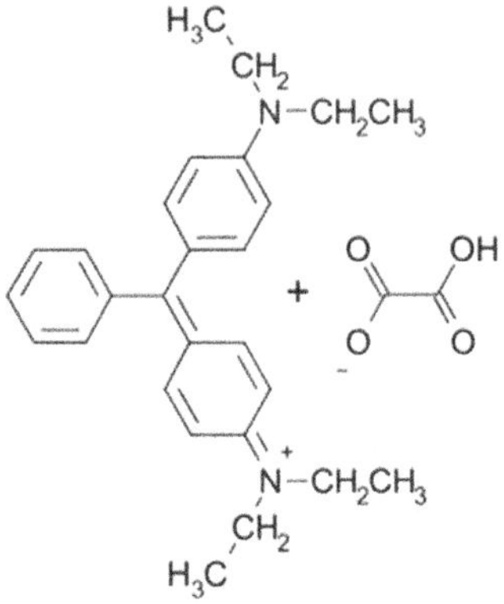

Figure 8.2 Structural formula of *Viride nitens* (brilliant green) $C_{27}H_{34}N_2O_4S$ (Source: (https://studbooks.net/2578424/tovarovedenie/brilliantovyy_zelenyy_tehnologiya_poluchen iya_brilliantovogo_zelenogo).

was added to the photocatalytic material. The layer thickness is 1.2 and 3 cm. At the same time, six different samples were placed under a UV lamp for 5 hours. Every hour, samples were taken from each container to study changes in color intensity.

(2) Study the effect of hydrogen peroxide (as a strong oxidizing agent, during the decomposition of which an OH radical is formed). The OH radical is responsible for the oxidation of organic matter in solution. In this experiment, a 3 or 30% solution of H_2O_2 was added to 100 mL of contaminated water, so that the ratio of H_2O_2 to organic matter concentration was 20:200 times.

(3) Study of the photocatalysis in a confined space. Peroxide as an oxidizing agent does not require atmospheric oxygen, therefore it can be conducted in an enclosed space. For this, a material transparent to UV radiation was placed above the surface of the liquid. The most common transparent coatings were used – glass, plexiglass, and fluoroplastic 1 mm thick. The height between the brilliant green (*Viride nitens*) solution and the closing surface was 4.5 cm.

(4) Comparison of the photocatalytic oxidation rate of organic compounds using light of different intensities using a UV lamp. The experimental work plan is presented in Table 8.3.

8.8.3.4 Properties of photocatalytic samples produced

An ultraviolet lamp was placed 10.5 cm from the solution surface and photocatalytic materials samples were placed under the lamp. At appropriate intervals, solution samples were taken to determine the color intensity (organic matter

Table 8.3 Experimental design.

Layer Height	**Samples**					
	Sample 1 (Polypropylene)	**Sample 2 (Polyethylene)**	**Sample 3 (PVC Film)**	**Sample 4 (Scotch Tape)**	**Sample 5 (Low Pressure Polyethylene)**	**Sample 6 (Silicone)**
1. Investigation of the effect of the brilliant green (*Viride nitens*) solution layer thickness						
1 cm	+	+	+	+	+	+
2 cm	+	+	+	+	+	+
3 cm	+	+	+	+	+	+
2. H_2O_2 addition						
1 cm	+	+	+	+	+	+
3. The study of the photocatalysis in a confined space						
Samples	Glass		Plexiglass		Fluoroplastic	
Sample 3 (PVC film)	+		+		+	
Sample 6 (silicone)	+		+		+	
4. The use of sunlight						
Samples	Under the lamp		50000 lux		500000 lux	
Sample 3 (PVC film)	+		+		+	
Sample 6 (silicone)	+		+		+	

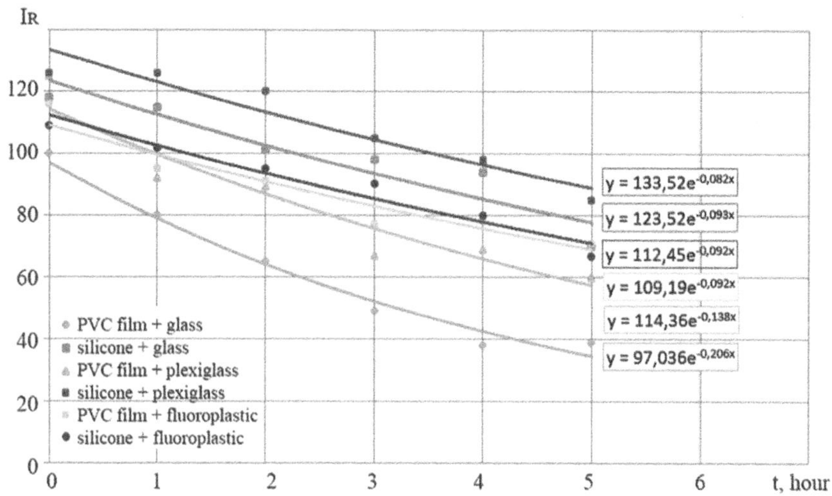

Figure 8.3 Intensity of red color wave, dependent on time.

concentration). The dependence of intensity of the red color wave on time was obtained for various photocatalytic materials (Figure 8.3).

The kinetics of the process can be described by an exponential dependence, where the value at the exponent determines the reaction rate and is independent of the initial concentration of the substance. Therefore, it was the parameter that was used to evaluate the effectiveness of the proposed technological solutions to compare them with each other.

8.8.3.5 Research results

In the first stage, sample 3 (PVC film) and sample 6 (silicone) saw the best performance. This may be due to their more developed surface. Decolorization of the solution by 99% for a layer thickness of 1 cm occurred in 5 hours and for a layer thickness of 3 cm in 13 hours (Figure 8.4). Hence, the cleaning time is directly related to the thickness of the solution layer.

In the second stage using H_2O_2, complete purification of the solution with a layer thickness of 2 cm occurred in 30 minutes (in contrast to 9 hours without using H_2O_2). In the third stage, glass was the most optimal coating for creating an enclosed space.

At the fourth stage, humic acid of the same concentration was used as a pollutant. For the ratio of H_2O_2 and organic substances 37:1, the time until the neutralization of the dye was 40 minutes, and the time until the bleaching of humates was 2 hours. Photocatalysis under the influence of sunlight occurs at a lower rate than under

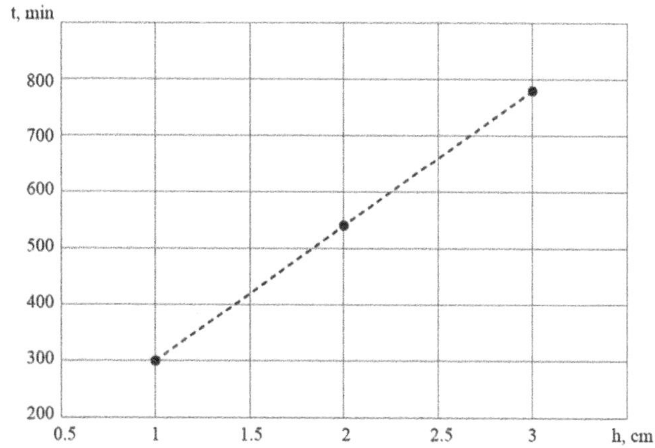

Figure 8.4 Complete water treatment, time dependent on layer thickness of solution.

the influence of an ultraviolet lamp. The experimental results are shown in Tables 8.4 and 8.5.

The photocatalytic material on PVC film had the best characteristics for creating an effective technology for the removal of organic substances from polluted waters. However, a UV lamp or sunlight should be used for lighting, where necessary, with the intensification of the cleaning process (10–20 times) requiring the addition of hydrogen peroxide.

Purification of contaminated water based on the combined effects of UV radiation and hydrogen peroxide is of great interest, and many scientific studies have demonstrated the effectiveness of this approach for the destruction of stable pharmaceutical compounds (Wu and Linden 2008; Kim *et al.*, 2008; Balcioglu and Otker 2004). Some studies also argue that the combination of UV radiation with one or more oxidizing agents (hydrogen peroxide, ozone, etc.) or catalysts (Fe_2, TiO_2) significantly increases the overall efficiency of the oxidative destruction process (Mahamuni and Adewuyi 2010; Liu *et al.*, 2012; Madhavan *et al.*, 2009). With the simultaneous exposure to several factors and/or substances, a synergistic effect is observed, which is explained by the influence on the process speed of intermediate products and other substances in contaminated water (Balcioglu and Otker 2004).

The oxidative destruction of 40 pharmaceutical preparations was carried out in a study (Wols *et al.*, 2013) using the combined effects of UV and H_2O_2. Furthermore, the effectiveness of the destruction of most drugs was more than 90% when the concentration of H_2O_2 in the treated solution was 5–20 mg/L. The rate of oxidative degradation is greatly reliant on the structure of the oxidized substance. So, for ketoprofen, prednisolone, the rate of destruction was significantly lower than for metformin, phosphamide.

Table 8.4 Effect of solution layer thickness on photocatalysis rate.

Carrier Material	The Decrease Rate in the Pollutant Concentration, Times/Hour	Time, Hour (to 90% Purification)
The solution layer thickness is 1 cm, H_2O_2 is absent, the covering surface is absent		
Sample 1 – Polypropylene	–0.1882	12.2361
Sample 2 – Polyethylene	–0.2781	8.2783
Sample 3 – PVC film	–0.0708	32.5114
Sample 4 – Scotch tape	–0.1163	19.7906
Sample 5 – Low pressure polyethylene	–0.1721	13.3803
Sample 6 – Silicone	–0.1346	17.1054
The solution layer thickness is 2 cm, H_2O_2 is absent, the covering surface is absent		
Sample 1 – Polypropylene	–0.130123143	17.6954
Sample 2 – Polyethylene	–0.085231303	27.0157
Sample 3 – PVC film	–0.067383306	34.1714
Sample 4 – Scotch tape	–0.074487403	30.9124
Sample 5 – Low pressure polyethylene	–0.082187881	28.0161
Sample 6 – Silicone	–0.088941873	25.8887
The solution layer thickness is 3 cm, H_2O_2 is absent, the covering surface is absent		
Sample 1 – Polypropylene	–0.072829215	31.6162
Sample 2 – Polyethylene	–0.070248522	32.7777
Sample 3 – PVC film	–0.04706078	48.9279
Sample 4 – Scotch tape	–0.054920417	41.9258
Sample 5 – Low pressure polyethylene	–0.053038689	43.4133
Sample 6 – Silicone	–0.053944763	42.6841
The solution layer thickness is 1 cm, the ratio of the concentration of H_2O_2 and humates is 180, the covering surface is absent		
Sample 1 – Polypropylene	–1.522426535	1.5124
Sample 2 – Polyethylene	–2.028148247	1.1353
Sample 3 – PVC film	–1.400087683	1.6446
Sample 4 – Scotch tape	–1.860752341	1.2374
Sample 5 – Low pressure polyethylene	–1.581786381	1.4557
Sample 6 – Silicone	–1.276293466	1.8041

Table 8.5 Photocatalysis in a confined space with solution thickness layer of 1 cm.

Carrier Material	Covering Material	The Decrease Rate in the Pollutant Concentration, Times/Hour		Time, Hour (to 90% Purification)	
		H_2O_2 is Absent	H_2O_2 is 1 ml*	H_2O_2 is Absent	H_2O_2 is 1 ml*
PVC film	Glass	–0.312009516	–4.356756987	7.3799	0.5285
	Plexiglass	–0.178912534	–4.208470989	12.8699	0.5471
	Fluoroplastic	–0.312009516	–3,295836866	7.3799	0.6986
Silicone	Glass	–0.116980649	–2.307399263	19.6835	0.9979
	Plexiglass	–0.099622475	–1.002893666	23.1131	2.2959
	Fluoroplastic	–0.118142771	–2.297948254	19.4899	1.0020

*The ratio of the concentration of H_2O_2 and humates is 180.

8.9 WATER TREATMENT METHODS FOR FIELD HOSPITALS

Understanding the usual conditions and effect of constantly operating sources of environmental pollution by pharmaceutical compounds, to reduce the burden on natural components, extreme situations like explosions, fires, military operations, and other happenings with mass people injury (disease) should be understood (Vambol *et al.*, 2019; Vambol *et al.*, 2019; Bar-On *et al.*, 2013; Sokolov *et al.*, 2018; Naor *et al.*, 2017). In situations with several victims, field hospitals should be organized to provide medical care. Many patients with various types of drug usage enhance the challenge of waste accumulation, including water contaminated with pharmaceutical compounds. The most important mobile hospital task is to assess the patient's health status and immediately provide emergency care, and then the patient can be sent to inpatient medical facilities.

Mobile hospitals extensively use antibiotics to treat patients; consequently, the extent of environmental pollution is significantly higher. To access the accurate strategy for wastewater treatment in mobile hospitals, parameters of drugs used should be taken into accounts, such as biological and physicochemical characteristics, toxicity, risk of circulating in the environment, and resistance to treatment (Kümmerer and Helmers 2000). The main conditions for the sewage treatment systems of hospitals in the field are simplicity, safety, and profitability. Since the field hospital is located temporarily, all life support systems must be mobile.

As stated in scientific studies on the single-stage treatment of waters contaminated with pharmaceutical compositions, their approximate effectiveness is as follows (Khan *et al.*, 2019, Göbel *et al.*, 2007; Lien *et al.*, 2016; Sim *et al.*, 2013; Gracepavithra *et al.*, 2017):

- membrane biological reactor (MBR) provides efficiency 39–60%;
- conventional activated sludge (CAS) and filtration – 59–76%;

- conventional activated sludge (CAS), filtration and activated carbon – 80%;
- ozonation process – 91%.

Hence, the most effective one-step method is the ozonation process which allows the removal of up to 91% of pharmaceutical compounds. However, the ozonation process has certain practical difficulties for field applications:

- it is expensive;
- ozone is a toxic gas that requires special monitoring and safety.

As such, the method is best used as one of the stages in the technology of combined purification of pharmaceutical compounds from contaminated waters.

According to research based on combined technologies for treatment of waters contaminated with pharmaceutical compositions, the basic biological phase of the HWW treatment is formed by one of four methods (Khan *et al.*, 2019; Batelaan *et al.*, 2013; Adamcza *et al.*, 2012; Grundfos BioBooster A/S 2015; Tambosi *et al.*, 2009; Martins *et al.*, 2008; Prado *et al.*, 2011; Shrestha *et al.*, 2001): septic tank, CAS, up flow anaerobic sludge layer (UASB) and MBR. Their effectiveness is as follows:

(1) Septic tank and:
  - wetlands – up to 77% (Shrestha *et al.*, 2001);
  - oxidation ponds – 54% (Beyene and Redaie 2011);
  - anaerobic filters – 65% (Martins *et al.*, 2008);

(2) CAS and:
  - chlorination – 65–92% (Göbel *et al.*, 2007; Shrestha *et al.*, 2001);

(3) UASB and:
  - anaerobic filters – 64% (Prado *et al.*, 2011);

(4) MBR and:
  - combined use ozonation process and ultraviolet – 50–90% (Bakker *et al.*, 2014);
  - combined use ozonation process and granulated activated carbon – up to 70% (Batelaan *et al.*, 2013);
  - Fenton process/ultraviolet – 12–100% (Tambosi *et al.*, 2009);
  - combined use powdered activated carbon and sand filtration – up to 80% (Adamcza *et al.*, 2012);
  - combined use ozonation process and powdered activated carbon and sand filtration – up to 80% (Adamcza *et al.*, 2012);
  - combined use granulated activated carbon and ultraviolet – up to 90% (Grundfos BioBooster A/S 2015);
  - chlorination – 95% (Liu *et al.*, 2010).

The extensive use of a variety of contaminated water treatment methods makes it possible to create an optimal treatment scenario to solve wastewater treatment of mobile hospitals. Additionally, they are applicable in the case of low values of water flow, which is characteristic of field hospitals (Khan *et al.*, 2019).

8.10 CONCLUSIONS

Because of the broad range of drugs available in wastewater, finding one standard method for removing them is complicated. Of the traditional methods, the largest group for wastewater treatment is chemical methods, which are based on the mineralization of organic pollutants under the influence of an oxidizing agent. Among them, the most applied approach is the use of oxygen as an oxidizing agent; however, a high temperature of about 1000°C and preliminary purification by sorption or membrane separation is required. Also, the process is complex and is environmentally and economically feasible only in cases of the high content of dissolved organic substances (more than 10% of the mass). Methods such as flotation and coagulation are only advisable at the pre-treatment stage. Since nanoparticles have a large surface area and can be synergistically associated with many chemical groups, as well as due to the increased stability and solubility, especially in water, many nanomaterials can be used to effectively remove or decompose pharmaceutical compounds. Without a doubt, nanotechnology has made significant advances in eliminating active pharmaceutical ingredients. On the other hand, the properties of the adsorbents must be improved to have higher selectivity and adsorption of PhAC, and the cost of their production should be reduced.

The photocatalytic oxidation of organic compounds is of particular interest; however, the practical application of this technology for wastewater treatment is not widespread due to the lack of an acceptable technical solution for the practical implementation of the method. An experimental study made it possible to establish the most effective photocatalytic material based on a PVC film. Greater efficiency of oxidative degradation is achieved by combined exposure to UV radiation with one or more oxidizing agents (hydrogen peroxide, ozone, etc.) or catalysts (Fe_2, TiO_2).

We should not forget the need for wastewater treatment for mobile hospitals since such hospitals widely use antibiotics to treat patients; therefore, the degree of environmental pollution is much higher. The multistage approach to removing pharmaceutical ingredients from the wastewater of mobile hospitals still needs to be improved.

However, the most effective method for preserving the environment from hazardous PhACs is achieved by using an integrated approach. In this case, an integrated approach implies:

- the development and use of engineering and technical measures and protective equipment (Vambol *et al.*, 2017; Bakker *et al.*, 2014; Zlarati *et al.*, 2020; Yousif and Haddad 2013);

- the use of a criminal law approach as an action to prevent criminal acts (misuse, spoilage, pollution) against natural resources (by adopting laws to protect natural resources and prohibiting criminal attacks on natural resources) (Khan *et al.*, 2019);
- a sociological approach, as the influence of the environment on social behavior in society (through mass media, education on contemporary environmental issues, informing the public and reminding them of the need to protect nature);
- a social approach, such as reducing tension due to social inequality, eliminating, or reducing greed for profit in any way, etc. (through media, education about the role of every inhabitant of the planet in ensuring a healthy environment).

REFERENCES

Adamcza K., Lyko S., Nafo S., Evenblij H., Cornelissen A., Igos E., Klepiszewski K., Venditti S., Kovalova L., McArdell C., Helwig K., Pahl O., Barraud O., Casellas M., Dagot C., Maftah C., Ploy M.-C. and Stalder T. (2012). Pharmaceutical residues in the aquatic system – A challenge for the future. Insights and activities of the European cooperation project PILLS. Available from: www.pills-project.eu/PILLS_summary_english.pdf.

Adams C., Wang Y., Loftin K. and Meyer M. (2002). Removal of antibiotics from surface and distilled water in conventional water treatment processes. *Journal of Environmental Engineering*, **128**, 253–260.

Adriano W. S., Veredas V., Santana C. C. and Gonçalves L. R. B. (2005). Adsorption of amoxicillin on chitosan beads: kinetics, equilibrium and validation of finite bath models. *Biochemical Engineering Journal*, **27**(2), 132–137.

Ajayan P. M., Schadler L. S. and Braun P. V. (2006). Nanocomposite Science and Technology. John Wiley & Sons, Germany.

Amouzgar P. and Salamatinia B. (2015). Short review on presence of pharmaceuticals in water bodies and the potential of chitosan and chitosan derivatives for elimination of pharmaceuticals. *Journal of Molecular and Genetic Medicine*, **S4**, 1–7.

Arami M., Limaee N. Y. and Mahmoodi N. M. (2006). Investigation on the adsorption capability of egg shell membrane towards model textile dyes. *Chemosphere*, **65**(11), 1999–2000.

Arvidsson R., Kushnir D., Sandén B. A. and Molander S. (2014). Prospective life cycle assessment of graphene production by ultrasonication and chemical reduction. *Environmental Science and Technology*, **48**(8), 4529–4536.

Attia T. M. S., Hu X. L. and Qiang Y. D. (2013). Synthesized magnetic nanoparticles coated zeolite for the adsorption of pharmaceutical compounds from aqueous solution using batch and column studies. *Chemosphere*, **93**, 2076–2085.

Bakker C., Wang F., Huisman J. and Den Hollander M. (2014). Products that go round: exploring product life extension through design. *Journal of Cleaner Production*, **69**, 10–16.

Balcioglu A. and Otker M. (2004). Pre-treatment of antibiotic formulation wastewater by O_3, O_3/H_2O_2, and O_3/UV processes. *Turkish Journal of Engineering and Environmental Sciences*, **28**, 325–331.

Bandala E. R., Quiroz Alfaro M. A., Cerro López M. and Méndez-Rojas M. A. (2014). Nanostructured metal oxides for waste-water treatment. In: Book of nanomaterials for environmental protection, first edition, B. I. Kharisov, O. V. Kharissova and H. V. Rasika Dias (eds), 1st edn. Wiley, Hoboken, pp. 27–40.

Bar-On E., Abargel A., Peleg K. and Kreiss Y. (2013). Coping with the challenges of early disaster response: 24 years of field hospital experience after earthquakes. *Disaster Medicine and Public Health Preparedness*, **7**(5), 491–498.

Batelaan M. V., van den Berg E. A., Koetse E., Wortel N. C., Rimmelzwaan J. and Vellinga S. (2013). Evaluation report Pharmafilter. STOWA, the Netherlands.

Bavastrello V., Jiang Z., Nicolini C., Schjødt-Thomsen J., Srivastava S. K. and Tripathy D. K. (2013). Nanocomposites: Materials, Manufacturing and Engineering. Walter de Gruyter, Germany, p. 2.

Belkheiri D., Fourcade F., Geneste F., Floner D., Aït-Amar H. and Amrane A. (2011). Feasibility of an electrochemical pre-treatment prior to a biological treatment for tetracycline removal. *Separation and Purification Technology*, **83**(1), 151–156.

Bellona C., Oelker G., Luna J., Filteau G., Amy G. and Drewes J. E. (2008). Comparing nanofiltration and reverse osmosis for drinking water augmentation. *Journal American Water Works Association*, **100**(9), 102–116.

Bernadiner I. M. (2010). Neutralization of dangerous outcomes: selection of the optimal technology. *Municipal Solid Waste*, **9**, 18–26.

Beyene H. and Redaie G. (2011). Assessment of waste stabilization ponds for the treatment of hospital wastewater: the case of Hawassa university referral hospital. *World Applied Sciences Journal*, **15**(1), 142–150.

Bickley R. I., Slater M. J. and Wang W.-J. (2005). Engineering development of a photocatalytic reactor for wastewater treatment. *Process Safety and Environmental Protection*, **83**, 205–216.

Brillas E. and Martínez-Huitle C. A. (2015). Decontamination of wastewaters containing synthetic organic dyes by electrochemical methods: An updated review. *Applied Catalysis B: Environmental*, **166–167**, 603–643.

Brinzila C. I., Pacheco M. J., Ciriaco L., Ciobanu R. C. and Lopes A. (2012). Electrodegradation of tetracyclin on BDD anode. *Chemical Engineering Journal*, **209**, 54–61.

Bruno J., DePablo J., Duro L. and Figuerola E. (1995). Experimental study and modeling of the U (VI)-Fe $(OH)_3$ surface precipitation/ coprecipitation equilibria. *Geochimica et Cosmochimica Acta*, **59**(2), 4113– 4123.

Bui T. X. and Choi H. (2009). Adsorptive removal of selected pharmaceuticals by mesoporous silica SBA-15. *Journal of Hazardous Materials*, **168**(2–3), 602–608.

Buser H.-R, Müller M. D. and Theobald N. (1998). Occurrence of the pharmaceutical drug clofibric acid and the herbicide mecoprop in various Swiss lakes and in the North Sea. *Environmental Science and Technology*, **32**(1), 188–192.

Buser H.-R, Poiger T. and Müller M. D. (1998). Occurrence and fate of the pharmaceutical drug diclofenac in surface waters: rapid photodegradation in a lake. *Environmental Science and Technology*, **32**(22), 3449–3456.

Buser H.-R, Poiger T. and Müller M. D. (1999). Occurrence and environmental behavior of the chiral pharmaceutical drug ibuprofen in surface waters and in wastewater. *Environmental Science and Technology*, **33**, 2529–2535.

Cai N. and Larese-Casanova P. (2014). Sorption of carbamazepine by commercial graphene oxides: A comparative studywith granular activated carbon and multiwalled carbon nanotubes. *Journal of Colloid and Interface Science*, **426**, 152–161.

Cai Z., Dwivedi A. D., Lee W.-N, Zhao X., Liu W., Sillanpää M., Zhao D., Huang C.-H and Fu J. (2013). Application of nanotechnologies for removing pharmaceutically active compounds in water: Development and future trends. *Environmental Science: Nano*, **00**, 1–20.

Calza V. A., Sakkas C., Medana C., Baiocchi A., Dimou E. and Pelizzetti T. (2006). Albanis Photocatalytic degradation study of diclofenac over aqueous TiO_2 suspensions. *Applied Catalysis*, **67**, 197–205.

Camilli L., Pisani C., Gautron E., Scarselli M., Castrucci P., D'Orazio F. and De Crescenzi M. (2014). A three-dimensional carbon nanotube network for water treatment. *Nanotechnology*, **25**(6), 065701.

Carabineiro S. A. C., Thavorn-Amornsri T., Pereira M. F. R., Serp P. and Figueiredo J. L. (2012). Comparison between activated carbon, carbon xerogel and carbon nanotubes for the adsorption of the antibiotic ciprofloxacin. *Catalysis Today*, **186**, 29–34.

Carbajo J. B., Petre A. L., Rosal R., Berná A., Letón P., García-Calvo E. and Perdigón-Melón J. A. (2016). Ozonation as pre-treatment of activated sludge process of a wastewater containing benzalkonium chloride and NiO nanoparticles. *Chemical Engineering Journal*, **283**, 740–749.

Cerro-Lopez M. and Méndez-Rojas M. A. (2017). Ecopharmacovigilance: Multidisciplinary Approaches to Environmental Safety of Medicines. Springer International Publishing, Cham.

Cerro-Lopez M., Meas-Vong Y., Méndez-Rojas M. A., Martinez-Huitle C. A. and Quiroz M. A. (2014). Formation and growth of PbO_2 inside TiO_2 nanotubes for environmental applications. *Applied Catalysis B Environmental*, **144**(1), 174–181.

Chamjangali M. A., Bagherian G., Javid A., Boroumand S. and Farzaneh N. (2015). Synthesis of Ag-ZnO with multiple rods (multipods) morphology and its applicationin the simultaneous photo-catalyticdegradation of methyl orange and methylene blue. *Spectrochimica Acta Part A*, **150**, 230–237.

Chao Y., Zhu W., Wu X., Hou F., Xun S., Wu P., Ji H., Xu H. and Li H. (2014). Application of graphene-like layered molybdenum disulfide and its excellent adsorption behavior for doxycycline antibiotic. *Chemical Engineering Journal*, **243**, 60–67.

Chen Y., Lan T., Duan L., Wang F., Zhao B., Zhang S. and Wei W. (2015). Adsorptive removal and adsorption kinetics of fluoroquinolone by nano-hydroxyapatite. *PLoS ONE*, **10**(12), e0145025.

Cincinelli A., Martellini T., Coppini E., Fibbi D. and Katsoyiannis A. (2015). Nanotechnologies for removal of pharmaceuticals and personal care products from water and wastewater. *Journal of Nanoscience and Nanotechnology*, **15**(5), 3333–3347.

Collado S., Quero D., Laca A. and Diaz M. (2013). Efficiency and sensitivity of the wet oxidation/biological steps in coupled pharmaceutical wastewater treatment. *Chemical Engineering Journal*, **234**, 484–490.

Collins P. G., Bradley K., Ishigami M. and Zettl A. (2000). Extreme oxygen sensitivity of electronic properties of carbon nanotubes. *Science*, **287**, 1801–1804.

Cui Y., Meng Q., Deng X., Ma Q., Zhang H., Cheng X., Li X., Xie M. and Cheng Q. (2016). Fabrication of platinum nano-crystallites decorated TiO_2 nano-tube array photoelectrode and its enhanced photoelectrocatlytic performance for degradation of aspirin and mechanism. *Journal of Industrial and Engineering Chemistry*, **43**, 177–184.

Da Silva W. and Dos Santos J. H. Z. (2017). Ecotechnological strategies in the development of alternative photocatalysis. *Current Opinion in Green and Sustainable Chemistry*, **6**, 63–68.

Deegan A. M., Shaik B., Nolan K. and Urell K. (2011). Treatment options for wastewater effluents from pharmaceutical companies. *International Journal of Environmental Science and Technology*, **8**(3), 649–666.

Dhingra A., Khan N. A., Ahmed S., Gautam S., Vambol S., Vambol V. and Kovalenko S. (2020). Investigation of medical institutions in India as a source of surface water pollution. *World Review of Science, Technology and Sustainable Development (in press)*. [to be updated]

Dirany A., Sirés I., Oturan N. and Oturan M. A. (2010). Electrochemical abatement of the antibiotic sulfamethoxazole from water. *Chemosphere*, **81**, 594–602.

Dirany A., Sirés I., Oturan N., Özcan A. and Oturan M. A. (2012). Electrochemical treatment of the antibiotic sulfachloropyridazine. Kinetics, reaction pathways and oxicity evolution. *Environmental Science and Technology*, **46**(7), 4074–4082.

Duczmal L. and Sobczynski A. (2007). Photocatalytic water purification: destruction of resorcinol by irradiated titania. *Reaction Kinetics and Catalysis Letters*, **66**(2), 289–295.

El-Ghenymy A., Centellas F., Garrido J. A., Rodríguez R. M., Sirés I., Cabot P. L. and Brillas E. (2014). Decolorization and mineralization of Orange Gazo dye solutions by anodic oxidation with a boron-doped diamond anode in divided and undivided tank reactors. *Electrochimica Acta*, **130**, 568–576.

Fang Z., Chen J., Qiu X., Cheng W. and Zhu L. (2011). Effective removal of antibiotic metronidazole from water by nanoscale zero-valent iron particles. *Desalination*, **268** (1–3) 60–67.

Flox C., Garrido J. A., Rodríguez R. M., Cabot P. L., Centellas F., Arias C. and Brillas E. (2007). Mineralization of herbicide mecoprop by photoelectro-Fenton with UVA and solar light. *Catalysis Today*, **129**(1–2), 29–36.

Fogarty C. (2013). Photocatalytic Oxidation of Ciprofloxacin under UV-LED Light. BSc degree. Worcester University, Worcester.

Fujishima A., Hashimoto K. and Watanabe T. (1999). TiO_2 Photocatalysis Fundamentals and Applications. A Revolution in Cleaning Technology. BKC, pp. 14–21. Available from: https://books.google.com/books?id=lDpxNAAACAAJ.

Gao Y., Li Y., Zhang L., Huang H., Shah S. M. and Su X. (2012). Adsorption and removal of tetracycline antibiotics from aqueous solution by graphene oxide. *Journal of Colloid and Interface Science*, **368**(1), 540–546.

Garza-Campos B., Brillas E., Hernandez-Ramirez A., El-Ghenymy A., Guzman-Mar J. L. and Ruiz-Ruiz E. J. (2016). Salicylic acid degradation by advanced oxidation processes. Coupling of solar photoelectro-Fenton and solar heterogeneous photocatalysis. *Journal of Hazardous Materials*, **319**, 34–42.

Geim A. K. and Novoselov K. S. (2007). The rise of graphene. *Nature Materials*, **6**, 183–191.

Göbel A., McArdell C. S., Joss A., Siegrist H. and Giger W. (2007). Fate of sulfonamides, macrolides, and trimethoprim in different wastewater treatment technologies. *Science of the Total Environment*, **372**, 361–371.

Golberg D., Bando Y., Huang Y., Terao T., Mitome M., Tang C. and Zhi C. (2010). Boron nitride nanotubes and nanosheets. *ACS Nano*, **4**(6), 2979–2993.

Gracepavithra K., Ponnusamy S. K., Sundarrajan P., Saravanan A. and Naushad M. (2017). Sources and impacts of pharmaceutical components in wastewater and its treatment process: A review. *Korean Journal of Chemical Engineering*, **34**(11), 2787–2805.

Grobert N. (2007). Carbon nanotubes – becoming clean. *Materials Today*, **10**(1–2), 28–35.

Grundfos BioBooster A/S. (2015). Private-Public Innovation Project Background and Project Objectives. Denmark.

Hayasi M. and Saadatjoo N. (2017). Preparation of magnetic nanoparticles functionalized with poly (styrene-2-acrylamido-2-methyl propanesulfonic acid) as novel adsorbents for removal of pharmaceuticals from aqueous solutions. *Advances in Polymer Technology*, **37**(6), 1941–1953.

Hua M., Zhang S., Pan B., Zhang W., Lv L. and Zhang Q. (2012). Heavy metal removal from water/wastewater by nanosized metal oxides: A review. *Journal of Hazardous Materials*, **211**, 317–31.

Islam M. S., Zhang Y., McPhedran K. N., Liu Y. and Gamal El-Din M. (2015). Granular activated carbon for simultaneous adsorption and biodegradation of toxic oil sands process-affected water organic compounds. *Journal of Environmental Economics and Management*, **152**, 49–57.

Jodeh S., Abdelwahab F., Jaradat N., Warad I. and Jodeh W. (2016). Adsorption of diclofenac from aqueous solution using Cyclamen persicum tubers based activated carbon (CTAC). *Journal of the Association of Arab Universities for Basic and Applied Sciences*, **20**, 32–38.

Kamrukov A. S., Kozlov N. P., Novikov D. O. and Seliverstov A. F. (2013). New technological scheme of sewage treatment with high concentration of organic pollutants. *Safety in Technosphere*, **5**(44), 35–41.

Khajeh M., Laurent S. and Dastafkan K. (2013). Nanoadsorbents: classification, preparation and applications (with emphasis on aqueous media). *Chemical Reviews*, **113**, 7728–7768.

Khan A., Wang J., Li J., Wang X., Chen Z., Alsaedi A., Hayat T., Chen Y. and Wang X. (2017). The role of graphene oxide and graphene oxide-based nanomaterials in the removal of pharmaceuticals from aqueous media: A review. *Environmental Science and Pollution Research*, **24**(9), 7938–7958.

Khan A. H., Khan N. A., Khan S. A., Dhingra A., Khan S. U., Mohammadi A. A., Changani F., Yousefi M., Alam S., Vambol S., Vambol V., Khursheed A. and Ali I. (2020). Application of advanced oxidation processes followed by different treatment technologies for hospital wastewater: A case study of multispecialty hospitals. *Journal of Cleaner Production*, **269**, 122411–122424.

Khan N. A., Ahmed S., Farooqi I. H., Ali I., Vambol V., Changani F., Yousefi M., Vambol S., Khan S. U. and Khan A. H. (2020). Occurrence, sources and conventional treatment techniques for various antibiotics present in hospital wastewaters: A critical review. *Trends in Analytical Chemistry*, **129**, 115921–115935.

Khan N. A., Ahmed S., Farooqi I. H., Vambol S., Vambol V. and Koloskov V. (2021). Operational parameters optimisation for simultaneous removal of two drugs in hospital wastewater – RSM approach. *International Journal of Environment and Waste Management* (in press).[to be updated]

Khan N. A., Ahmed S., Vambol S., Vambol V. and Farooqi I. (2019). Field hospital wastewater treatment scenario. *Ecological Questions*, **30**, 57–69.

Khan N. A., Khan S. U., Ahmed S., Farooqi I. H., Hussain A., Vambol S. and Vambol V. (2019). Smart ways of hospital wastewater management, regulatory standards and conventional treatment techniques: A short review. *Smart and Sustainable Built Environment*, [to be updated] https://doi.org/10.1108/SASBE-06-2019-0079

Khataee A., Gholami P. and Sheydaei M. (2016). Heterogeneous Fenton process by natural pyrite for removal of a textile dye from water: effect of parameters and intermediate identification. *Journal of the Taiwan Institute of Chemical Engineers*, **58**, 366–373.

Khin M. M., Nair A. S., Babu V. J., Murugan R. and Ramakrishna S. (2012). A review on nanomaterials for environmental remediation. *Energy and Environmental Science*, **5**(8), 8075–8109.

Kim S. D., Cho J., Kim I. S., Vanderford B. J. and Snyder S. A. (2007). Occurrence and removal of pharmaceuticals and endocrine disruptors in South Korean surface, drinking, and waste waters. *Water Research*, **41**(5), 1013–1021.

Kim H. J., Choi K., Baek Y., Kim D. G., Shim J., Yoon J. and Lee J. C. (2014). High-performance reverse osmosis cnt/polyamide nanocomposite membrane by controlled interfacial interactions. *ACS Applied Materials and Interfaces*, **6**(4), 2819–2829.

Kim H., Hwang Y. S. and Sharma V. K. (2014). Adsorption of antibiotics and iopromide onto single-walled and multi-walled carbon nanotubes. *Chemical Engineering Journal*, **255**, 23–27.

Kim I. H., Tanaka H., Iwasaki T., Takubo T., Morioka T. and Kato Y. (2008). Classification of the degradability of 30 pharmaceuticals in water with ozone UV and H_2O_2. *Water Science and Technology*, **57**, 195–200.

Larsson D. G. J., Adolfsson-Erici M., Parkkonen J., Pettersson M., Berg A. H., Olsson P. E. and Förlin L. (1999). Ethinyloestradiol – an undesired fish contraceptive? *Aquatic Toxicology*, **45**, 91–97.

Lens P., Virkutyte J., Jegatheesan V., Kim S. H. and Al-Abed S. (2013). Nanotechnology for water and wastewater treatment, IWA Publishing, London, UK. https://doi.org/10.2166/9781780404592

Lien L. T. Q., Hoa N. Q., Chuc N. T. K., Thoa N. T. M., Phuc H. D., Diwan V., Dat N. T., Tamhankar A. J. and Lundborg C. S. (2016). Antibiotics in wastewater of a rural and an urban hospital before and after wastewater treatment, and the relationship with antibiotic use – A one year study from Vietnam. *International Journal of Environmental Research and Public Health*, **13**, 1–13.

Lin B. and Liaw S.-L. (2015). Simultaneous removal of volatile organic compounds from cooking oil fumes by using gas-phase ozonation over $Fe(OH)_3$ nanoparticles. *Journal of Environmental Chemical Engineering*, **3**, 1530–1538.

Linsebigler A. L., Lu G. Q. and Yates Y. T. (1995). Photocatalysis on TiO_2 surfaces: principles, mechanisms, and selected results. *Chemical Reviews*, **95**, 735–758.

Liu D., Lei W., Qin S., Klika K. D. and Chen Y. (2016). Superior adsorption of pharmaceutical molecules by highly porous BN nanosheets. *Physical Chemistry Chemical Physics*, **18**(1), 84–88.

Liu L., He D. and Zhao D. (2012). Study on photocatalysis degradation of phenol by using Tourmaline/TiO2 system as catalyst. *Advanced Materials Research*, **399–401**, 1337–1341.

Liu M., Chen C., Hu J., Wu X. and Wang X. (2011). Synthesis of magnetite/graphene oxide composite and application for cobalt(II) removal. *The Journal of Physical Chemistry C*, **115**(51), 25,234–25,240.

Liu M., Wen T., Wu X., Chen C., Hu J., Li J. and Wang X. (2013). Synthesis of porous Fe_3O_4 hollow microspheres/graphene oxide composite for Cr(vi) removal. *Dalton Transactions*, **42**, 14,710–14,717.

Liu Q., Zhou Y., Chen L. and Zheng X. (2010). Application of MBR for hospital wastewater treatment in China. *Desalination*, **250**(2), 605–608.

Liu Q. S., Zheng T., Wang P., Jiang J. P. and Li N. (2010). Adsorption isotherm, kinetic and mechanism studies of some substituted phenols on activated carbon fibers. *Chemical Engineering Journal*, **157**(2–3), 348–356.

Liu R., Wilding A., Whibberd A. and Zhou J. L. (2005). Partition of endocrine–disrupting chemicals between colloids and dissolved phase as determined by cross–flow ultrafiltration. *Environmental Science and Technology*, **39**, 2753–2761.

Liu W., Zhao X., Cai Z., Han B. and Zhao D. (2016). Aggregation and stabilization of multiwalled carbon nanotubes in aqueous suspensions: influences of carboxymethyl cellulose, starch and humic acid. *RSC Advances*, **6**(71), 67,260–67,270.

Lu L., Li J., Yu J., Song P. and Ng D. H. L. (2016). A hierarchically porous $MgFe_2O_4/\gamma\text{-}Fe_2O_3$ magnetic microspheres for efficient removals of dye and pharmaceutical from water. *Chemical Engineering Journal*, **283**, 524–534.

Kümmerer K. (2009). Antibiotics in the aquatic environment – A review – Part I. *Chemosphere*, **75**, 417–434.

Kümmerer K., Al-Ahmad A., Wiethan J., Hertle W. and Henninger A. (2000). Effects and fate of antibiotics in sewage treatment plants. Proceedings of the Third SETAC World Congress, May, pp. 21–25.

Kümmerer K. and Helmers E. (2000). Hospital effluents as a source of Gadolinium in the aquatic environment. *Environment Science and Technology*, **34**, 573–577.

Machulek A., Jr, Oliveira S. C., Osugi M. E., Ferreira V. D., Quina F. H., Dantas R. F., Oliveira S. L., Casagrande G. A., Anaissi F. J., Silva V. O., Cavalcante R. P., Gozzi F., Ramos D. D., Da Rosa A. P. P., Santos A. P. F., De Castro D. C. and Nogueira J. A. (2013). Application of different advanced oxidation processes for the degradation of organic pollutants. Organic Pollutants – Monitoring, Risk and Treatment. IntechOpen, London, p. 238.

Madhavan J., Grieser F. and Ashok kumar M. (2009). Kinetics of the sonophotocatalytic degradation of orange G in presence of Fe^{3+}. *Water Science and Technology*, **60**, 2195–1202.

Madrakian T., Afkhami A., Ahmadi M. and Bagheri H. (2011). Removal of some cationic dyes from aqueous solutions using magnetic-modified multi-walled carbon nanotubes. *Journal of Hazardous Materials*, **196**, 109–114.

Mahamuni N. N. and Adewuyi Y. G. (2010). Advanced oxidation processes (AOPs) involving ultrasound for wastewater treatment: a review with emphasis on cost estimation. *Ultrasonics Sonochemistry*, **17**, 990–1003.

Marselli B., García-Gomez J., Michaud P. A., Rodrigo M. A. and Comninellis C. (2003). Electrogeneration of hydroxyl radicals on boron-doped diamond electrodes. *Journal of the Electrochemical Society*, **150**, 79–83.

Martins A. F., Vasconcelos T. G., Henriques D. M., Frank C., König A. and Kümmerer K. (2008). Concentration of ciprofloxacin in Brazilian hospital effluent and preliminary risk assessment: A case study. CLEAN – Soil, Air, Water, **36**, 264–269.

Masciangioli T. and Zhang W. X. (2003). Peer reviewed: Environmental technologies at the nanoscale. *Environmental Science and Technology*, **37**(5), 102a–108a.

Mazhar M. A., Khan N. A., Ahmed S., Khan A. H., Hussain A., Changani F. and Vambol V. (2020). Chlorination disinfection by-products in Municipal drinking water–A review. *Journal of Cleaner Production*, **273**, 123159–123171.

Mestre A. S., Pires J., Nogueira J. M. F. and Carvalho A. P. (2007). Activated carbons for the adsorption of ibuprofen. *Carbon*, **45**(10), 1979–1988.

Michael I., Rizzo L., McArdell C. S., Manaia C. M., Merlin C., Schwartz T., Dagot C. and Fatta-Kassinos D. (2013). Urban wastewater treatment plants as hotspots for the release of antibiotics in the environment: A review. *Water Research*, **47**(3), 957–995.

Mohle E., Kempter C., Kern A. and Metzger J. (1999). Examination of the degradation of drugs in municipal sewage plants using liquid chromatography-electrospray mass spectrometry. *Acta Hydrochimica et Hydrobiologica*, **27**(6), 430–436.

Nam S. W., Jung C., Li H., Yu M., Flora J. R., Boateng L. K., Her N., Zoh K. D. and Yoon Y. (2015). Adsorption characteristics of diclofenac and sulfamethoxazole to graphene oxide in aqueous solution. *Chemosphere*, **136**, 20–26.

Naor M., Heyman S. N., Bader T. and Merin O. (2017). Deployment of field hospitals to disaster regions: Insights from ten medical relief operations spanning three decades. *American Journal of Disaster Medicine*, **12**(4), 243–256.

Ni X., Zhao Y. and Song Q. (2015). Electrochemical reduction and in-situ electrochemiluminescence detection of nitroaromatic compounds. *Electrochimica Acta*, **164**, 31–37.

Nidheesh P. V. and Gandhimathi R. (2012). Trends in electro-Fenton process for water and wastewater treatment: An overview. *Desalination*, **299**, 1–15.

Novoselov K. S., Geim A. K., Morozov S. V., Jiang D., Zhang Y., Dubonos S. V., Grigorieva I. V. and Firsov A. A. (2004). Electric field effect in atomically thin carbon films. *Science*, **306**, 666–669.

O'Brien P. (2012). Nanoscience: Nanostructures through Chemistry. Royal Society of Chemistry, Cambridge.

Oller I., Malato S. and Sánchez-Pérez J. A. (2011). Combination of Advanced Oxidation Processes and biological treatments for wastewater decontamination – A review. *Science of the Total Environment*, **409**, 4141–4166.

Önal Y., Akmil-Başar C. and Sarıcı-Özdemir Ç. (2007). Elucidation of the naproxen sodium adsorption onto activated carbon prepared from waste apricot: kinetic, equilibrium and thermodynamic characterization. *Journal of Hazardous Materials*, **148**(3), 727–734.

Panizza M. and Cerisola G. (2009). Direct and mediated anodic oxidation of organic pollutants. *Chemical Reviews*, **109**, 6541–6569.

Park H.-S., Lee B.-J. and Kim S.-J. (2005). Medical waste treatment using plasma. *Journal of Industrial and Engineering Chemistry*, **11**(3), 353–360.

Perreault F., Fonseca de Faria A. and Elimelech M. (2015). Environmental applications of graphene-based nanomaterials. *Chemical Society Reviews*, **44**, 5861–5896.

Prachi P. G., Madathil D. and Brijesh Nair A. N. (2013). Nanotechnology in waste water treatment: A review. *International Journal of ChemTech Research*, **5**(5), 2303–2308.

Prado T., Silva D. M., Guilayn W. C., Rose T. L., Gaspar A. M. C. and Miagostovich M. P. (2011). Quantification and molecular characterization of enteric viruses detected in effluents from two hospital wastewater treatment plants. *Water Research*, **45**, 1287–1297.

Prieto-Rodnguez L., Oller I., Klamerth N., Aguera A., Rodnguez E. M. and Malato S. (2013). Application of solar AOPs and ozonation for elimination of micropollutants in municipal wastewater treatment plant effluents. *Water Research*, **47**, 1521–1528.

Qu X., Alvarez P. J. J. and Li Q. (2013). Applications of nanotechnology in water and wastewater treatment. *Water Research*, **47**(12), 3931–3946.

Qu X., Brame J., Li Q. and Alvarez P. J. J. (2013). Nanotechnology for a safe and sustainable water supply: enabling integrated water treatment and reuse. *Accounts of Chemical Research*, **46**(3), 834–843.

Rahardjo A. K., Susanto M. J. J., Kurniawan A., Indraswati N. and Ismadji S. (2011). Modified Ponorogo bentonite for the removal of ampicillin from wastewater. *Journal of Hazardous Materials*, **190**, 1001–1008.

Ramírez-Sánchez I. M., Méndez-Rojas M. A. and Bandala E. R. (2016). Photocatalytic degradation of natural and synthetic estrogens with semiconducting nanoparticles. Book of advanced environmental analysis: applications of nanomaterials. Royal Society of Chemistry, London, pp. 153–177.

Richardson M. L. and Bowron J. M. (1985). The fate of pharmaceutical chemicals in the aquatic environment. *Journal of Pharmacy and Pharmacology*, **37**(1), 1–12.

Rodriguez J., Parra C., Contreras D., Freer J. and Baesa J. (2001). Dihydroxybenzenes: drive Fenton reactions. *Water Science and Technology*, **44**, 251–256.

Santhosh C., Velmurugan V., Jacob G., Jeong S. K., Grace A. N. and Bhatnagar A. (2016). Role of nanomaterials in water treatment applications: A review. *Chemical Engineering Journal*, **306**, 1116–1137.

Särkkä H., Vepsäläinen M. and Sillanpää M. (2015). Natural organic matter (NOM) removal by electrochemical methods – A review. *Journal of Electroanalytical Chemistry*, **755**, 100–110.

Schiavon M., Scapinello M., Tosi P., Ragazzi M., Torretta V. and Rada E. C. (2015). Potential of non-thermal plasmas for helping the biodegradation of volatile organic compounds (VOCs) released by waste management plants. *Journal of Cleaner Production*, **104**, 211–219.

Schwarzenbach R. P., Gschwend P. M. and Imboden D. M. (2003). General introduction and sorption processes involving organic matter. Environmental Organic Chemistry, 2nd edn. Wiley-Interscience, New York, pp. 275–330.

Shrestha R. R., Haberl R. and Laber J. (2001). Constructed wetland technology transfer to Nepal. *Water Science and Technology*, **43**, 345–350.

Silva E. J., Rocha e Silva N. M. P., Rufino R. D., Luna J. M., Silva R. O. and Sarubbo L. A. (2014). Characterization of a biosurfactant produced by Pseudomonas cepacia CCT6659 in the presence of industrial wastes and its application in the biodegradation of hydrophobic compounds in soil. *Colloids and Surfaces B: Biointerfaces*, **117**, 36–41.

Sim W. J., Kim H. Y., Choi S. D., Kwon J. H. and Oh J. E. (2013). Evaluation of pharmaceuticals and personal care products with emphasis on anthelmintics in human sanitary waste, sewage, hospital wastewater, livestock wastewater and receiving water. *Journal of Hazardous Materials*, **248–249**, 219–227.

Singh K. P., Singh A. K., Singh U. V. and Verma P. (2012). Optimizing removal of ibuprofen from water by magnetic nanocomposite using Box–Behnken design. *Environmental Science and Pollution Research*, **19**(3), 724–738.

Singh R. K., Patel K. D., Kim J. J., Kim T. H., Kim J. H., Shin U. S., Lee E. J., Knowles J. C. and Kim H. W. (2014). Multifunctional hybrid nanocarrier: magnetic CNTs ensheathed with mesoporous silica for drug delivery and imaging system. *ACS Applied Materials and Interfaces*, **6**(4), 2201–2208.

Sirés I. and Brillas E. (2012). Remediation of water pollution caused by pharmaceutical residues based on electrochemical separation and degradation technologies: A review. *Environment International*, **40**, 212–229.

Sokolov D., Sobyna V., Vambol S. and Vambol V. (2018) Substantiation of the choice of the cutter material and method of its hardening, working under the action of friction and cyclic loading. *Archives of Materials Science and Engineering*, **94**(2), 49–54.

Stamatelatou K., Frouda C., Fountoulakis M., Drillia P., Kornaros M. and Lyberatos G. (2003). Pharmaceuticals and health care products in wastewater effluents: the example of carbamazepine. *Water Science and Technology Water Supply*, **3**(4), 131–137.

Strikalenko T. V. (2005). Some problems of wastewater toxicology. IV International Congress on Waste Management (WasteTech-2005), Moscow, May 31–June 3, 2005: collection of documents. SIBIKO Int., Moscow, pp. 647–648.

Strlic M., Radovic T., Kolar J. and Pihlar B. (2002). Anti- and prooxidative properties of gallic acid in Fenton-type systems. *Journal of Agricultural and Food Chemistry*, **50**, 6313–6317.

Stumpf M., Ternes T. A., Wilken R. -D, Rodrigues S. V. and Baumann W. (1999). Polar drug residues in sewage and natural waters in the state of Rio de Janeiro, Brazil. *Science of the Total Environment*, **225**, 135–141.

Suchikova Y., Bogdanov I., Vambol S., Vambol V. and Kondratenko O. (2018). Investigation of the porous GaP layers' chemical composition and the quality of the tests carried out. *Journal of Achivements in Materials and Manufacturing Engineering*, **2**(86), 49–60.

Suhikova Y., Vambol S., Vambol V., Mozaffari N. and Mozaffari N. (2019). Justification of the most rational method for the nanostructures synthesis on the semiconductors surface. *Journal of Achievements in Materials and Manufacturing Engineering*, **92**(1–2), 19–28.

Sun Y. and Pignatello J. J. (1993). Photochemical reactions involved in the total mineralization of 2,4-D by iron (3)/hydrogen peroxide/UV. *Environmental Science and Technology*, **27**, 304–310.

Sun Y., Yang S., Zhao G., Wang Q. and Wang X. (2013). Adsorption of polycyclic aromatic hydrocarbons on graphene oxides and reduced graphene oxides. *Chemistry: An Asian Journal*, **8**(11), 2755–2761.

Taghipour S., Hosseini S. M. and Ataie-Ashtiani B. (2019). Engineering nanomaterials for water and wastewater treatment: review of classifications, properties and applications. *New Journal of Chemistry*, **43**(21), 7902–7927.

Tambosi J. L., de Sena R. F., Gebhardt W., Moreira R. F. P. M., José H. J. and Schröder H. F. (2009). Physicochemical and advanced oxidation processes – A comparison of elimination results of antibiotic compounds following an MBR treatment. *Ozone: Science and Engineering*, **31**, 428–435.

Tan X., Ren X., Chen C. and Wang X. (2014). Analytical approaches to the speciation of lanthanides at solid-water interfaces. *Trends in Analytical Chemistry*, **61**, 107–132.

Tang Y. and Reeder R. J. (2009). Uranyl and arsenate cosorption on aluminum oxide surface. *Geochimica et Cosmochimica Acta*, **73**(10), 2727–2743.

Ternes T. A. (1998). Occurrence of drugs in German sewage treatment plants and rivers. *Water Research*, **32**, 3245–3260.

Ternes T. A., Meisenheimer M., Mcdowell D., Sacher F., Brauch H.-J., Haist-Gulde B., Preuss G., Wilme U. and Zulei-Seibert N. (2002). Removal of pharmaceuticals during drinking water treatment. *Environmental Science and Technology*, **36**(17), 3855–3863.

Thomason T. B. and Modell M. (1984). Supercritical water destruction of aqueous wastes. *Journal of Hazardous Waste*, **1**(4), 453–467.

Tijani J. O., Fatoba O. O., Madzivire G. and Petrik L. F. (2014). A review of combined advanced oxidation technologies for the removal of organic pollutants from water. *Water, Air, and Soil Pollution*, **225**(2102), 1–30.

Upadhyayula V. K. K., Ruparelia J. P. and Agrawal A. (2011). Use of Carbon Nanotubes in Water Treatment. Nanoscale Multifunctional Materials: Science and Applications. Wiley, Hoboken, NJ, pp. 321–368.

Urase T., Kagawa C. and Kikuta T. (2005). Factors affecting removal of pharmaceutical substances and estrogens in membrane separation bioreactors. *Desalination*, **178**, 107–113.

Vambol S., Bogdanov I., Vambol V., Suchikova Y., Lopatina H. and Tsybuliak N. (2017). Research into effect of electrochemical etching conditions on the morphology of porous gallium arsenide. *Eastern-European Journal of Enterprise Technologies*, **6/5** (90), 22–31.

Vambol S., Shakhov Y., Vambol V. and Petukhov I. (2016). A mathematical description of the separation of gas mixtures generated by the thermal utilization of waste. *Eastern-European Journal of Enterprise Technologies*, **1**(2/79), 35–41.

Vambol S., Vambol V. and Al-Khalidy K. A. H. (2019). Experimental study of the effectiveness of water-air suspension to prevent an explosion. *Journal of Physics: Conference Series*, **1294**(7), 72009.

Vambol S., Vambol V., Kondratenko O., Koloskov V. and Suchikova Y. (2018). Substantiation of expedience of application of high-temperature utilization of used tires for liquefied methane production. *Archives of Materials Science and Engineering*, **87**(2), 77–84.

Vambol S., Vambol V., Sobyna V., Koloskov V. and Poberezhna L. (2018). Investigation of the energy efficiency of waste utilization technology, with considering the use of low-temperature separation of the resulting gas mixtures. *Energetika*, **64**(4), 186–195.

Vambol S., Vambol V., Sundararajan M. and Ansari I. (2019). The nature and detection of unauthorized waste dump sites using remote sensing. *Ecological Questions*, **30**(3), 43–55. DOI: 10.12775/EQ.2019.018.

Verlicchi P., Galletti A. and Masotti L. (2010). Management of hospital wastewaters: The case of the effluent of a large hospital situated in a small town. *Water Science and Technology*, **61**, 2507–2519.

Vogna D., Marotta R., Andreozzi R., Napolitano A. and d'Ischia M. (2004). Kinetic and chemical assessment of the UV/H_2O_2 treatment of antiepileptic drug carbamazepine. *Chemosphere*, **54**, 497–505.

Wang C., Li H., Liao S., Zheng H., Wang Z., Pan B. and Xing B. (2013). Coadsorption, desorption hysteresis and sorption thermodynamics of sulfamethoxazole and carbamazepine on graphene oxide and graphite. *Carbon*, **65**, 243–251.

Wang H. (2014). Nanocomposites in Wastewater Treatment, A. K. Mishra (ed.). CRC Press Book, Soil Science Society of America, New York.

Wang S. and Peng Y. (2010). Natural zeolites as effective adsorbents in water and wastewater treatment. *Chemical Engineering Journal*, **156**(1), 11–24.

Wei H. R., Deng S. B., Huang Q., Nie Y., Wang B., Huang J. and Yu G. (2013). Regenerable granular carbon nanotubes/alumina hybrid adsorbents for diclofenac sodium and carbamazepine removal from aqueous solution. *Water Research*, **47**(12), 4139–4147.

Wols B. A., Hofman-Caris C. H. M., Harmsen D. J. H. and Beerendonk E. F. (2013). Degradation of 40 selected pharmaceuticals by UV/H_2O_2. *Water Research*, **47**, 5876–5888.

Wu C. and Linden K. G. (2008). Degradation and byproduct formation of parathion in aqueous solutions by UV and UV/H_2O_2 treatment. *Water Research*, **42**, 4780–4790.

Yahiaoui I., Aissani-Benissad F., Fourcade F. and Amrane A. (2013). Removal of tetracycline hydrochloride from water based on direct anodic oxidation (Pb/PbO_2 electrode) coupled to activated sludge culture. *Chemical Engineering Journal*, **221**, 418–425.

Yang W., Lu Y., Zheng F., Xue X., Li N. and Liu D. (2012). Adsorption behavior and mechanisms of norfloxacin onto porous resins and carbon nanotube. *Chemical Engineering Journal*, **179**, 112–118.

Yin X.-T., Dastan D., Wu F.-Y. and Li J. (2019). Facile synthesis of SnO_2 /$LaFeO_3$ -XNX composite: photocatalytic activity and gas sensing performance. *Nanomaterials*, **9**, 1163.

Yousif E. and Haddad R. (2013). Photodegradation and photostabilization of polymers, especially polystyrene. *SpringerPlus*, **2**(1), 398.

Yu S., Wang X., Chen Z., Wang J., Wang S., Hayat T. and Wang X. (2017). Layered double hydroxide intercalated with aromatic acid anions for the efficient capture of aniline from aqueous solution. *Journal of Hazardous Materials*, **321**, 111–120.

Yuan Y., Lai B. and Tang Y.-Y. (2016). Combined Fe^0/air and Fenton process for the treatment of dinitrodiazophenol (DDNP) industry wastewater. *Chemical Engineering Journal*, **283**, 1514–1521.

Zaib Q., Mansoor B. and Ahmad F. (2013). Photo-regenerable multi-walled carbon nanotube membranes for the removal of pharmaceutical micropollutants from water. *Environmental Science: Processes and Impacts*, **15**, 1582–1589.

Zanta C. L. P. S., Friedrivh L. C., Machulek A., Jr., Higa K. M. and Quina F. H. (2010). Surfactant degradation by a catechol-driven Fenton reaction. *Journal of Hazardous Materials*, **178**, 258–263.

Zazo J. A., Casas J. A., Mohedano A. F., Gilarranz M. A. and Rodriguez J. J. (2005). Chemical pathway and kinetics of phenol oxidation by Fenton's reagent. *Environmental Science and Technology*, **39**, 9295–9320.

Zhang N., Liu S. and Xu Y. J. (2012). Recent progress on metal core@semiconductor shell nanocomposites as a promising type of photocatalyst. *Nanoscale*, **4**(7), 2227–2238.

Zhang S. J., Shao T., Kose H. S. and Karanfil T. (2010). Adsorption of aromatic compounds by carbonaceous adsorbents: a comparative study on granular activated carbon, activated carbon fiber, and carbon nanotubes. *Environmental Science and Technology*, **44**(16), 6377–6383.

Zhang Y., Wu B., Xu H., Wang M., He Y. and Pan B. (2016). Nanomaterials-enabled water and wastewater treatment. *NanoImpact*, **3–4**, 22–39.

Zhao G., Jiang L., He Y., Li J., Dong H., Wang X. and Hu W. (2011). Sulfonated graphene for persistent aromatic pollutant management. *Advanced Materials*, **23**(34), 3959–3963.

Zhi J., Wang Y. and Luo G. (2005). Adsorption of diuretic furosemide onto chitosan nanoparticles prepared with a water-in-oil nanoemulsion system. *Reactive and Functional Polymers*, **65**(3), 249–257.

Zhi M. J., Xiang C. C., Li J. T., Li M. and Wu N. Q. (2013). Nanostructured carbon–metal oxide composite electrodes for supercapacitors: A review. *Nanoscale*, **5**, 72–88.

Zhou Q., Li Z., Shuang C., Li A., Zhang M. and Wang M. (2012). Efficient removal of tetracycline by reusable magnetic microspheres with a high surface area. *Chemical Engineering Journal*, **210**, 350–356.

Zhuo N., Lan Y., Yang W., Yang Z., Li X., Zhou X., Liu Y., Shen J. and Zhang X. (2017). Adsorption of three selected pharmaceuticals and personal care products (PPCPs) onto MIL-101(Cr)/natural polymer composite beads. *Separation and Purification Technology*, **177**, 272–280.

Ziarati P., Vambol V. and Vambol S. (2020). Use of inductively coupled plasma optical emission spectrometry detection in determination of arsenic bioaccumulation in *Trifolium pratense* L. from contaminated soil. *Ecological Questions*, **31**(1),15–22.

Zwiener C. and Frimmel F. (2003). Short-term tests with a pilot sewage plant and biofilm reactors for the biological degradation of the pharmaceutical compounds clofibric acid, ibuprofen, and diclofenac. *Science of the Total Environment*, **309**, 201–211.

Chapter 9

Effect on treatment schemes pharmaceutical wastewater on conventional treatment plants

Ahmed Kettab[1] and Fatima Zohra Batana[2]

[1]International Consultant-Expert, Associate Professor University of Bouira – Algeria; Professor/Research Director of the National Polytechnic School – Algiers; Member of French Water Academy

[2]Engineer, PhD student at Amar Telidji – Laghouat University, Algeria, North Africa

ABSTRACT

New environmental contaminants from industrial, agricultural and human activities, called emerging pollutants, include human and veterinary pharmaceuticals, personal care products, household chemicals, pesticides and fertilizers. The presence of pharmaceuticals in water represents an environmental problem due to their therapeutic and biological activity, even at low levels. To combat this environmental scourge linked to the presence and persistence of pharmaceutical products in water and wastewater, new advanced water treatment techniques are currently the subject of a thorough investigation to pay particular attention to methods based on adsorption and degradation processes.

Keywords: Environmental impact, Pharmaceutical products, Pollution, Treatment, Wastewater.

Contents

9.1 Introduction 222
9.2 Water Pollution Worldwide 223
9.3 Pharmaceuticals in Water 224

Editor(s): Nadeem A Khan, Sirajuddin Ahmed, Viola Vambol and Sergij Vambol
doi: 10.2166/9781789061338_0221

9.3.1 History 224
9.3.2 Definition 224
9.4 Medication Use 225
9.5 Drug Circuit After Administration 225
9.6 Diversity of Compounds and Varieties of Concentrations 226
9.7 Frequency of Detection and Contamination of Aquatic Environments 227
9.7.1 River waters 227
9.7.2 Surface water 228
9.7.3 Deep waters 228
9.7.4 Coastal and estuarine waters 228
9.7.5 Drinking water 229
9.8 Origin of Pharmaceutical Liquid Effluents 229
9.8.1 Domestic waste 229
9.8.2 Discharges from healthcare establishments and hospitals 229
9.8.3 Medical analysis laboratories 230
9.8.4 Farm waste 230
9.8.5 Industrial discharges 230
9.9 Environmental Impacts and Risks Related to Pharmaceutical Products 231
9.9.1 Potential impacts on human health 231
9.9.2 Impact on ecosystems 231
9.10 Pharmaceuticals Removal from Water 232
9.10.1 Biodegradation 233
9.10.2 Membrane separation processes 233
9.10.3 Adsorption 233
9.10.4 Urban WWTP flow reduction 233
9.10.5 Activated sludge 234
9.10.6 Bacterial beds 235
9.11 Conclusion 236
References 237

9.1 INTRODUCTION

Pharmaceutical medicine plays a key role in life expectancy improvement and the quality of life. Thousands of tons of pharmaceuticals are used every year in human and veterinary medicine in order to treat diseases, bacterial infections and to stimulate the growth of agricultural and aquaculture farms.

Since the 1980s, several pharmaceutical molecules have been detected in effluents and aquatic environments, but it is only recently that studies have highlighted the presence of more than 80 pharmaceutical products in water (Pépin, 2006).

Currently, several pharmaceutical products are found in different environmental media: groundwater, at the outlet of treatment plants, in activated sludge, sediment, dams, rivers and even in drinking water, in river waters and groundwater, because the treatment plants are not sufficiently equipped to eliminate them (Kettab *et al.*, 2008). Precisely, these are anti-cancer drugs, antibiotics, feminizing hormones, antidepressants, antiepileptic drugs, and pain killers. So far, there are no studies which associate the presence of these contaminants with human diseases. However, several of these pollutants have a biological impact on fish, especially at the hormonal level (mainly feminization, but also carcinogenic effects and on the immune system). In recent years, their impact on the environment has become a subject of public and global interest. Previously, the fate of drugs in the environment has been relatively little studied due to analytical chemistry methods which were not yet effective in detecting this type of pollutant. Another major problem is the number of pharmaceutical compounds consumed worldwide each year.

The environmental issue concerning drugs is linked to the fact that they can present a significant environmental risk, even at low concentrations. It is even a concern because of their direct toxicity and their likely cumulative or synergistic effects with other micropollutants. Certain drugs have biological effects on non-target organisms, a classic example being that of certain synthetic hormones on fish (Stülten *et al.*, 2008).

Analgesic drugs are the most widely used products followed by antibiotics, anti-inflammatory drugs and drugs for chronic diseases such as diabetes and hypertension.

The question that arises is on the effects that drug residues can have on human health and the environment in the long term, and it is in this context that this bibliographical synthesis is carried out on the substances of drugs discharged in wastewater.

9.2 WATER POLLUTION WORLDWIDE

Water is considered to be one of the most important and essential components of the ecosystem. The demand for water has increased considerably with the increase of the world population, industrialization and agricultural development, which has caused significant water stress due to the large quantities of wastewater discharged which contains high levels of chemicals (organic and inorganic), a complex mixture of toxins and pollutants that require attention in order to meet environmental standards (Hilton *et al.*, 2003; Jurado *et al.*, 2012).

This global problem requires immediate management of water quality, which is a real challenge due to the discovery of a huge number of artificial chemicals that occur in water bodies around the world (Yu *et al.*, 2006).

Global environmental concerns regarding water quality have focused on pH, turbidity, electrical conductivity, total dissolved solids (TDS), total suspended solids (TSS), the presence of heavy metals, biochemical oxygen demand (BOD), chemical oxygen demand (COD), microbial and nutrient load and total coliforms. This is why the majority of global pollution control regulations mention release limits for organic materials, toxic metals and trace elements in surface water (Kettab *et al.*, 2017).

A list of 129 pollutants that can pose a threat to receiving waters has been drawn up by the European Union (EU) and the United States Environmental Protection Agency (US EPA) with a limited technical contribution composed of chemicals selected according to stability, structure, chemical composition, hazardous nature and presence in wastewater (Geissen *et al.*, 2015; Jurado *et al.*, 2012).

9.3 PHARMACEUTICALS IN WATER

9.3.1 History

In 1976, in Kansas City, Missouri, the first drug residues in the waters were detected, where salicylic acid and clofibric acid had been quantified in the waters of the Big Blue Rivers treatment plant.

Currently, drug residues in waters have become a fact proven and recognized by scientists which attracts the interest of communities (Rodriguez-Narvaez *et al.*, 2017).

9.3.2 Definition

Pharmaceutical substances are molecules which have curative, preventive or administered properties in order to establish a diagnosis. They are classified according to their effect they are designed for (antibiotics, analgesics, etc.), their chemical structure or their mode of action (antimetabolites or alkylating agents). They have physicochemical properties which give them the capacity to cross biological membranes (Saptarshi *et al.*, 2019).

Medicines are made up of two kinds of substances: the first is the active substance or the active principle having a pharmacological effect and therapeutic interest demonstrated clinically and the second is a pharmacologically inert substance called an excipient. Drugs can be grouped into six broad categories as listed in Table 9.1.

Table 9.1 Main pharmaceutical categories.

	Anesthetic	Antalgic	Sedative
Psychic	Antidepressant	Anxiolytic	Psychotropic
Infections	Antibiotic	Antiviral	Antiretroviral
	Diuretic	Antidiuretic	Laxative
Strong reactions	Antipyretic	Antihistamine	Anti-inflammatories
Other	Bronchodilator	Vasopressor	

Table 9.2 Evolution of the distribution of drug purchase by pharmacy worldwide in 2006.

Medicine Purchase Worldwide		
Continent	**Country**	**Million USD**
North America	USA	197,802
	Canada	13,719
Europe	Germany	27,668
	France	25,630
	UK	15,666
	Italy	14,942
	Spain	11,629
Asia	Japan	56,675
Latin America	Brazil	8366
	Mexico	8096
	Argentina	2285
Australia	New-Zeeland	5803

9.4 MEDICATION USE

Medicines or pharmaceutical substances have curative and preventive properties and are currently necessary for the health of humans and animals (Kettab, 2014). However, the residues of these substances are considered to be emerging pollutants and are frequently classified in the category of micropollutants given their concentrations in the order of nanograms or micrograms per liter in aquatic environments (Saptarshi *et al.*, 2019).

The drugs are taken following a prescription by a doctor, advice given by a pharmacist and self-medication. These uses are more concentrated in public and private hospitals.

On a global scale, industrialized countries are the largest consumers of pharmaceutical products followed by Europe. The consumption of North America and Japan represent about 80% of the global market for less than 15% of the world population, as shown in Table 9.2.

The use of medicines in veterinary medicine for the treatment of domestic and farm animals comes from the veterinarians themselves and from the pharmacists. Several prescription drugs for humans are also used for pets, including tranquilizers and antidepressants in the United States (Arpin-Pont, 2015).

9.5 DRUG CIRCUIT AFTER ADMINISTRATION

After its administration, whether by humans or animals, the drug diffuses into the body and then undergoes a transformation or metabolization to modify its active molecule and facilitate its elimination from the body; these reactions will produce secondary derivatives called metabolites. Generally, drug transformations are dependent on enzymes, but some of them are carried out by hydrolysis in an acid or alkaline medium. According to the transformation processes induced by enzymes, there are two phases of metabolization:

- Phase I includes the oxidation, reduction and hydrolysis reactions which give derivatives whose functional groups are most often carboxylates (–COOH), hydroxyls (–OH), amines (–NH_2). These reactions are generally localized in the hepatic microsomes and at the tissue and plasma level.
- Phase II reactions to make the molecule water-soluble in order to accelerate its elimination by conjugation of the functional groups resulting from phase I reactions. These metabolites are found in bile, urine or stool, in more or less significant quantities according to their physico-chemical characteristics.

The original molecules and their metabolites are released into wastewater for various physico-chemical treatments or will be released directly into the environment (Shaw *et al.*, 2005).

9.6 DIVERSITY OF COMPOUNDS AND VARIETIES OF CONCENTRATIONS

More than 600 pharmaceutical compounds belonging to various therapeutic classes have been detected worldwide in all aquatic compartments: surface water, groundwater and drinking water, but also in the effluents of treatment plants or even marine waters at concentrations ranging from nanograms per liter (ng/L^{-1}) to micrograms per liter ($\mu g/L^{-1}$).

Among the therapeutic classes most found in the environment (Table 9.3), there are antibiotics, anti-inflammatories and beta-blockers for human and veterinary use.

Due to the complexity of the research and the lack of suitable analytical techniques, many pharmaceutical substances and their transformation products are little cited and are present in the environment.

Table 9.3 Example of drugs found in the environment.

Therapeutic Category	Molecules
Antibiotics	Ofloxacin, chlortetracycline, streptomycin, flumequine, penicillin, amoxicillin, erythromycin, ciprofloxacin, N4-acetyl-sulfamethazine
Anti-inflammatory/ analgesic	Diclofenac, ibuprofen, acetaminophen, codeine, paracetamol, acetylsalicylic acid, metramizol, indomethacine, naproxen, fenopren, phenazon
Beta-blockers	Metoprolol, propranolol, nadolol, aténolo, sotalol, betaxolol

Furthermore, the quantification of drug residues present in waters differs enormously between countries. Indeed, the concentrations are variable due to socio-economic development, human activity, regulations, healthcare accessibility, the receiving environment, as well as the means used for water treatment and for the quantitative analytical approach.

In developing countries, water pollution by drugs is very important despite the limited access to care, this is due to the persistence of the compounds in the environment, untreated discharges and the high number of industries compared to the number of inefficient water treatment plants.

The residues of the rejected drugs continue to deteriorate. Despite the fact that the residues of rejected drugs continue to deteriorate, they remain harmful for long periods which is often neglected.

The levels of drugs found in the waters are very low. However, we cannot know what the effect of very long-term exposure is on the body, nor the synergistic effects of different drug residues or the possible cross-effects between drugs and micropollutants. On the other hand, this risk has already been proven in fish, and therefore suggests potential dangers for humans (Rodriguez-Narvaez *et al.*, 2017).

9.7 FREQUENCY OF DETECTION AND CONTAMINATION OF AQUATIC ENVIRONMENTS

The waters affected by the contamination of aquatic environments by pharmaceutical products are of several types, as discussed in the following.

9.7.1 River waters

River waters are primarily affected by pollution in pharmaceutical products, furthermore, a large part of the treatment plants discharge their effluents directly into rivers, thus making them sensitive to typically anthropogenic contaminants, such as, for example, the active ingredients.

Indeed, the most frequently detected contaminants are both the most consumed (paracetamol, aspirin) and the most resistant (carbamazepine, diclofenac) (Kools *et al.*, 2008).

This pollution is progressive due to the location of the withdrawals in the rivers. Indeed, the concentrations are maximum at the junction between the river and the effluent while the concentration systematically tends to decrease if one moves away from the effluent discharge points.

River waters are therefore the natural aquatic compartments most affected by anthropogenic contamination, such as pharmaceutical products, and occurrences of concentrations greater than $100\,ng/L^{-1}$ are frequent, indicating the great sensitivity of these waters (Merceron, 2016).

9.7.2 Surface water

Surface water is the final receptacle for discharges of wastewater from domestic, industrial or hospital sources, so the environmental environment is most affected by these discharges (Saptarshi *et al.*, 2019).

The abundance of drug residues in surface waters depends on the nature of the molecule, the level of consumption, the dilution factor in the receiving medium and the time of sampling (year, season, etc.) (Loos *et al.*, 2010).

The concentrations vary from a few ng/L^{-1} to a few μg/L^{-1}, but depending on inputs and degradation phenomena, these concentrations will vary in aquatic ecosystems. Non-steroidal anti-inflammatory drugs are systematically found at rates of the order of μg/L^{-1} because of their high consumption (Acuña *et al.*, 2015).

9.7.3 Deep waters

Knowledge of the pollution present in deep waters is important to limit the risks. On the one hand, these waters are used for the irrigation of crops which could have consequences for human health as in the case of market gardening, or contamination of agricultural soils in general, and on the other hand, drinking water is often taken from groundwater, which is quite shallow and sensitive in certain regions (Loos *et al.*, 2010).

In addition, deep waters are not very sensitive to pollution from pharmaceuticals and drug residues, only 13% can be affected by heterogeneous composites of pharmaceuticals (Saptarshi *et al.*, 2019).

In some cases, contamination of the aquifers follows the percolation of rainwater via old recycling centers, which leads to contamination of the deep aquifers. Certain compounds which are slightly degraded in the treatment plant resist the percolation step and thereafter will be found in the aquifer where their decay will remain very weak (Loos *et al.*, 2010).

9.7.4 Coastal and estuarine waters

Estuarine and coastal waters are strongly influenced by the amount of pollutants crossing rivers and large rivers.

There is a very significant difference in the maximum pollutant concentrations of around 100 ng/L^{-1} for the estuaries of large rivers, on the other hand, this content is less important in the coastal zones strongly influenced by large metropolises and is of the order of 1 ng/L^{-1}. However, due to a very strong dilution effect, data on ocean waters are still lacking.

Currently, the pollution discovered in fish farming is linked to pharmaceutical products, which present significant risks for marine microfauna.

Thus, the widespread use of antibiotics which are administered to farmed fish via food increases the bio-resistance to antibiotics. About 70–80% of antibiotics administered to fish are not consumed or are excreted and are found in the aquatic environment.

Even if the concentrations of pharmaceutical products remain rare in marine waters, their detection is limited because of the variability of the waters and the dilution of the coastal contributions which require methods of trace analysis (Loos *et al.*, 2010; Merceron, 2016).

9.7.5 Drinking water

A third of the water intended for human consumption comes mainly from surface water and the remaining two-thirds from groundwater. The presence of pharmaceutical residues in these two aquatic compartments can thus generate contamination of water intended for human consumption, which is a major concern for public authorities (Loos *et al.*, 2010).

So, these waters should have our special attention and be the subject of rigorous surveillances, because of its direct impact on public health due to the rapid transmission of contaminations to humans.

At present, the toxicity of pharmaceuticals at such low concentrations in humans can be considered very limited. However, some hormonal disturbances are attributed to endocrine disruptors to which humans can be exposed by various routes, including drinking water (Merceron, 2016).

9.8 ORIGIN OF PHARMACEUTICAL LIQUID EFFLUENTS

9.8.1 Domestic waste

The main source of releases is the consumption of drugs by the population. Once the drug is administered, it will be absorbed and metabolized by the body, then excreted, and then released to wastewater. Then, the residue reaches urban wastewater treatment plants which only degrade a part of it, their treatment is inefficient in removing all residue compounds.

All drugs are more or less metabolized and then excreted by humans. Generally, hormonal drugs are eliminated by more than 90% by the treatment of wastewater treatment plants; other pharmaceutical residues such as beta-blockers have a reduction rate of less than 20% (Destrieux, 2018), as shown in Figure 9.1.

9.8.2 Discharges from healthcare establishments and hospitals

These establishments and hospitals present a high risk, due to the concentration of their patients and the massive use of drugs; they are considered primarily responsible for the elimination of radiopharmaceuticals and anti-cancer drugs.

The increase in the incidence of cancers will only amplify their presence in the environment with risks for the immune and genetic systems of fauna and flora. In addition, the detection and evaluation of residues originating from the establishment is very complex and is generally not taken into account.

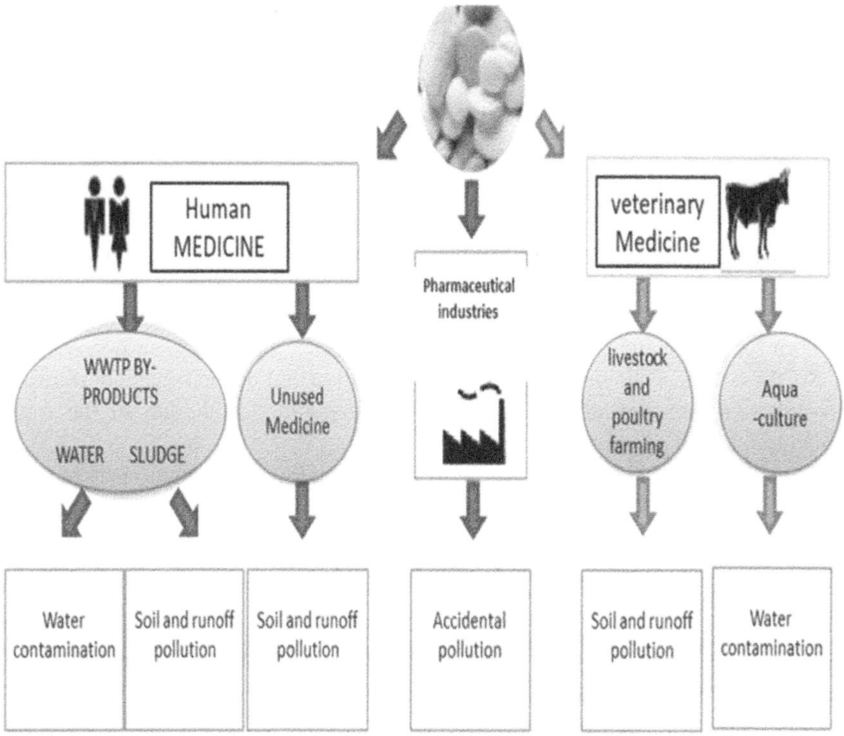

Figure 9.1 Sources and routes of contamination of the aquatic environment by drug residues (Blondel, 2013).

9.8.3 Medical analysis laboratories

Discharges from medical analysis laboratories may be considered harmful waste due to their multiple analysis activities. However, due to their limitation; they represent a low source of pollution compared to those generated by hospitals.

9.8.4 Farm waste

Antibiotics, anti-parasites and hormones (growth promoters) are the main classes of drugs used in livestock. Veterinary medicinal products can be dispersed directly in the environment when they are used in the treatment of farm animals (poultry, cattle, goats, horses, etc.) or in aquaculture, or indirectly in the event of spreading of liquid manure and liquid manure in agricultural soils (Shaw *et al.*, 2005).

9.8.5 Industrial discharges

The pharmaceutical industries are the first link in the drug production chain. By synthesizing drug molecules and pharmaceuticals, they can release not only the molecules of the drugs produced, but also the chemicals used in the synthesis of drugs as well as the products of side reactions into the environment.

In fact, discharges can occur, leading to localized peaks of contamination in the event of poor treatment of effluents or accidental pollution, particularly in developing countries. It is not impossible that releases of pharmaceutical substances may occur even if the pharmaceutical industry complies with good manufacturing practices and the legislation on classified installations (Blondel, 2013).

9.9 ENVIRONMENTAL IMPACTS AND RISKS RELATED TO PHARMACEUTICAL PRODUCTS

9.9.1 Potential impacts on human health

Once released in the environment, pharmaceutical products can have a negative impact on flora and fauna and subsequently alter ecosystems and human health when they persist despite drinking water treatments (Loos *et al.*, 2010).

Endocrine disruptors disrupt people's hormonal systems and are toxic to reproduction. However, certain molecules (antibiotics, lipid regulators, diuretics, beta-blockers) have effects at the molecular level through alterations of gene expression, on the modification of behaviors that could affect reproduction and growth, or on the individual by an increase in cardiac activity.

Also, certain synthetic hormones act on fertility, while other molecules could be carcinogenic, mutagenic and toxic for reproduction and cause pathogenic effects such as tumors which appear not long after exposure (Rodriguez-Narvaez *et al.*, 2017).

9.9.2 Impact on ecosystems

Increasingly, the presence of pharmaceutical residues and their impact on the environment are being taken into account (Loos *et al.*, 2010). There are two types of toxicity of pharmaceutical products to living beings:

- *Acute toxicity*. It is induced by the administration of a single and massive dose of toxic substance(s). Generally, it is expressed in the form of LD50 (Lethal Dose 50%), an acute toxicity dose is called a dose of organic compound which causes 50% of the death of an exposed biological population.
- *Chronic toxicity*. It is a long-term effect that can be linked to a change in growth, metabolism, reproduction and resistance to diseases or that leads to the death of an organism due to the presence of one or more substances in its environment. It helps to understand the environment of the pollutants impact on living organisms.

One of the characteristics of organic micropollutants, and in particular endocrine disruptors, is that the addition of several compounds could increase the toxicity of the whole combination exponentially.

Ecosystems are exposed to a concoction of substances. When there are several molecules of the same class, it is possible that their effects are cumulative or that there is a synergy between certain compounds. It is also necessary to consider the eco-toxicity of metabolites which remains little studied (Acuña *et al.*, 2015; Rodriguez-Narvaez *et al.*, 2017).

In the environment, the metabolites could be more toxic than the parent chemical itself. The physico-chemical properties of pharmaceutical products allow them to be persistent and non-biodegradable in aqueous environments (Destrieux, 2018; Kolpin *et al.*, 2004).

Many pharmaceuticals are bioactive compounds that can directly affect non-target organisms.

Fish are very vulnerable to contamination by pollutants and in particular by residues of pharmaceutical products found in water because they are linked to an environment where the mobility of these contaminants is maximal. Chronic effects can limit the chances of survival and cause reproductive failure (Geissen *et al.*, 2015; Rodriguez-Narvaez *et al.*, 2017).

Generally, algae are phototrophic organisms that regulate the energy and nutrient cycles in water, which makes their activity crucial for the survival of the ecosystem as a whole (Gabet-Giraud, 2014).

The presence of certain pharmaceutical molecules such as non-steroidal anti-inflammatory drugs and beta-blockers has greatly damaged the algae chloroplasts. Thus, prolonged exposure to antibiotics leads to chronic toxicity in algae, interfering with photosynthesis (Saptarshi *et al.*, 2019).

The presence of antibiotics even at very low levels, can develop resistance in natural bacteria.

The presence of hormones in aqueous medium tends to disturb the activity of the endocrine system causing a disturbance in the homeostasis of non-target organisms. Some inhibit the activity of natural hormones and others could induce physiological changes, such as damage to the reproductive mechanism and even induce female traits in male aquatic species (Blondel, 2013).

It is impossible to understand all the chronic effects of the exposure of aquatic and terrestrial organisms to pharmaceutical residues which can induce biological changes over long periods, sometimes for an entire community of organisms over several generations (Boumalek *et al.*, 2019).

9.10 PHARMACEUTICALS REMOVAL FROM WATER

Over the years, a wide variety of wastewater treatment technologies have been developed. They can be classified as physical treatment involving the use of adsorbents, chemical treatment using advanced oxidation processes and biological treatment based on the biodegradation of pollutants by selected microbes in treatment chambers or specially designed reactors. Treatment methods can be further classified as conventional and unconventional (Cherifi *et al.*, 2008; Geissen *et al.*, 2015).

At present, where hazardous drug residues from pharmaceutical industries are increasing considerably, several techniques are developing, such as coagulation-flocculation, adsorption, chemical oxidation and foam flotation (Janah *et al.*, 2017).

9.10.1 Biodegradation

Biodegradation is considered among the main routes for the degradation of drugs. Some pharmaceuticals biodegrade more easily than others. This is particularly the case for radio contrast.

For biodegradable molecules, degradation products can escape treatment: this is the case, for example, of clofibric acid, but also of estrone, which corresponds to the biodegradation of a natural hormone (estrogen).

9.10.2 Membrane separation processes

Membrane separation processes are mainly used to concentrate molecules or ions, or to separate microorganisms or particles that are suspended in a liquid.

Membrane retention treatments use filters with very low porosity. The membranes are hollow and porous fibers based on metallic powder or cellulose. The two most important processes in the disposal of pharmaceuticals are nanofiltration and reverse osmosis (Blondel, 2013; Karef *et al.*, 2013).

Reverse osmosis is based on the use of semi-permeable membranes that allow the passage of water without the passage of drug residues (Gabet-Giraud, 2014).

9.10.3 Adsorption

The adsorption process on the various adsorbents, and in particular activated carbon, makes it possible to eliminate several kinds of pollutants existing in water such as dissolved organic pollutants, certain pesticides, hydrocarbons and drugs, as shown in Figure 9.2.

Activated carbon can be used either in the form of powder during flocculation, or in grains in thick filtration beds. In the latter case, the drug molecules penetrate and become fixed in the pores of the activated carbon.

9.10.4 Urban WWTP flow reduction

The majority of residues of pharmaceutical products for human use, consumed by patients, pass through urban wastewater treatment plants (WWTP). The degree of elimination of drug residues in efficient WWTPs is evaluated at 80%. This percentage varies according to the molecules: some are more effectively degraded than others. Certain residues, depending on their lipophilicity, pass into the sludge phase unlike hydrophilic molecules.

Current treatment plants such as adsorption and biodegradation plants are not sufficient for the elimination of pharmaceutical molecules and their metabolites.

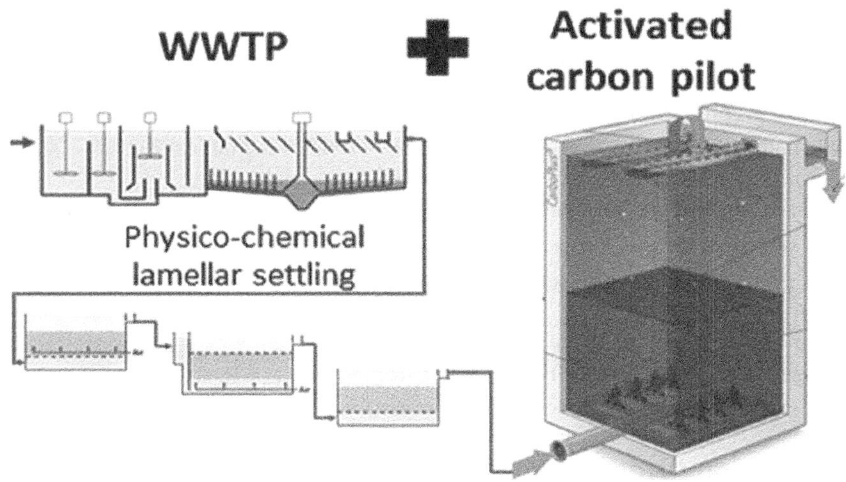

Figure 9.2 Wastewater treatment by adsorption (Guillossou *et al.*, 2019).

Some molecules pass through different barriers during the process and end up in the environment. Thus, complementary treatments such as ozone oxidation and photolysis pose certain environmental impacts resulting from reaction and/or degradation products.

Currently, technologies are moving towards sorting solutions at source and, also, improving the effectiveness of pollution control which can involve different techniques and in particular the optimization of biological treatment (residence time, level of aeration).

In certain particular cases, one can also envisage the addition of a complementary stage of depollution implementing techniques similar to that of the refining in drinking water (membrane filtration, activated carbon, ozonation, ultraviolet rays).

It is better to carry out the purification at the source to avoid hampering the proper purification of the water and to avoid dilution of the waste and the establishment of specific treatment sites for occasional pollution for high risk sites (healthcare establishments), releases from animal husbandry and releases from the pharmaceutical industry (Blondel, 2013).

9.10.5 Activated sludge

Activated sludge is the most widespread process for treating urban wastewater and one can imagine that this supremacy will last for many years. It is a set of processes which have in common the development of a bacterial culture arranged in the form of activated sludge in an aeration tank, supplied with water to be purified.

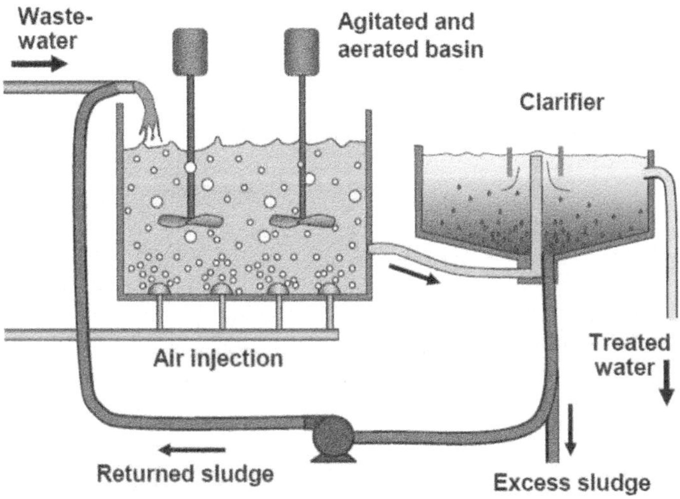

Figure 9.3 Activated sludge wastewater treatment system (Hubbe *et al.*, 2016).

The purpose of stirring is to avoid deposits and to homogenize the mixture of bacterial flocs and wastewater while aeration is intended to provide aerobic bacteria with the oxygen, they need to purify water. After a sufficient contact time, the mixed liquor is sent to a clarifier intended to separate the purified water from the sludge. This major work very often conditions the good or bad functioning of a station and makes it possible to minimize and better distribute the residence time of the sludge, as shown in Figure 9.3.

The sludge deposited at the bottom of the secondary settling tank is partly recycled in the aeration tank in order to maintain a sufficient concentration of purifying bacteria inside, and it can be extracted and evacuated to the sludge treatment line for another stage (Cherifi *et al.*, 2008).

9.10.6 Bacterial beds

This process consists of supplying already decanted water, a structure containing a mass of material (pozzolan or plastic) which serves as a support for the purifying microorganisms which have the form of a biological film responsible for the assimilation of pollution. The biological film is released as the water percolates. At the exit from the bacterial bed, a mixture of treated water and biofilm is collected. The latter is trapped at a secondary decanter in the form of sludge and the treated water rejoins the natural environment. The recirculation of sludge to the digestion tank is essential.

Pollutants are assimilated by the microorganisms forming the biological film. This consists of aerobic bacteria on the surface and anaerobic bacteria in depth, as shown in Figure 9.4.

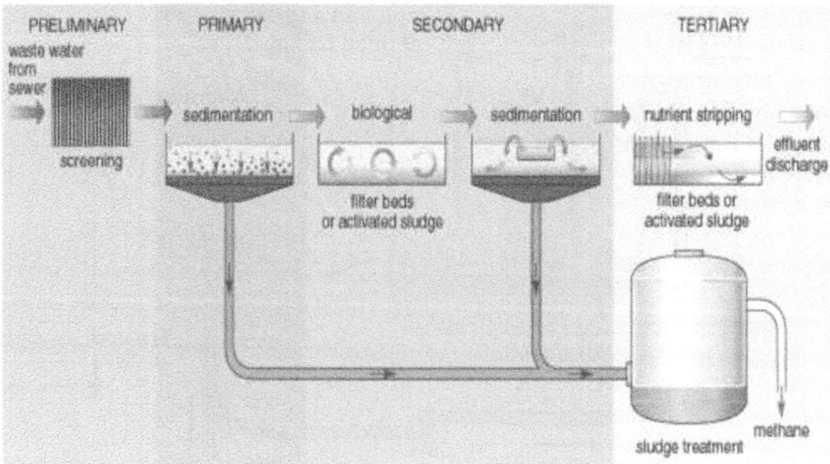

Figure 9.4 Diagram of a treatment plant with a bacterial bed (Guillossou *et al.*, 2019).

The by-products and carbon dioxide produced by the purification are evacuated in liquid or gaseous fluids. Excess sludge that naturally falls off the support under the effect of hydraulic load is separated by secondary settling (Karef *et al.*, 2013).

9.11 CONCLUSION

Now, global environmental concerns about water quality are increasing, given the challenges posed by pollutants and emerging contaminants released to the environment from a variety of sources that have the potential to cause serious negative impacts for a long period of time.

Their concentration in the environment may be low, but their persistence and continuous addition to the environment is a major preoccupation considering the unprecedented side effects they can have.

The majority of pollutants have complex chemical structures and are found in the environment via wastewater treatment plants. Certain pharmaceutical products are not completely metabolized by the human or animal body and are therefore excreted in a biologically active form, the latter has the potential to have an impact on non-target organisms living in aquatic systems such as fish.

These pharmaceuticals can also undergo biotransformation in the environment, forming transformation products more toxic than the original compound.

With the growing concern for the elimination of pollutants, different treatment technologies must be considered to find the best solution to this problem. Conventional techniques do not seem very favorable to their removal; thus, a combination of conventional and unconventional treatment systems must be explored and envisaged to obtain a solution to this ever-increasing problem.

REFERENCES

Acuña V., Von Schiller D., García-Galán M. J., Rodríguez-Mozaz S. and Corominas L. (2015). Occurrence and in-stream attenuation of wastewater-derived pharmaceuticals in Iberian rivers. *Science of the Total Environment*, **503**, 133–141.

Arpin-Pont L. (2015). Les produits pharmaceutiques et de soin personnel en milieu marin : prédiction des concentrations environnementales et étude des effets sur le métabolisme endogène d'organismes exposés (Pharmaceutical and personal care products in the marine environment: predicting environmental concentrations and studying the effects on the endogenous metabolism of exposed organisms). PhD thesis, University of Montpellier, France.

Blondel B. (2013). Résidus médicamenteux dans les eaux : états des lieux, risques sanitaires et solutions d'avenir. Medicated residues in waters (inventory, health risks and solutions for the future). PhD thesis. University of Nantes, France.

Boumalek W., Kettab A., Bensacia N., Bruzzoniti M., Ben Othman D., Mandi L. M., Chabaca N. and Benziada S. (2019). Specification of sewage sludge arising from a domestic wastewater treatment plant for agricultural uses. *Desalination and Water Treatment*, **143**, 178–183.

Cherifi H., Hanini S., Bentahar F. and Kettab A. (2008). Traitement des effluents aqueux par adsorption sur des fibres naturelles (Treatment of aqueous effluents by adsorption on natural fibers). *Algerian Journal of Technology*. Special Number.

Destrieux D. (2018). Résidus de medicaments d'un cours d'eau urbain – Constitution d'une base de données pour la gestion des risques écotoxicologiques (Drug residues in an urban watercourse – Creation of a database for the management of ecotoxicological risks). PhD thesis. University of Toulouse, France.

Gabet-Giraud V. (2014). Distribution d'estrogènes et de bétabloquants dans les stations d'épuration des eaux résiduaires et dans l'eau de surface (Distribution of estrogens and beta-blockers in wastewater treatment plants and in surface water). PhD thesis. University of Lyon, France.

Geissen V., Mol H., Klumpp E., Umlauf G., Nadal M., Van der Ploeg M., Van de Zee S. E. and Ritsema C. (2015). Emerging pollutants in the environment: A challenge for water resource management. *International Soil and Water Conservation*, **3**, 57–65.

Guillossou R., Le Roux J. et al. 2019. Organic micropollutants in a large wastewater treatment plant: what are the advantages of an advanced treatment by activated carbon adsorption compared to conventional treatment? *Chemosphere*, **218**, 1050–1060.

Hilton M. J., Thomas K. V. and Ashton D. (2003). Targeted monitoring programme for pharmaceuticals in the aquatic environment. R&D Technical Report P6-012/06/TR, UK Environment Agency, Bristol, UK.

Hubbe MA., Metts J. R. et al. (2016). Wastewater treatment and recovery: review of practices and opportunities in the pulp and paper industry. *BioRessources*, **11**(3), 7953–8091.

Janah Y., Ait Nouh F. and Kettab A. (2017). Desalination techniques and renewable energies. European University Editions.

Jurado A., Vazquez- Sune E., Carrera J., Lopez de Alda M., Pujades E. and Barcelo D. (2012). Emerging organic contaminants in groundwater in Spain: a review of sources, recent occurrence and fate in a European context. *Science of the Total Environment*, **440**, 82–94.

Karef S., Nakib M., Kettab A., Benmaamar S., Benziada S. and Bouguerra K. (2013). Study of performance indicators of a sewage treatment plant (Algeria), 3rd International Water Conference, Algiers.

Kettab A., Metiche R. and Bennacar N. (2008). De l'eau pour un développement durable: enjeux et stratégie-(Water for sustainable development: encounters and strategies). *Science de l'Eau*, **21**(2), 247–256.

Kettab A., Chabaca N., Karef S., Nakib M., Boumalek W., Bouanani H., Djillali Y., Benziada S. and Benmamar S. (2017). Traitement des eaux usées urbaines et leurs réutilisations en Agriculture (Urban wasrewater treatment and agricultural reuse). Technical days on treated waste water reuse. National Polytechnic School of Algiers, Algiers.

Kettab A. (2014). Water for all with quality and quantity: it is the concern of all. *Desalination and Water Treatment*, **52**, 1965–1966.

Kolpin D. W., Skopec M., Meyer M. T., Furlong E. T. and Zaugg S. D. (2004). Urban contribution of pharmaceuticals and other organic wastewater contaminants to streams during differing flow conditions. *Science of the Total Environment*, **328**, 119–130.

Kools S., Moltmann J. F. and Knacker T. (2008). Estimating the use of veterinary medicines in the European Union. *Regulator Toxicology Pharmacology*, **50**, 59–65.

Loos R., Locoro G., Comero S. and Contini S. (2010). Pan-European survey on the occurrence of selected polar organic persistent pollutants in ground. *Water Research*, **44**, 4115–4126.

Merceron A. (2016). Les résidus médicamenteux dans l'environnement : Origines, réglementation et risques (Drug residues in the environment: origins, regulations and risks). PhD thesis. University of Nantes, France.

Pépin J. M. (2006). Impacts éco-toxicologiques de certains médicaments dans l'environnement (Eco-toxicological impacts of certain drugs in the environment). PhD thesis. Université of Sherbrooke, Canada.

Rodriguez-Narvaez O., Peralta- Hernandez J. M., Goonetilleke A. and Bandala E. R. (2017). Treatment technologies for emerging contaminants in water. *Chemical Engineering Journal*, **323**, 361–380.

Saptarshi D., Farhat B. and Anushree M. (2019). Pharmaceuticals and personal care product (PPCP) contamination – A global discharge inventory. *Emerging Contaminants and Micro Pollutants*, 1–26.

Shaw J., Horrace W. and Vogel R. (2005). The determinants of life expectancy: an analysis of the OECD health data. *Southern Economic Journal*, **71**, 768–783.

Stülten D., Zühlke S., Lamshöft M. and Spiteller M. (2008). Occurrence of diclofenac and selected metabolites in sewage effluents. *Science of the Total Environment*, **405**, 310–316.

Yu J., Bouwer E. and Coelhan M. (2006). Occurrence and biodegradability studies of selected pharmaceuticals and personal care products in sewage effluent: a review. *Agricultural Water Management*, **86**, 72–80.

Chapter 10

Urban wastewater treatment plants

Ahmed Kettab[1] and Hanane Bouanani[2]

[1]International Consultant-Expert, Associate Professor University of Bouira – Algeria; Professor/Research Director of the National Polytechnic School-Algiers; Member of French Water Academy

[2]Engineer, PhD student at the National Polytechnic School of Algiers

ABSTRACT

Wastewater cannot join the natural aquatic environment without prior management of the flow and the pollutant load originating from the various discharges of the current urban society. The itinerary of the hydraulic flow leading to a treatment plant that is designed for the pollution eradication is called "the sewer system" because it contributes to the sanitization of urban communities. The aim of wastewater treatment plants is to offer the purest water quality to the aquatic environment. Sanitization and treatment techniques have not stopped to evolve along with the sophistication of hydraulic and environmental technologies.

Keywords: Aquatic, Environment, Environmental technologies, Eradication, Sanitization.

Contents

10.1 Introduction 241
10.2 Definition and Characterization of the Pollution Concept 242
10.3 Sources of Wastewater 243
  10.3.1 Domestic wastewater 243
  10.3.2 Industrial wastewater 243

Editor(s): Nadeem A Khan, Sirajuddin Ahmed, Viola Vambol and Sergij Vambol
doi: 10.2166/9781789061338_0239

10.3.3 Urban wastewater 243
10.4 Pollutants Classification 244
10.4.1 Organic/mineral nature 244
10.4.2 Solubility 244
10.4.3 Toxicity 244
10.4.4 Bioactivity 245
10.5 Characterization Parameters 245
10.5.1 Physico-chemical parameters 245
10.5.1.1 Temperature 245
10.5.1.2 pH (hydrogen potential) 245
10.5.1.3 Dissolved oxygen (DO) and percentage (%) of oxygen saturation 245
10.5.1.4 Conductivity 246
10.5.2 Indicators of particulate pollution 246
10.5.2.1 Total suspended solids (TSS) 246
10.5.2.2 Volatile suspended solids (VSS) 247
10.5.2.3 Turbidity 247
10.5.3 Parameters of organic pollution 247
10.5.3.1 Chemical oxygen demand (COD) 247
10.5.3.2 Biochemical oxygen demand (BOD) 247
10.5.3.3 Total oxygen demand (TOD) 248
10.5.3.4 Total organic carbon (COT) 248
10.5.4 Nutritive elements 248
10.5.5 Microbiological pollutants 248
10.5.5.1 Viruses 249
10.5.5.2 Bacteria 249
10.5.5.3 Protozoa 249
10.5.5.4 Helminths 250
10.5.6 Toxic elements 250
10.5.6.1 Trace, mineral or organic elements 250
10.5.6.2 Heavy metals 250
10.5.7 Emerging pollutants 250
10.5.7.1 Pharmaceutical micropollutants 251
10.5.7.2 Pesticides 251
10.6 Urban Sanitation 252
10.7 Treatment Stages 252
10.7.1 Pre-treatment 253
10.7.1.1 Screening 253
10.7.1.2 Desanding 253
10.7.1.3 Degreasing 254
10.7.2 Primary sedimentation 254
10.7.3 Biological treatment 254
10.7.3.1 Intensive process 255
10.7.3.2 Trickling filter 255

10.7.3.3 Biodisks 256
10.7.3.4 Activated sludge 257
10.7.3.5 Extensive process 261
10.7.3.6 Infiltration-percolation 261
10.7.3.7 Planted filters with vertical flow 261
10.7.3.8 Reed filter beds with horizontal flow 261
10.7.3.9 Natural lagoons 262
10.7.3.10 Aerated lagooning 263
10.7.4 Tertiary treatment techniques for disinfection 263
10.7.4.1 Chlorination 263
10.7.4.2 Ozonation 264
10.7.4.3 Ultraviolet 264
10.7.5 Quaternary treatment techniques for micro-pollution removal 264
10.8 Sludge Treatment 264
10.8.1 Mineralization 265
10.8.1.1 Biological stabilization 265
10.8.1.2 Composting 266
10.8.1.3 Chemical stabilization 266
10.8.1.4 Thermal drying 266
10.8.2 Physical conditioning 267
10.8.3 Volume reduction 267
10.8.3.1 Thickening 267
10.8.3.2 Dehydration 267
10.8.4 Sterilization 268
10.9 Conclusion 268
References 268

10.1 INTRODUCTION

Water is a precious resource that faces several types of pollution and degradation: ecosystems and human health are directly impacted. Water pollution is caused by various origins such as industrial, domestic or agricultural (Bassompierre, 2007). The figures published by the World Health Organization (WHO) in 2004 reveal that every year 1.8 million people, 90% of whom are children under five years old, mostly living in developing countries, die from diarrheal diseases. However, globally, 88% of diarrheal diseases are attributable to poor quality of drinking water and inadequate sanitation (Bixio *et al.*, 2005). Water management has become a global strategic issue that must imperatively be integrated into a political perspective of sustainable development. It is indeed claimed that it will be, in the third millennium, a military issue such as oil has been during the last decades (Anctil, 2008; Kettab, 2001).

Water scarcity in Algeria is increasing significantly and the resource is less and less renewable. It is currently being exploited competitively between human, agricultural and industrial needs, all striving for limited availability. In addition, in Algeria, everyone remembers the consequences of recent and recurrent drought years and the resulting restricting measures of water consumption (Kettab, 2001).

Wastewater treatment is therefore essential to preserve the quality of natural resources and protection of the environment, especially surface and groundwater. Sanitized water is currently mostly released. Widespread reuse of this water can therefore help to compensate water deficiency (Kettab, 2001). Water reuse in industries and recreational areas can help to reduce the pressure on conventional resources. The use of unconventional water in agriculture can contribute to the creation of new irrigated lands or constitute a back-up resource for the existing perimeters. It is practiced in developed and developing countries (Anctil, 2008).

10.2 DEFINITION AND CHARACTERIZATION OF THE POLLUTION CONCEPT

Water pollution is a qualitative alteration, which makes its use dangerous and disturbs the aquatic ecosystem (Kettab, 2014). It can concern surface water (rivers, water bodies) and groundwater (Blinda, 2011). Water pollution manifests itself mainly in surface waters by:

- *Dissolved oxygen decrease*: organic matter which is essential to aquatic life as a nutrient can become a disturbing element when the quantity is high. Among the substances that cause a significant consumption of oxygen, we should particularly note the by-products rejected by the dairy industry, the blood rejected by the meat industry, the waste contained in domestic wastewater, etc. This phenomenon is likely to generate significant fish mortality.
- *The presence of toxic products*: these substances cause effects which can come in two forms: immediate or short-term effect leading to brutal toxic effects fast eradication of different organisms; and the delayed or long-term effect, by cumulative effect of some substances in some microorganisms (Kettab, 2014). Most toxic products originate from chemical and metal industries, agricultural activity and domestic or industrial waste landfills (Kettab *et al.*, 2008).
- *Algae proliferation*: although the presence of algae in aquatic environments is beneficial for the production of dissolved oxygen, these organisms can significantly proliferate and inconveniently generate eutrophication.
- *A physical alteration of the receiving environment*: the environment can be disturbed by contributions with various effects, water turbidity rise (e.g. washing of sand or quarry materials), salinity alteration (e.g. dewatering of salt mines), temperature raise (e.g. cooled water from nuclear power plants).
- *The presence of dangerous bacteria or viruses*: domestic households, hospitals, farms and certain food industries reject germs that can present a health hazard.

10.3 SOURCES OF WASTEWATER

Wastewater is usually defined as water that has been used for domestic, industrial or even agricultural uses, which consequently constitutes a polluted effluent and is discharged into a sewer outfall (Kettab *et al.*, 2008). Wastewater is generally loaded with undesirable elements, which, depending on their quantity and composition, represent a real danger for the receiving environments or their users. The elimination of these toxic elements requires the design of a treatment chain. However, before designing any treatment process, it is imperative to characterize the effluent, quantitatively and qualitatively.

10.3.1 Domestic wastewater

Domestic wastewater includes household water (toilet water, washing water, cooking water) and sewage water (urine and feces) (Kettab, 2014). Domestic wastewater contains mineral and organic substances. Mineral materials (chlorides, phosphates, sulfates, etc.) and organic materials that mostly represent ternary compounds, such as sugars and fats (containing carbon, oxygen and hydrogen, but also of nitrogen and, in some cases, other components such as sulfur, phosphorus, iron, etc.) (Kettab *et al.*, 2008).

10.3.2 Industrial wastewater

All discharges resulting from the use of water other than domestic are classified as industrial discharges. This definition concerns rejects from factories, but also rejects from craft or commercial activities: laundry, restaurant, medical laboratories, etc. The quality spectrum is very large (Kettab, 2014). Some of these waters are toxic to aquatic flora and fauna, or/and humans. We must distinguish between wastewater and waste liquids from certain industries. Wastewater is water that has been used in refrigeration circuits as well as the cleaning of devices, equipment, raw materials or products of a factory. Another industrial use is to retain dust smoke. They may contain chemicals used during manufacturing. Waste liquids, however, are liquids resulting from manufacturing, which is usually the case with chemical solutions, by-product solutions, and tanks discharge (Ecosse, 2001; Kettab, 2014).

Three options are possible (Andreadakis, 2003; Kettab, 2014):

- Direct release into the urban sewer systems,
- Pre-treatment before joining the urban sewer systems,
- Complete local treatment before joining surface waters.

10.3.3 Urban wastewater

Urban wastewater includes domestic wastewater and runoff (rainwater, the cleaning of roads, gutters, markets and yards). These waters that run off on roofs, courtyards, gardens, green spaces, public roads and markets carry all kinds of mineral and organic waste: clay, silt, sludge, sand, vegetable waste (grass, straw, leaves, seeds, etc.) (Andreadakis, 2003; Kettab, 2014). They also contain all kinds of micropollutants (hydrocarbons, pesticides from gardens, detergents used for washing yards, public roads, cars, microscopic rubber debris from wear and tear on vehicle tires, lead from tetraethyl, lead in gasoline, various fallouts from the atmosphere, notably from domestic chimneys and factory chimneys) (Ecosse, 2001).

10.4 POLLUTANTS CLASSIFICATION

The classification of the polluting substances allows the selection of processes that will guarantee maximal eradication (Blinda, 2011; Karef *et al.*, 2013). The nature of the pollutants in the water, of course, depends on the origin of the wastewater. They are classified according to the characteristics below.

10.4.1 Organic/mineral nature

Organic matter is the matter that mainly originates from living species (plants, animals, etc.) and from the chemical industry. Its composition is structured around carbon. There are sugars, proteins, organic acids (lactic, acetic), fatty acids and macromolecules such as starch, cellulose, etc. (Kettab *et al.*, 2017). Mineral matter is generally the matter that does not contain carbon. It contains heavy metals, ammonia, nitrates and phosphates (Karef *et al.*, 2013). Practically, wastewater always contains these two types of pollution in varying amounts depending on its origin.

10.4.2 Solubility

Soluble matter is dissolved in water and is therefore found in the form of a simple chemical unit as a molecule or a macromolecule, such as proteins and colloids (Karef *et al.*, 2013). Insoluble matter is an aggregate of matter that is found in particulate form as solid particles that can either float or settle, depending on their densities (Kettab *et al.*, 2017).

10.4.3 Toxicity

Among the various materials in polluted waters, some have a high toxicity for the living world. That is to say that at very low concentrations, they have a significant impact on the balance of natural ecosystems (Karef *et al.*, 2013; Kettab, 2014). For instance, cyanide in very small quantities can have a devastating effect on an ecosystem. This is also the case for heavy metals such as cadmium and mercury, which, in very small quantities, strongly modify the balance of ecosystems (Kettab, 2014; Kettab *et al.*, 2017).

10.4.4 Bioactivity

Polluted water contains organic and/or mineral matter which therefore does not have the characteristics of "living things" and which can be described as "inert matter". Nevertheless, there are often microorganisms (bacteria for example), which represent living matter. These develop as soon as the water is soiled. They can be pathogenic.

10.5 CHARACTERIZATION PARAMETERS

10.5.1 Physico-chemical parameters

10.5.1.1 Temperature

The temperature of the water plays an important role in the solubility of salts and gases, as well as oxygen availability for aquatic life. In addition, temperature increases the rates of chemical and biochemical reactions by a coefficient of 2–3 for a temperature increase of 10°C. The metabolic activity of aquatic organisms is therefore also accelerated when water temperature increases (Andreadakis, 2003; Kettab *et al.*, 2017).

The value of this parameter is influenced by the ambient temperature but also by possible discharges of hot wastewater. Sudden changes in temperature of more than 3°C are often harmful (Ecosse, 2001; Kettab *et al.*, 2017).

10.5.1.2 pH (hydrogen potential)

pH estimates water acidity, or chemically, the concentration of hydrogen ions (H+). The pH scale practically ranges from 0 (very acidic) to 14 (very alkaline); the median value of 7 corresponds to a neutral solution at 25°C. The pH value of natural water can vary from 4 to 10 depending on the acidic or basic nature. Low pH (acidic water) in particular increases the risk of the presence of metals in a more toxic ionic form. High pH values increase concentrations of ammonia NH_3, which is a toxic form to aquatic life. It is generally acknowledged that a natural pH between 6.5 and 8.5 characterizes aquatic environments where life expectancy is optimal (Kettab *et al.*, 2008, 2017).

10.5.1.3 Dissolved oxygen (DO) and percentage (%) of oxygen saturation

Dissolved oxygen concentrations, along with pH values, are one of the most important parameters of water quality for aquatic life. Dissolved oxygen in surface waters comes primarily from the atmosphere and from the photosynthetic activity of algae and aquatic plants. The dissolved oxygen concentration varies daily and seasonally as it depends on many factors such as the partial oxygen pressure of the atmosphere, the water temperature, the salinity, the penetration of light, the agitation of the air and nutrient availability. It is also a function of the rate of depletion of the oxygen environment by the activity of aquatic organisms and the processes of oxidation and decomposition of organic matter present in water (Karef *et al.*, 2013).

Overall, the closer the dissolved oxygen concentration (OD) is to saturation, the greater the ability of the river to absorb pollution: a value of less than 1 mg of O_2 per litre indicates a state close to anaerobic. This state occurs when the oxidation processes of mineral waste, organic matter and nutrients consume more oxygen than that available. A low dissolved oxygen content causes an increase in the solubility of toxic elements, which are released from the sediments. A value of 1–2 mg of O_2 per litre indicates a heavily polluted but reversible river; a content of 4–6 mg of O_2 per litre characterizes good quality water; higher amounts than the natural oxygen saturation point indicate eutrophication, resulting in intense photosynthetic activity. Sensitive aquatic species can be disturbed by an oxygen content of less than 4 mg/L (Karef *et al.*, 2013).

10.5.1.4 Conductivity

Conductivity measurements in water allow assessment of the amount of dissolved salts. It mainly depends on the quality of the used drinking water and the industrial activities of the population. This parameter is more informative on the eventual pollutants than other parameters (Blinda, 2011). Standard electrical conductivity is generally expressed in millisiemens per meter (mS/m) at 20°C. The conductivity of natural water is between 50 and 1500 μS/cm (Karef *et al.*, 2013).

Estimation of the total amount of dissolved matter can be obtained by multiplying the value of the conductivity by an empirical factor depending on the nature of the dissolved salts and water temperature. Assessment of the amounts of dissolved salts is important since each aquatic organism has specific requirements regarding this parameter. Aquatic species generally do not tolerate significant variations, which can be observed, for example, in the case of wastewater spills (Blinda, 2011; Karef *et al.*, 2013).

10.5.2 Indicators of particulate pollution

10.5.2.1 Total suspended solids (TSS)

Suspended solids are non-solubilized materials, of organic and mineral nature. They are generally volatile and they can be treated by biological degradation but, since they are essentially inert in nature, they can be largely eliminated by sedimentation (Blinda, 2011; Kettab, 2014).

The quantities of TSS are particularly variable according to the seasons and the water flow regime. These particles affect the limpidity of water and decrease light penetration potential and, subsequently, photosynthesis. They can also interfere with fish breathing. Besides, suspended solids can accumulate high quantities of toxic materials (metals, pesticides, mineral oils, polycyclic aromatic hydrocarbons, etc.) (Kettab, 2014).

10.5.2.2 Volatile suspended solids (VSS)

The organic fraction of TSS is called "volatile" because it is estimated by volatilizing the totality of suspended solids in an oven. The estimation of this parameter allows distinguishing the organic part from the mineral part (Kettab *et al.*, 2017). VSS represent only the particles of biomass, whether the organisms are living or dead, as well as certain organic particles not involved in the depollution process (Kettab *et al.*, 2017):

- Its value has consequences on the biological reactor dimensioning (optimal volume requires working with high but limited values of this parameter due to the hydraulic capacity of the clarifier).
- The suspended biomass of activated sludge can easily be estimated by measuring volatile suspended matter.
- The volume of the aeration tank that is necessary for the treatment of a daily demand can accordingly be calculated using the mass loading value and the concentration of MVS in the aeration tank.

10.5.2.3 Turbidity

It is an alteration of water limpidity due to the presence of undissolved matter (TSS). It gives a first indication of the content of colloidal matter of mineral or organic origin (Blinda, 2011; Karef *et al.*, 2013).

10.5.3 Parameters of organic pollution

10.5.3.1 Chemical oxygen demand (COD)

It represents the quantity of oxygen consumed for the chemical oxidation of all the organic and mineral matter dissolved in the water. The potassium dichromate acting in the medium of sulfuric acid for 2 hours at boiling point, at 150°C, almost completely oxidizes the reducing materials. This test is particularly useful for assessing the plant performance (Ecosse, 2001; Kettab *et al.*, 2017).

10.5.3.2 Biochemical oxygen demand (BOD)

It represents the quantity of oxygen necessary to decompose by oxidation (with the intervention of bacteria) the organic matter contained in wastewater. It indicates the quantity of organic matter, which is one of the most important criteria in order to assess the metabolic pollution degree (Kettab *et al.*, 2017):

Organic matter + bacteria = Sludge + gas + water

Generally, this type of pollution is significantly degraded over a period of 5 days. Beyond 5 days, oxygen consumption decreases enormously, so we adopted the concept of BOD5 obtained after 5 days of incubation at 20°C and in dark conditions (Kettab *et al.*, 2017).

10.5.3.3 Total oxygen demand (TOD)

The TOD is the quantity of oxygen (expressed in mg/L) that is necessary for the total combustion in an oxidizing atmosphere of 1 L of a wastewater sample.

10.5.3.4 Total organic carbon (COT)

COT represents, like COD or BOD5, an important composition parameter in the determination of organic contamination of water. The measurement of total organic carbon makes it possible to give a direct indication of the organic load. It provides information on the type and origin of the contamination (Andreadakis, 2003).

Organic compounds are on the one hand proteins, lipids, carbohydrates and humus, as well as organic carbonaceous substances developed or used by chemical, pharmaceutical, petroleum industry, etc. (Andreadakis, 2003; Karef *et al.*, 2013).

10.5.4 Nutritive elements

Nutrients are found in large quantities in wastewater, and constitute an important quality parameter for the development of these waters in agriculture and in landscape management. The most common elements in wastewater are nitrogen, phosphorus and sometimes potassium, zinc, boron and sulphur. These elements are found in appreciable quantities, but in very variable proportions. In addition, the presence of organic matter in different forms in wastewater (suspended solids, colloidal elements and dissolved matter) can, by its long-term effect on soil fertility, also contribute to the structural stability of the soil (Karef *et al.*, 2013).

10.5.5 Microbiological pollutants

These pollutants originate from the environment, human feces, enteric flora and may contain pathogen microorganisms. They have various effects on health: they are the cause of mild infections (gastroenteritis for example) as well as fatal diseases (cholera for instance) (Catherine *et al.*, 2008). The pathogenicity of microorganisms (or pathogenicity) depends on several factors. Microorganisms in the environment or in water will not systematically trigger disease if absorbed. Pathogenicity depends on several factors that can be grouped under two categories: the physiology of the microorganism and that of the infected host (Catherine *et al.*, 2008).

First, latency is the time it takes a pathogen to become infectious. It differs according to microorganisms. Thus, it is low (from 0 to 48 h) for the majority of viruses, bacteria and protozoa which are immediately infectious as soon as they enter the host. On the other hand, the incubation can take several weeks due to the required period for eggs to hatch (Kettab *et al.*, 2017). Furthermore, under favourable conditions (pH, temperature, sunshine, etc.) and depending on the nature of the microorganism, the pathogens can survive for several weeks, even several months on the ground, on plants or in the water, which poses a significant risk to public health (Catherine *et al.*, 2008; Kettab *et al.*, 2017).

Besides, another important parameters is the infectious minimal dose (IMD), which is proper to the host and which represents the amount of pathogens that must be absorbed in order to observe any symptoms. Its value is highly variable depending on the biological type of the agent.

Microorganisms in wastewater are divided into four main groups (Catherine *et al.*, 2008), as discussed below.

10.5.5.1 Viruses

Viruses are very small intracellular parasites (10–350 nm) that can only multiply in a host cell. Their concentration in urban wastewater is between 103 and 104 particles per litre. Their isolation and enumeration in wastewater is difficult, which probably leads to an underestimation of their real number.

Viruses are not naturally present in metabolic waste, unlike bacteria. They are present either intentionally (after an intramuscular vaccination against polio for example), or in an infected individual. The mode of infection is, in most cases, ingestion, but there can also be cases of inhalation. Viruses appear to be more resistant in the environment than bacteria and their small dimensions are the main cause of dissemination.

10.5.5.2 Bacteria

Bacteria are simple, cell-free, single-celled organisms. Their size is between 0.1 and 10 μm. Urban wastewater contains approximately 107–108 bacteria/litre, including 106 enterococci and entero-bacteria, 104–105 fecal streptococci and 103–104 Clostridium. The majority of these organisms do not present a health hazard and the concentration of pathogenic bacteria can reach around 104/litre.

However, in an infected host, the number of pathogenic bacteria can be very large. Enteric bacteria are adapted to living conditions in the intestine, i.e. a large amount of carbonaceous matter and nutrients, and a relatively high temperature (37°C). Their survival time in the external environment, where the conditions are very different, is therefore limited. In addition, pathogen bacteria will find themselves in competition with native bacteria of environmental origin, which will limit their development. The main contamination way is ingestion. Water pathogens are responsible for the deaths of 3–10 million people a year worldwide (Catherine *et al.*, 2008).

10.5.5.3 Protozoa

Protozoa are single-celled organisms with a nucleus, more complex and larger than bacteria (size between 1 and 200 μm). Most pathogenic protozoa are parasitic organisms that develop at the expense of their host. Some protozoa adopt a form of resistance, called a cyst, during their life cycle. This form can generally resist wastewater treatment processes. Among the most "known" protozoa, there is *Entamoeba histolytica*, responsible for amoebic dysentery (Catherine *et al.*, 2008; Kettab *et al.*, 2008).

10.5.5.4 Helminths

Helminths are multicellular worms. Like protozoa, they are mostly parasitic organisms. Helminth eggs are very resistant and can survive for several weeks or even months on soils or cultivated plants. The concentration of helminth eggs in wastewater is in the range of 10–103 eggs/L.

10.5.6 Toxic elements

10.5.6.1 Trace, mineral or organic elements

Organic and inorganic micropollutants result from multiple and complex pollution. Micropollutants are elements that are present in infinitesimal quantities in wastewater. The main way of contamination, in the case of reuse, is ingestion. Indirect contamination is generally a worrying point. Thus, certain micropollutants, such as heavy metals or pesticides, can accumulate in the tissues of living beings, and particularly in cultivated plants (Kettab *et al.*, 2017). There can therefore be contamination of the food chain. The volatile scientific opinion about their toxicity, and the lack of knowledge about their long-term effects, requires careful analysis of their presence in wastewater (Wu *et al.*, 2018).

10.5.6.2 Heavy metals

The spectrum of these elements in urban wastewater is very large. The most abundant elements are iron, zinc, copper and lead. The other metals (manganese, aluminium, chromium, arsenic, selenium, mercury, cadmium, molybdenum, nickel, etc.) are usually present in trace amounts (Kettab *et al.*, 2017). The origins are various: they come from "individual product consumption, material corrosion in sewer systems, rainwater in the case of a unitary system, community services (health, automotive) and possibly industrial waste. The mentioned elements in the literature as being the most dangerous are lead (Pb), arsenic (As), mercury (Hg), cadmium (Cd) and nickel (Ni) (Kettab *et al.*, 2008, 2017).

10.5.7 Emerging pollutants

An emerging micropollutant is a polluting substance (of mineral, biological, organic or radioactive nature) with infinitesimal concentration in an aqueous environment. The presence of these elements has recently been detected by the scientific community as a consequence of methodological enhancement of water quality monitoring (Kettab *et al.*, 2017; Wu *et al.*, 2018). Toxicity has however only been identified for some of these elements and has not been fully studied or included in aquatic environment monitoring programs and its behaviour, fate and (eco) toxicological effects are not yet fully understood (Wu *et al.*, 2018).

The spectrum of emerging pollutants covers: hormones, additives for plastics, new pesticides and their metabolites, pharmaceutical compounds, body care products or any other new compound where the risk assessment is not sufficiently studied by the scientific community (Ramya *et al.*, 2019).

Several micropollutants have been identified as Endocrine Disruptors (PE). They are also mostly present in every-day products. There is no environmental persistence, the continuous introduction into aqueous environments defines them as pseudo-persistent. This is mainly the case for body care products and pharmaceutical residues. However, additives can also have this continuous aspect through plastic leaching for instance (Amy *et al.*, 1996; Ramya *et al.*, 2019).

10.5.7.1 Pharmaceutical micropollutants

These are the active substances present in small quantities in any composition presented as having curative or preventive properties regarding human or animal diseases with the aim of establishing a medical diagnosis or to restoring, correcting or modifying their physiological functions by exercising a pharmacological, immunological or metabolic action (Wu *et al.*, 2018).

Pharmaceutical products are mainly found in wastewater following their metabolic excretion by humans and animals, in particular via hospital effluents or agricultural effluents (veterinary drugs). In addition, pharmaceuticals assimilated by humans and animals can degrade in living organisms, and degradation products from metabolism, also excreted in the urine or faeces, can also be considered micropollutants (Wu *et al.*, 2018).

10.5.7.2 Pesticides

The term pesticide is a generic designation covering all substances (molecules) or products (formulations, commercial preparations) eliminating harmful organisms, whether used in the agricultural sector or for other applications. About 10% of the tonnage of pesticides marketed in France is used outside agriculture for the maintenance of green spaces, private gardens, roads and railways. They are also designated in France by the terms "phytosanitary products" or "anti-parasitic products". They are used mainly for the protection of plants in the agricultural or non-agricultural sector in a wide range of applications such as agriculture, landscaping or the maintenance of areas around transport axes, forestry and amateur gardening. These products are mainly divided into the following families: fungicides acting on fungi, herbicides acting on "weeds" and insecticides that aim for the elimination insects and mites (Ramya *et al.*, 2019).

A pesticide is generally composed of one or more active substances, which are the main cause of toxicity: a solvent, used to dilute the active substance in order to reduce its concentration and adjuvants to simplify the use and the assimilation of the pesticide by plants. They contain more than 1000 chemicals belonging to almost 150 different chemical families. Pesticides consequently constitute a set of heterogeneous substances and products regarding physico-chemical properties, environmental fate and eco-toxicological properties (Kettab *et al.*, 2017).

10.6 URBAN SANITATION

While it is relatively easy to predict and control the volumes of domestic wastewater, the approach is more complicated from rainwater. Therefore, the discharged water comes from two different sources that need to be managed (Kettab *et al.*, 2017). Eliminating pollution requires not only an efficient treatment process but also resistant and well-dimensioned sewer systems that can safely vehicle wastewater to the plant. When specific sewer systems are installed for rainwater and wastewater is separately carried to the plant, the system is called "separate sewerage" whereas the system that regroups both of these waters is called "combined sewerage" (Audic, 2002; Kettab *et al.*, 2008).

The combined sewer system is cost-effective, besides, run-off water contributes to dilute pollution of urban wastewater. Nevertheless, in the event of heavy rain, the storage capacity of treatment plants does not allow the entire volume of water to be treated. The excess is released without treatment into the natural environment. Besides, abundant precipitation leads to a resuspension of deposits in sewer pipes.

The separate sewer system avoids the risk of wastewater overflowing into natural environment when it rains. It allows better control of the flow and its pollution concentration and better adaptation of the capacity of treatment plants. However, this system is costlier and is difficult to set up in neighbourhoods that already have a combined system.

10.7 TREATMENT STAGES

Because of urbanization and industrialization, wastewater discharges have evolved and become more difficult to treat. Domestic waste has been supplemented with complex and polluting products. The sewerage networks also collect industrial, commercial and artisanal waste, loaded with all kinds of polluting products. Rainwater is loaded with various mineral and organic products and therefore increases the amount of wastewater to be treated (Catherine *et al.*, 2008). If no treatment is proposed the rivers are polluted, wastewater treatment therefore aims to reduce the pollution load in order to obtain suitable water quality for future use.

The decontamination of wastewater requires a series of stages using physical, physico-chemical and biological treatments. Apart from solid waste that is accidentally present in wastewater, treatment must at least allow the total elimination of carbonaceous pollution (Catherine *et al.*, 2008).

10.7.1 Pre-treatment

Pre-treatments allow the elimination of the coarsest solid pollutants (sieving, sand and oil removal) through three stages using simple physical mechanisms.

10.7.1.1 Screening

The principle of screening consists of the insertion of a grid across the wastewater in order to eliminate bulky waste and protect the treatment plant. The screen can be straight or curved, placed in a vertical or tilted position. The space between the bars makes it possible to fix the size of the waste to be eliminated; for this, there are three types of screening, depending on the size of the litter to be removed (Audic, 2002):

- 30–100 mm: pre-screening;
- 10–30 mm: medium screen;
- 3–10 mm: fine screen.

To avoid clogging of the installation, a cleaning operation is necessary. The stations can be either manual or automatic according to the volume. The flow speed must allow retention by the screen without causing a significant pressure drop or clogging at the bottom of the bars. The average speed value must be in the interval: $0.6 < V < 1$ m/s (Bassompierre, 2007).

10.7.1.2 Desanding

The main objective of sand removal is to capture particles in order to protect further installations. In theory, sand removal aims to trap dense (density >2) and grainy materials (diameter >0.2–0.25 mm, even 0.5 mm), i.e. mineral particles such as gravel and sand. However, a small proportion of organic matter is inevitably retained along with those materials, which engenders anoxic fermentation problems at the bottom of the tank and that is why sand washing processes are usually applied (Bassompierre, 2007).

Basically, the particles are trapped by reducing their speed, which makes it possible to distinguish rectangular from tangential desanders. The dimensions of longitudinal or rectangular desanders allow a particle speed of usually between 0.2 and 0.5 m/s. These structures may or may not be aerated in order to ensure separation of the organic fraction fixed on the sand and the grease (grit separators are then used in the treatment plant for flows greater than about 50 L/s). The efficiency of the structures is very variable and depends in part on the design conditions (hydraulic) and especially on the operating conditions, that is to say the admitted flows. The collected sand volume varies between 5 and 12 L per inhabitant and per year. It depends on the nature of the effluent, the sewerage type and condition, etc. (Kettab *et al.*, 2017).

10.7.1.3 Degreasing

The degreasing and de-oiling processes are very complex. Actually, these two operations usually occur together despite belonging to two different physical phenomena. De-oiling corresponds to a liquid–liquid separation whereas degreasing corresponds to a solid–liquid separation operation (if the water temperature is low enough to allow the fat jellification) (Ecosse, 2001; Kettab *et al.*, 2017).

The objective of oil removal is to retain fats and oils (particles with lower density than water) in order to protect the environment. In fact, these components can form a thin layer on the surface of calm waters, which reduces the gas and oxygen exchange between the water and atmosphere. This separation operation is also essential to protect sensitive installations in the plant (in wastewater treatment plants, greasy substances can interfere with biological treatment efficiency (Bassompierre, 2007).

These installations are generally designed for ascending particles with a speed that is close to 15 m/h, and there are different types: independent or combined with desanders. The first type is often used for rainwater retention basins with siphoid bulkheads. The second type is more compact; it is generally used in treatment plants and uses air blowing in order to scum the floating particles (Kettab *et al.*, 2017; Nakib *et al.*, 2015a, b).

10.7.2 Primary sedimentation

The primary sedimentation aims for the elimination of the particulate matter (essentially TSS) by physical or physico-chemical processes. Among the technologies usually encountered at this stage of treatment, the sedimentation basin allows the physical process to occur, and the coagulation-flocculation eventually allows accelerating the process through physico-chemical action (Catherine *et al.*, 2008). Forty to 60% of suspended matter is consequently removed along with 25–40% of organic matter. Coagulation and flocculation can even ensure the sedimentation of finer suspended particles and a part of the colloidal fraction reaching a removal percentage of 60–90%. Organic load is also reduced in that case with higher efficiency than simple primary decantation (Karef *et al.*, 2013).

10.7.3 Biological treatment

Biological treatment of wastewater allows the degradation of pollutants through the action of microorganisms. This process spontaneously occurs in natural environments with low pollution loads. The rich microflora can adapt to various types of pollutants that it consumes as substrates to proliferate. It is therefore possible to use this microflora in a controlled process in order to eradicate organic matter (Amy *et al.*, 1996).

The different biological processes can be classified according to the nature of the involved microorganisms (aerobic and anaerobic). In big cities, anaerobic processes are the most used. Anaerobic processes are either reserved for the stabilization of excessive sludge production or for some industrial effluents (Audic, 2002).

10.7.3.1 Intensive process

The most developed techniques in urban wastewater treatment plants are intensive biological processes. The common criteria of these processes are reduced surfaces and intensified phenomena of organic matter transformation and destruction of organic matter that can be observed in the natural environment (Bassompierre, 2007).

Two main types of process are used:

- *treatments on a fixated support*: bacterial beds and biological disks;
- *treatments without fixated support*: activated sludge.

10.7.3.2 Trickling filter

This process consists of supplying water, primarily treated, to a structure containing a mass of material (pozzolan for instance) serving as a support for the purifying microorganisms which form a biological film responsible for the assimilation of pollution. The biological film is released as the water percolates. As a result, a mixture of treated water and biofilm is collected. The biofilm is trapped in a secondary sedimentation basin in the form of sludge and the treated water joins the natural environment. The sludge reuse as a biomass is essential as shown in Figure 10.1.

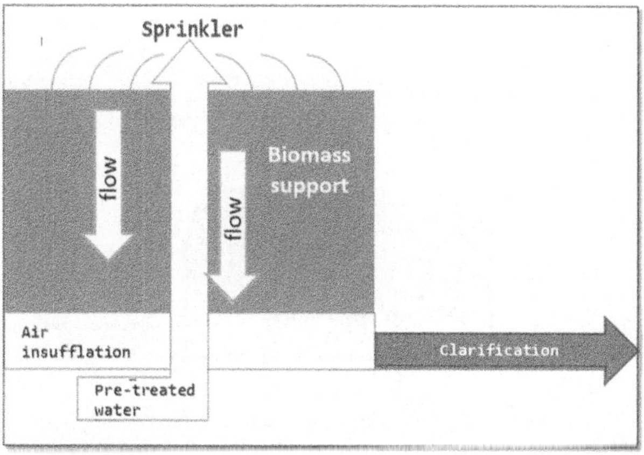

Figure 10.1 Trickling filter (Kettab *et al.*, 2017).

Wastewater is distributed over the upper part of the bed in most cases, by means of a rotary distributor (sprinkler). The satisfaction of oxygen requirements is obtained either by natural or artificial ventilation. The microorganisms forming the biological film that consists of aerobic bacteria on the surface and anaerobic bacteria in depth assimilate pollutants. The produced by-products and carbon dioxide are evacuated in liquid or gaseous fluids (Kettab *et al.*, 2008, 2017).

10.7.3.3 Biodisks

The support disks of the purifying microflora are partially immersed in the effluent and activated by a rotational movement which ensures both mixing and aeration. Microorganisms develop and form a purifying biological film on the surface of the disks. The disks are semi-submerged; their rotation allows the oxygenation of the fixed biomass. The effluent is primarily treated before this process in order to avoid clogging the support material. The dropping sludge is separated from the treated water through clarification.

The bio-disk unit consists of rotating plastic disks mounted on an axis in an open basin filled with wastewater. The disks rotate slowly in the basin and when they pass into the wastewater, the organic matter is absorbed by the biofilm fixed on the rotating disk. The accumulation of biological material on the disks increases its thickness and forms a layer of sludge. When the disks are in the aerated part, oxygen is absorbed, which promotes the growth of this biomass. When the latter is thick enough (about 5 mm), a certain amount comes off and settles at the bottom of the unit. During the submerged phase, the biomass absorbs organic matter which it degraded by aerobic fermentation thanks to atmospheric oxygen from the emerged phase. The used material is progressively lighter (generally expanded polystyrene) and the actual surface area is progressively developing (flat or alveolar disk) as shown in Figure 10.2.

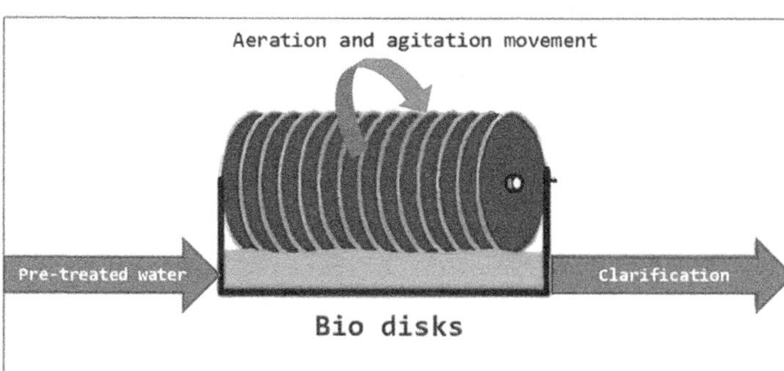

Figure 10.2 Biodisks (Kettab *et al.*, 2017).

10.7.3.4 Activated sludge

It is currently the most widespread process of wastewater treatment and research has confirmed its long-lasting feature. It is in fact a set of processes which all have in common the development of a bacterial culture arranged in the form of flakes (activated sludge) in a stirred and aerated tank (aeration tank or bioreactor), supplied with wastewater which may or may not have undergone a more or less extensive primary treatment.

In the aeration tank, the purpose of mixing is to avoid deposits and to homogenize the mixture of bacterial flocs and wastewater. The purpose of aeration is to provide aerobic bacteria with the oxygen they need to purify water. Aeration is generally carried out using oxygen from the air either via surface aerator with vertical axis (slow turbines) or horizontal axis (brushes) or by air blowing devices (porous disks and more recently micro-perforated membranes). In the first cases, stirring and aeration are ensured by the same device, while for the blowing of air, dissociated mixing devices (propeller at slow speed) are immersed to ensure stirring when the aeration stops. In some cases, it may be necessary to use an oxygen-enriched gas, or pure oxygen (Audic, 2002).

After a sufficient contact time, the mixed liquor is sent to a clarifier, also called a secondary sedimentation basin, in order to separate pure water from sludge. The deposited sludge at the bottom of the secondary basin is partly recycled in the aeration tank in order to maintain a sufficient concentration of purifying bacteria there (Audic, 2002).

One of the criteria frequently used to characterize the different activated sludge systems is the mass loading, which is the ratio between the daily mass of pollution and the used biomass as shown in Figure 10.3:

- high mass load >0.5 kg BOD5/kg of sludge/day;
- medium mass load: 0.2 < cm <0.5;

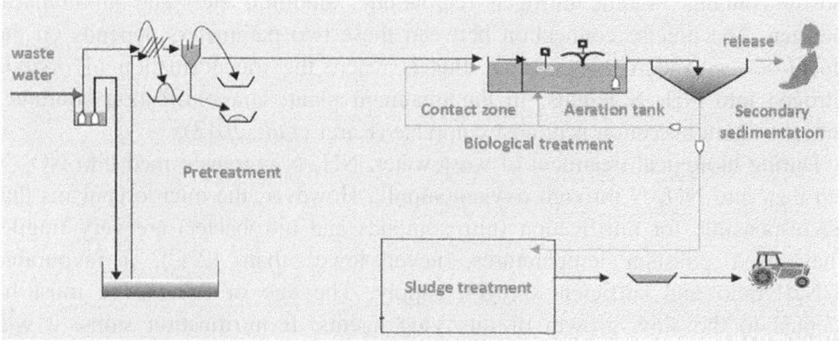

Figure 10.3 Activated sludge (Kettab *et al.*, 2017).

- low mass load: $0.07 < cm < 0.2$;
- low mass load with prolonged ventilation: $cm < 0.07$.

Each of these systems has very specific characteristics (efficiency, sludge production, energy consumption). These figures, given for information purposes, are understood for an urban waste water moderately charged (BOD5 = 250 mg/L) without industrial inputs, with perfect final clarification (Audic, 2002).

For high-load processes, the removal efficiency of BOD5 drops to less than 85%, which in conventional urban wastewater is no longer sufficient to produce an effluent containing 25 mg/L of BOD5. It is then necessary to precede the activated sludge stage with a primary sedimentation. On the other hand, for lower load processes primary sedimentation is not essential to obtain high removal efficiency (Karef *et al.*, 2013), as shown in Table 10.1.

These different permissible sludge concentrations in the aeration tanks are explained by their different sedimentation capacities. These high-load processes generally generate more voluminous sludge (high Mohlman index) than the lower-load processes. This consequently requires lower sludge concentrations in order to keep a reasonable volume of sludge in the clarifiers. Specific oxygen consumption is higher in the extended aeration type processes than in the high load type processes. The selection of the right process must take into account (Audic, 2002; Amy *et al.*, 1996):

- investment costs;
- operating costs;
- the desired pollution removal efficiency;
- operating constraints;
- the presumed fate of the sludge excess.

Today, aerobic biological processes are the most competitive and best suited for the treatment of nitrogen in urban and industrial effluents, because of their ease of implementation, their moderate cost and the possibility of combining the elimination of carbon and nitrogen pollution (Karef *et al.*, 2013). The effluent mostly contains organic nitrogen (carbamide, albumin, etc.) and ammoniacal nitrogen. The precise connection between these two parameters depends on the global sewer system length, since this is where the transformation of organic nitrogen into NH_4-N begins. In the treatment plant, ammonification continues until the ammonification is almost complete (Karef *et al.*, 2013).

During biological treatment of wastewater, NH_4-N is transformed into NO_2-N and then into NO_3-N through oxygen supply. However, the microorganisms that are responsible for nitrification (nitrosomonas and nitrobacter) are very fragile. They need constant temperatures (never lower than 12°C), a favourable C/N/P ratio and sufficient oxygen supply. The age of the sludge must be adapted to the slow growth of nitrifying agents. If nitrification stops, it will take several days for the process to stabilize (Karef *et al.*, 2013; Kettab *et al.*, 2017).

Table 10.1 Mass loading impacts on the treatment process (Amy *et al*., 1996).

Process Parameters	**Mass Loading**			
	$<$**0.07**	**Between 0.07 and 0.2**	**Between 0.2 and 0.5**	$>$**0.5**
	Low with Extended Aeration	**Low**	**Medium**	**High**
BOD removal rate (%)	95%	95%	90%	85%
Nitrification	+	+	–	
Excess sludge production (kg of sludge/kg of eliminated BOD)	0.8	0.9	0.9–1.1	$>$1.2
Maximal admissible sludge quantity (g/L)	5	4–5	3–4	2
Contact time (hour)	17–18	10–15	4–8	5
Oxygen consumption (kg of O2/kg of eliminated BOD)	1.5–1.7	1.3–1.5	0.9–1	0.7–0.8

Under anoxic conditions, the transformation of NO_3-N into elemental nitrogen takes place via the intermediate products NO_2-N and NO/N_2O. For this step, it is important to have a sufficient amount of carbon (mainly methanol) easily degradable. No trace of dissolved oxygen should remain (Karef *et al.*, 2013; Kettab *et al.*, 2017).

In a single tank, nitrification and denitrification are carried out alternately by sequencing the aeration. The separation of the brewing and aeration functions is mandatory here to optimize the transformation of nitrates. The single basin must be properly dimensioned to allow the successive realization of the two reactions; it is a full-mix hydraulic basin, classic rectangular basin type or better oxidation channel. Obviously, a regulation of the operating ranges of the aerators must be implemented (use of redox probes).

However, when the anoxic zone is set upstream, the process is based on the concept of the anoxic conditions. The installation relies on the following conditions (Amy *et al.*, 1996):

- a brewed and unventilated anoxic zone into which the raw effluent, recirculated sludge and mixed liquor are introduced;
- the aeration tank operating at a daily load of 0.15 kg of BOD/kg of VSS in order to ensure nitrification;
- the clarifier ensuring the separation of the biomass from the purified water.

The produced nitrates are returned via the recycling of sludge and mixed liquor to the anoxic zone, where they are then transformed to gaseous nitrogen by using the carbon contained in raw water. The degree of denitrification will therefore be dependent on the recycling rate (a 100% yield results in infinite recycling). Practically, a part of the denitrification occurs in the aeration tank during the aerator shutdown sequences. This additional endogenous denitrification will occur if the dimensions are adjusted, to obtain efficiency rates that are greater than 90% (Karef *et al.*, 2013).

For phosphorous removal, the principle consists of an over-accumulation of phosphorus in a biomass. If phosphorus constitutes 2–3% in the sludge under normal conditions, the mechanism of over-accumulation requires placing the biomass alternately in the anaerobic and aerobic phase (Amy *et al.*, 1996).

For the processes ensuring nitrogen elimination, the denitrification zone is therefore assimilated to an aerobic zone. Put under these conditions, microorganisms accumulate phosphorus in the form of polyphosphate granules. The basic scheme must include at least one anaerobic zone where the release of phosphorus will occur and at least one aerated zone where the reactions of over-assimilation will be carried out.

From conventional urban wastewater, we obtain, on average, 50 and 65% of phosphorus elimination by biological means. This limitation means that, in order to reach higher rates, the remaining phosphorus has to sediment by the addition of reagent. The physico-chemical process is particularly suitable for low flow rates. The main reagents used are metal salts and/or lime Fe (SO_4): ferrous sulphate, Al_2SO_4: alumina sulphate, $FeCl_3$: ferric chloride, $FeClSO_4$: ferric chlorosulphate, and $Ca(OH)_2$: lime (Kettab *et al.*, 2017).

10.7.3.5 Extensive process

Extensive techniques are processes which carry out treatment using fixed or unfixed biomass and usually use solar energy to produce oxygen by photosynthesis. This type of process without electricity is possible, except for aerated lagoons, which require an energetic supply to supply the aerators or air blowing devices (Kettab *et al.*, 2017).

10.7.3.6 Infiltration-percolation

Wastewater infiltration-percolation is a treatment process by aerobic biological filtration on fine granular support. Water is successively spilled over several infiltration units. Hydraulic loads are some hundred litres per square meter. Wastewater is uniformly distributed over the surface of the filter, which is not covered. The water distribution range is kept open and visible.

Another interesting enhancement consists of horizontal or vertical buried sand filters. These techniques are very practical to populations of a few hundred inhabitants. For a buried vertical sand filter, an area of 3.5 m^2/inhabitant is necessary and low-pressure supply is recommended (Karef *et al.*, 2014; Kettab *et al.*, 2017).

10.7.3.7 Planted filters with vertical flow

The filters are excavations, sealed from the ground, filled with successive layers of gravel or sand of variable particle size according to the quality of the wastewater to be treated. Unlike the infiltration-percolation previously mentioned, the effluent is distributed directly, without prior sedimentation, on the surface of the filter. It flows within it undergoing physical (filtration), chemical (adsorption, complexation) and biological (biomass fixed on a fine support) treatment. The purified water is drained. The filters are supplied with raw sewage by tarpaulins. For the same stage, the filtration surface is separated into several units allowing the establishment of feeding and rest periods (Karef *et al.*, 2013; Kettab *et al.*, 2017).

The treatment principle is based on the development of an aerobic biomass fixed on a reconstituted soil. Oxygen is supplied by convection and diffusion. The supply of oxygen by the rootlets of plants is, here, negligible compared to the needs. The sector consists of a screening grid and two sequential installations of vertical filters as shown in Figure 10.4.

10.7.3.8 Reed filter beds with horizontal flow

In horizontal flow filters, the filter block is almost completely saturated with water. The effluent is distributed over the entire width and height of the bed by a

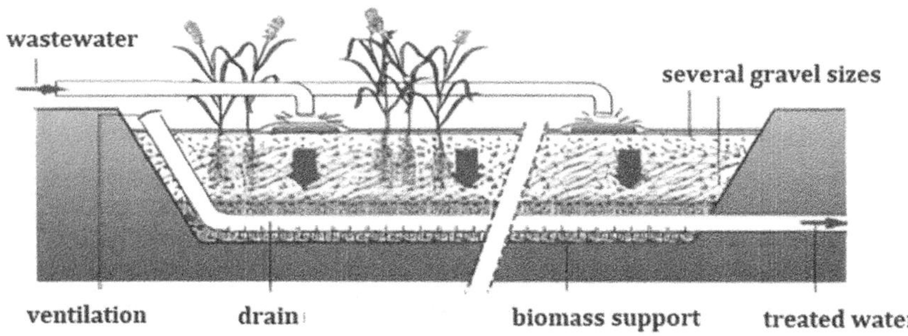

Figure 10.4 Vertical percolation filter (Kettab *et al.*, 2017).

distributing system that is located at one extreme of the basin. It then flows in a horizontal direction through the substrate. A drain ensures water evacuation at the opposite extreme of the bed, at the bottom and buried in a trench of draining stones. This pipe is connected to a siphon in order to adjust the height of the overflow (Kettab *et al.*, 2017).

The water level should be kept approximately 5 cm below the surface of the material. In fact, water must not circulate above the surface so as not to short-circuit the treatment chain, as shown in Figure 10.5.

10.7.3.9 Natural lagoons

The treatment is ensured through long contact periods, through a series of basins. The number of basins most commonly encountered is three. However, using a configuration with four or even six basins allows for more thorough disinfection.

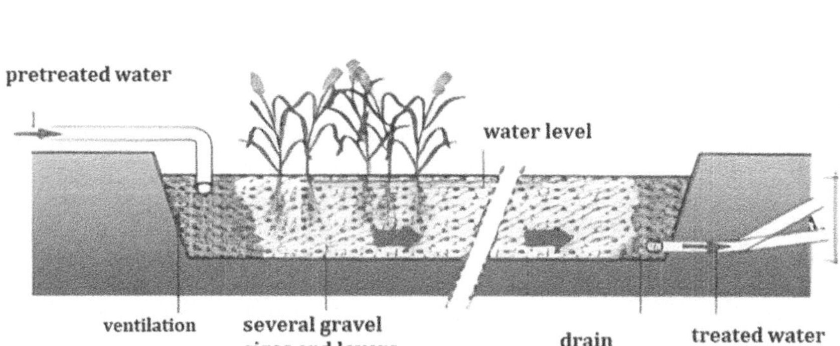

Figure 10.5 Horizontal percolation filter (Kettab *et al.*, 2017).

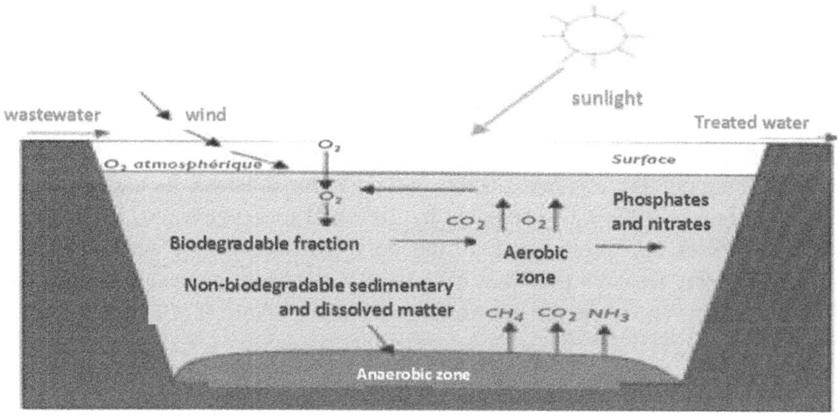

Figure 10.6 Natural lagooning (Kettab *et al.*, 2017).

The basic mechanism is photosynthesis. The upper layer of water is exposed to sunlight. This allows the existing algae to produce the required oxygen for the development and maintenance of aerobic bacteria. The carbon dioxide formed by bacteria, as well as the mineral salts contained in wastewater, allow algae to multiply. There is consequently a proliferation of two interdependent populations: bacteria and planktonic algae, also called "microphytes". This cycle is self-sustaining as long as the system receives solar energy and organic matter. At the bottom of the basin, where light does not penetrate, anaerobic bacteria degrade the sedimented organic matter. Carbon dioxide and methane are released at this level, as shown in Figure 10.6.

10.7.3.10 Aerated lagooning

Oxygenation in this case is brought mechanically by surface aerators or air blowing devices. This principle differs from activated sludge only by the absence of the sludge recycling system. The energy consumption of these two processes is similar (1.8–2 kW/kg of daily eliminated BOD).

10.7.4 Tertiary treatment techniques for disinfection

10.7.4.1 Chlorination

Chlorine is a powerful oxidant, which reacts with organic molecules and microorganisms. Chlorine works in the following order: it reacts first with reducing minerals like iron, manganese or cyanide. Then it reacts with organic molecules, which gives halogenated derivatives that may be carcinogenic. Subsequently, nitrogen compounds, which gives chloramines with a germicidal power. Finally, it directly attacks microorganisms by disrupting enzymatic activity. Therefore, the secondary sedimentation is very important to allow efficient treatment, and avoid having to use too much chlorine (Karef *et al.*, 2013; Nakib *et al.*, 2015a, b).

10.7.4.2 Ozonation

Ozone is produced by the impact of an electrical flow through oxygen. It is active on mineral compounds, organic matter and microorganisms. It allows colour elimination up to 90%, COD removal by 20% and re-oxygenation. It eliminates bacteria, viruses and protozoa. It is the only radical effective method against viruses. Toxicity research tests that have been carried out on aquatic species like fish, crustaceans and algae did not reveal any toxicity (Amy *et al.*, 1996).

10.7.4.3 Ultraviolet

UV systems disinfect by inactivating microorganisms like bacteria, viruses, and parasites that could cause epidemics. In the broad spectrum of electromagnetic radiation, the effectiveness of UV-C wavelengths (200–280 nm) in inactivating these microorganisms is recognized. They damage nucleic acids (DNA or RNA) and stop the reproduction of these organisms (Amy *et al.*, 1996; Kettab *et al.*, 2017).

The germicidal effect of UV systems varies depending on several factors: quality and flow of water to be treated, as well as the UV dose that is applied. The dose is a function of the UV intensity delivered by the system and the exposure time in the reactor (Amy *et al.*, 1996; Karef *et al.*, 2013).

10.7.5 Quaternary treatment techniques for micro-pollution removal

Domestic treatment plants were not originally designed to treat micropollutants. The efficiency of a treatment system strongly depends on the properties of the substances.

In general, the main physical elimination mechanisms in aquatic ecosystems are respectively:

- adsorption on activated carbon;
- chemical or photochemical degradation or transformation like zonation;
- reverse osmosis, which is a costly process that retains the totality of micropollutants because of the membranes that are adapted to their molecular sizes and weights. It is so effective that it even requires the addition of minerals (Nakib *et al.*, 2015a, b).

10.8 SLUDGE TREATMENT

Depending on the wastewater treatment processes, three main categories of sludge can be distinguished: primary sedimentation sludge, physico-chemical sedimentation sludge and biological treatment sludge (Karef *et al.*, 2014):

- The primary treatment sludge is produced by simple decantation of suspended solids (TSS) contained in wastewater.
- Physico-chemical treatment sludge comes from the agglomeration of particulate or colloidal organic matter contained in water by the addition of a coagulating reagent. The sludge is then separated by decantation. Almost 90% of the suspended solids can therefore be retained. Sludge is mainly produced in seaside or mountain resorts, which are designed to treat significant seasonal variations in the pollution load (Nakib *et al.*, 2015a, b).
- The residues of "cultivated" bacteria in the purification works essentially form what we call biological treatment sludge. These bacteria digest organic matter in wastewater.

There are two categories of sludge:

- Mixed sludge: made up of a mixture of primary and biological sludge, it originates from most treatment stations.
- Sludge originating from prolonged aeration without primary sedimentation with polluting materials intensively aerated. The sludge is less organic and therefore safer for reuse.

Wastewater treatment processes produce purified water and a large quantity of by-products: sludge. In the same way that the treated waters return permanently in the natural environment, the resulting sludge, which represents a considerable volume every day, requires a safe destination. That is why sludge treatment techniques are progressively developed. The goal of this stage is to transform produced waste within the plant into a stabilized and reusable product (Karef *et al.*, 2014).

10.8.1 Mineralization

This treatment mode consists of inducing natural mineralization through digestion or the interruption of life in the sludge through physico-chemical stabilization. The notion of stabilization provides information on the olfactive aspect (Djillali *et al.*, 2020).

The stabilization treatments are either biological, chemical or thermodynamic. They are applied to fresh mixed sludge or biological sludge in order to reduce their fermentability and olfactory inconveniences (Djillali *et al.*, 2020; Karef *et al.*, 2014).

10.8.1.1 Biological stabilization

It reduces the amount of fermentable matter in the sludge through aerobic (oxygenated) digestion. The energy consumption of this process does not allow systematic sludge reuse of urban origin. However, in anaerobic digesters, methane biogas is produced and eventually recycled without any presence of oxygen. The extraction of a portion of the biomass is always necessary to ensure the continuous activity of the plant. Composting is a special aerobic biological stabilization process. It is preferably carried out on already dewatered sludge so as to save the supply of composting medium, the sludge not being self-compostable (Karef *et al.*, 2014; Kettab *et al.*, 2017).

10.8.1.2 Composting

Composting is a special aerobic biological stabilization process. It is preferably applied on dehydrated sludge in order to have a good composting support since sludge is not self-compostable. The composted sludge has a soil-like appearance and a solid and stable structure. There is currently a strong revival of interest in this technique due to new regulatory and economic data regarding waste management. This process is usual in medium-sized stations and represents only 2% of the tonnages of sludge (Karef *et al.*, 2014; Kettab *et al.*, 2017).

10.8.1.3 Chemical stabilization

It simply inhibits biological activity, by adding a significant amount of lime (10–50% of the dry matter, generally 30%) raising the pH above 12. This treatment provides extra calcium, which can be beneficial in the case of agricultural reuse (Nakib *et al.*, 2015a, b). Liming generally requires prior sludge dewatering, except for the case of the filter press where limewater is mixed with liquid sludge. The limed sludge can have a pasty or solid structure. Liming techniques have been developed for years because of their efficiency on olfactory inconveniences (Boumalek *et al.*, 2019).

The major disadvantage is the cost of this treatment, approximately 100 g of CaO per kg of dry matter is required, to reduce costs, ferric chloride is often added which improves the flocculation of the sludge and therefore their ability to decantation or filtration. Other chemical techniques, based on blocking biological activity, are starting to be proposed (example: nitrites treatment process at acid pH) (Boumalek *et al.*, 2019).

10.8.1.4 Thermal drying

Thermal drying has a temporary stabilizing effect as long as the sludge is dry. Drying involves two techniques (Kettab *et al.*, 2017):

- the action of the sunlight on sludge placed in a greenhouse, mechanically dehydrated in advance. This process mainly concerns small and medium treatment plants;
- the action of direct or indirect heating depending on device. This kind of treatment is reserved for larger capacity treatment plants.

10.8.2 Physical conditioning

This process induces liquid–solid separation. It uses synthetic organic flocculants (called polyelectrolytes) or minerals (lime, iron or aluminium salts). Conditioning can also be done thermally (autoclaving technique) and consists of sludge heating at 180–220°C for between 30 minutes and two hours in order to break the colloidal aspect that is specific to water retention (Nakib *et al.*, 2015a, b). This process is, however, rarely used because of the encountered management difficulties and the prohibitive cost.

10.8.3 Volume reduction

10.8.3.1 Thickening

Thickening aims to increase the dryness percentage of liquid sludge without altering its liquid state. It can simply be done by gravity in a concentrator or by mechanical means (draining, flotation or centrifugation). The dryness of the thickened sludge does not usually exceed 7% on average and is rather around 5–6%. Generally, gravity thickened sludge is not conditioned and its dryness is limited to 3 or 4%.

This process is frequent in rural areas and concerns small treatment plants, with less than 2000 capita. Dynamic (or mechanical) thickening becomes more frequent in plants between 2000 and 5000 capita. The benefits of thickening are various (Nakib *et al.*, 2015a, b):

- equal capacity, increased residence time in storage tanks of aerobic and anaerobic digesters;
- improvement of operational safety, in particular in urban waste water installations, the sludge from the primary settling tanks can be pumped in a very liquid form;
- efficiency improvement for dehydration devices, drying beds, vacuum filters, filter press, centrifuge.

10.8.3.2 Dehydration

The purpose of this technique is to eliminate most of the water that is contained in the sludge and to obtain a solid product that is easier to manage (Nakib *et al.*, 2015a, b).

Drying by mechanical processes requires belt filters and centrifuges, which give a pasty sludge due to dewatering performance which peaks at 18–20% for the first category, and 20–25% for the second. Filter presses are also an enhanced dehydration device which produce sludge with a solid structure (dryness reaches 30–35%) because it combines limewater conditioning and high physical pressure. These materials are reserved for the most important installations, because they are more expensive and costlier than the previous devices (Djillali *et al.*, 2020).

Drying beds in the open air combine natural evaporation and drainage through a filtering layer of sand and gravel. The individual section is 1 m^2 for 4–5 capita. This extensive system gives solid sludge with a dryness rate that reaches to 35–40%

dryness but remains highly dependent on weather conditions and landscaping. The obtained dry sludge is shovelled and can be used for agriculture. It is very likely to witness the growth beginning while the sludge is still in the drying phase (Boumalek *et al.*, 2019).

Thermal drying reaches a dryness rate that is close to 95%. The obtained sludge is granular. Due to energy costs, this process remains very little used, despite its obvious efficiency in volume reduction. Thermal drying should experience technological enhancement and evolution in the years to come since the other management options, including incineration, are becoming more expensive due to new regulatory and economic conditions.

10.8.4 Sterilization

Sanitized sludge has undergone a treatment that reduces pathogenic colonies in the sludge. Certain treatments destroy pathogenic germs contained in the sludge: liming by raising the pH, composting can also destroy the pathogenic germs through the high temperatures (up to 70°C) that are reached during composting and the selection of saprophytic microorganisms (Boumalek *et al.*, 2019; Kettab *et al.*, 2017).

Thermal drying is certainly the safest way to destroy pathogenic germs because of the very high temperatures and the dryness rise up to 95%, just like thermal conditioning (autoclaving at 195°C and high pressure) (Kettab *et al.*, 2008; Nakib *et al.*, 2015a, b).

10.9 CONCLUSION

The techniques for treating wastewater and the resulting by-products are diverse and present advantageous and attractive aspects but also generate inconvenience to the management according to the quantitative and qualitative parameters of the by-product, climate and topography, of the receiving environment, etc.

It is imperative that there is an optimization of the cost of treatment according to the chosen destination and a regular and careful monitoring of the by-product and the receiving environment. This will allow environmental and health risk management, cost minimization and maximization of water and agricultural amendments.

REFERENCES

Amy G., Debroux J. F., Arnold R. and Wilson L. G. (1996). Preozonation for enhancing the biodegradability of wastewater effluent in a potable-recovery soil aquifer treatment (SAT) system. *Revue des sciences de l'eau*, **9**(3), 365–380.

Anctil F. (2008). L'eau et ses enjeux (Water encounters). University of Laval, Quebec, p. 228.

Andreadakis A. (2003). Wastewater reuse criteria in Grèce. *Global Nest, the International Journal*, **5**(1), 9–14.

Audic J. M. (2002). Présentation des divers éléments des filières de traitement des eaux usées, Traitement des eaux usées urbaines (Urban wastewater treatment stages). Lyonnaise des eaux, Lyon, France, 523 pp.

Bassompierre C. (2007). Procédé à boues activées pour le traitement d'effluents papetiers: de la conception d'un pilote à la validation de modèles (Activated sludge process for wastewater originating from paper industries: Pilot station). Thèse Doctorat Institut National Polytechnique De Grenoble, pp. 25–42.

Bixio D., De Heyder B., Cikurel H., Miska V., Joksimovic D. and Ravazzini A. 2005. Municipale wastewater reclamation, where do we stand? An overview of treatment technology and management practice. *Water Science and Technology*, **5**(1), 77–85.

Blinda M. (2011). Stratégie méditerranéenne pour le développement durable, efficience d'utilisation de l'eau (Mediterranean strategy for sustainable development, efficient use, Blue plan). Athena, Greece, 58 p.

Boumalek W., Kettab A., Bensacia N., Concetta Bruzzoniti M., Ben Othman D., Mandi L., Nacer Chabaca M. and Benziada S. (2019). Specification of sewage sludge arising from a domestic wastewater treatment plant for agricultural uses. *Desalination and Water Treatment*, **143,** 178–183.

Catherine B., Alain H. and Jean-M H. (2008). Technologies d'épuration en vue d'une réutilisation des eaux usées traitées Rapport final, Convention de partenariat ONEMA (Treatment technologies for reuse. Final report, ONEMA partnership). Cemagref, National Centre of Management of Fauna Reserves, Beijing, 100 pp.

Djillali Y., Nacer Chabaca M., Benziada S., Bouanani H., Mandi L., Concetta Bruzzoniti M., Boujelbene N. and Kettab A. (2020). Effect of treated wastewater on strawberry. *Desalination and Water Treatment*, **181**, 338–345.

Ecosse D. (2001). Techniques alternatives en vue de subvenir à la pénurie d'eau dans le mond (Alternative techniques to fight against water scarcity). Graduation Thesis, Faculty of Sciences, University of Amiens, France, p. 15, 17, 21, 26.

Karef S., Nakib M., Kettab A., Benmaamar S., Benziada S. and Bouguerra K. (2013). A study of performance indicators of a sewage treatment plant (Algeria), 3rd International Water Conference, Algiers, November 2013.

Karef S., Kettab A. and Nakib M. (2014). Characterization of byproducts from wastewater treatment of Medea (Algeria) with a view to agricultural reuse. *Desalination and Water Treatment*, **52**, 2201–2207.

Kettab A. (2001). Water resources in Algeria: strategies, investments and vision. *Desalination*, **136**(1–3), 25–33.

Kettab A. (2014). Water for all with quality and quantity: it is the concern of all. *Desalination and Water Treatment*, **52**, 1965–1966.

Kettab A., Metiche R. and Bennacar N. (2008). De l'eau pour un développement durable: enjeux et stratégie (Water for sustainable development: encounters and strategies). *Revue des Science de l'Eau*, **21**(2), 247–256.

Kettab A., Chabaca N. M., Karef S., Nakib M., Boumalek W., Bouanani H., Djillali Y., Benziada S. and Benmamar S. (2017). Traitement des eaux usées urbaines et leurs réutilisations en Agriculture (Urban wastewater treatment and agricultural reuse). EUE Publishing (European University Publications), Greece.

Nakib M., Kettab A., Berreksi A. and Mandi L. (2015a). Study of the prospects for agricultural utilization of sludge produced from WWTPs in North Central Algeria. *Desalination and Water Treatment*, **55,** 1152–1166.

Nakib M., Kettab A., Bouanani H., Tebbal S. and Berreksi A. (2015b). A study of the fertilizing potential of the treated wastewater of the Kolea wastewater treatment plant (Algeria). *Desalination and Water Treatment*, **57**, 1–5.

Ramya T., Premkumar P., Thanarasu A., Velayutham K., Dhanasekaran A. and Subramanian S. (2019). Degradation of pesticide-contaminated wastewater (coragen) using electrocoagulation process with iron electrodes. *Desalination and Water Treatment*, **165**, 103–110.

Wu G., He K., Miao J., Yin Q. and Zhao Y. (2018). Strategies for sustainable wastewater treatment based on energy recovery and emerging compounds control: a mini review. *Desalination and Water Treatment*, **127**, 4–11.

Chapter 11

Comparison among pharmaceutical wastewater treatment schemes for the removal of particular type of drugs and their residues

Afzal Husain Khan

Civil Engineering Department, Jazan University, 114 Jazan, Saudi Arabia

ABSTRACT

The treatment of pharmaceutical wastewater (PWW) is a challenging task for health and environmental authorities since pharmaceutical effluent from pharmaceutical manufacturing industries and healthcare facilities have been found to adversely affect the aquatic environment and increase the load on sewage treatment plants (STPs). Usually, PWW contains drug residues, chemicals, cultures, toxins etc., which needs to be pre-treated before discharge to domestic sewers. Because of this, a dedicated PWW treatment system should be developed for the efficient removal of various types of, or specific, pharmaceuticals. In recent decades, the rapid development of advanced treatment technologies assured the efficient elimination of pharmaceuticals from industries or healthcare facilities (HCFs), moreover, pharmaceutical residues (discussed in Chapter 8) which can be processed and recovered in sustainable processes leading to wastewater revalorization. Hence, for the development of dedicated treatment schemes, priority selection and risk assessment of pharmaceuticals is much needed. The objective of this chapter is to provide a comprehensive review of the current state of the technologies in use or under development and comparison among available technologies is taken into account for the efficient treatment of PWW. Due to the uneven nature of PWW, different removal mechanisms would be needed and various combinations or couplings can be adopted for specific compound removal. Here, we present various treatment schemes for the treatment of PWW as a whole, or certain pharmaceutical compounds specifically.

 Pharmaceutical Wastewater Treatment Technologies: Concepts and Implementation Strategies
Editor(s): Nadeem A Khan, Sirajuddin Ahmed, Viola Vambol and Sergij Vambol
doi: 10.2166/9781789061338_0271

Keywords: On-site treatment, Pharmaceutical wastewater, Wastewater treatment plant.

Contents

11.1 Introduction 272
11.2 Treatment Technologies for PWW Treatment 274
  11.2.1 Primary treatment 278
  11.2.2 Secondary or biological treatment 279
    11.2.2.1 Treatment scheme integrated with activated sludge process 279
    11.2.2.2 Treatment scheme integrated with anaerobic digestion 281
    11.2.2.3 Treatment scheme integrated with constructed wetlands (CWs) 281
    11.2.2.4 Treatment scheme integrated with electro bioreactor (EBR) technology 282
    11.2.2.5 Treatment scheme integrated with aerobic or anaerobic fluidized bed reactor (FAB) technology 284
    11.2.2.6 Treatment scheme integrated with submerged aerobic fixed film (SAFF) technology 285
    11.2.2.7 Treatment scheme integrated with sequencing batch reactor (SBR) technology 285
    11.2.2.8 Treatment scheme integrated with upflow anaerobic sludge blanket UASB reactors 286
    11.2.2.9 Advanced treatment 286
    11.2.2.10 Membrane technology 287
    11.2.2.11 Adsorption 289
    11.2.2.12 Advance oxidation 290
    11.2.2.13 Hybrid treatments 291
11.3 Comparison of Technologies for Treatment of Hospital Effluents 292
11.4 Conclusion 294
References 294

11.1 INTRODUCTION

Globally, different reference guidelines and treatment technologies are being utilized for treating industrial effluents before release into the ecosystem. Tragically, the same cannot be said about hospital effluents, since there is not much unanimity about how pharmaceutical wastewater (PWW) ought to be dealt with (Verlicchi *et al.*, 2012a). A plausible explanation behind such a situation may be on the grounds that most nations consider hospital effluents to be on comparable lines as residential wastewater. Subsequently, such PWW is in effect straightforwardly released into the sewer framework, with no pre-treatment. Not many nations are considering hospital effluents as industrial effluent and adopting pre-treating before releasing into urban sewage treatment plants (Deegan *et al.*, 2011). After publication of the WHO guidelines "Safe Management of Wastes from Healthcare Activities" in 1999, competent authorities in regional and municipal levels in many countries formed protocols for pre-treatment or on-site treatment of PWW. In 2002 the US Geological Survey studied 139 streams in 30 states, 80% contained organic wastewater contaminants, mostly trace amounts of pharmaceuticals including drugs such as analgesics, antidepressants, lipid regulators, antibiotics and steroids, as well as substances such as disinfectants, nicotine, caffeine, food supplements, cosmetics, sunscreen, insecticides and detergent soaps (Kolpin *et al.*, 2002).

After almost 15 years, the WHO second edition guidelines mention the methods for the treatment and disposal of healthcare wastewaters, in which there is a segment that concerns the collection and the disposal of wastewater from healthcare activities. In this report, WHO define the wastewater from healthcare facilities as "A large part of the wastewater from healthcare facilities is of a similar quality to domestic wastewater and poses the same risks. Just as domestic wastewater is considered to be potentially infectious, wastewater from healthcare facilities must also be considered in a similar manner and precautions taken", but "A proportion of the generated wastewater from healthcare facilities will pose a higher risk than domestic wastewater. Depending on the service level and tasks of the healthcare facility, the wastewater might contain chemicals, pharmaceuticals and contagious biological agents, and might even contain radioisotopes" (WHO, 2014).

Nowadays, a major concern with pharmaceutical compounds such as anti-inflammatory agents, antibiotics, chemical residues, radio-elements, antibiotic resistance strains, and pathogens etc. is that they are commonly not secured by any administrative agency in many developing countries. Consequently, there is almost no comprehension of the antagonistic effect of these substances on human health and the environment (Luo *et al.*, 2014). Actually, very little investigation has been carried out on how wastewater treatment plants can lessen the presence of these pharmaceutical compounds. It must be understood that the release of huge amounts of wastewater, stacked with such pharmaceutical compounds, may deteriorate the aquatic environment as well as human beings (Quinn *et al.*, 2008; Sui *et al.*, 2015) In this manner, adopting onsite pre-treatment for the reduction/removal of pharmaceutical compounds from PWW will help to diminish the worry relating to the environment and public health too. Despite the fact that there is very little worldwide awareness towards the pre-treatment of PWW before discharge to STPs, some developed countries are adopting various technologies for on-site treatment of PWW. Only a couple of countries have reference standards and explicit treatment techniques to deal with these effluents. For pharmaceutical effluents, explicit reference measures and treatment strategies have been developed at provincial or municipal levels with respect to direct discharge (in surface waters) or indirect discharge (in STPs). Some countries consider hospital effluents to be domestic and therefore they are discharged directly into the drainage systems, rivers, and lakes without prior treatment.

In only a few countries is on-site treatment or pre-treatment of PWW mandatory before release to STPs.

The high priority drug categories detected in urban wastewaters include beta-blockers, sedatives/psychiatric drugs, analgesics/anti-inflammatories, antibiotics, diuretics, cytotoxic, and lipid regulators (as shown in Table 11.1). Conventional STP techniques have facilitated these pharmaceuticals removal to a certain extent, though these schemes are not deliberately designed to remove these compounds. Hence, while designing wastewater treatment plants for treating pharmaceuticals some factors should be considered including wastewater characteristics, flow volume, area requirement, erection, operation and maintenance costs. Based on these factors, various treatment technologies can be effectively integrated with existing units. If the industrial site is restricted in space, the plants should be installed with small units and operated without the continuous (few times a year) supervision of highly qualified operators. Constructed wetlands, oxidation and polishing ponds are usually preferred if shortage of funds for construction and electric supply power is not continuous. However, this treatment scheme requires large areas and moderate temperatures. In addition, the selection technique for pharmaceutical wastewater treatment should be designed on the basis of simple operation, efficiency, economical, durability and environmental friendliness.

Based upon an extensive literature survey of the research carried out on pharmaceutical wastewater treatment, a listing has been made of the available treatment technologies with respect to drug category and its compounds in Table 11.1. This chapter attempts to summarize the literature concerning the potential technologies and schemes for degrading or removing pharmaceuticals from wastewater, along with hybrid technologies.

11.2 TREATMENT TECHNOLOGIES FOR PWW TREATMENT

Urban STPs are usually designed to treat domestic effluents that are received on a regular basis and of the same characteristics (mostly nutrients, such as phosphorus and potassium, and organics). The context of pharmaceuticals is entirely different, they are individual compounds with distinctive characteristics that, at very trace amounts, cause detrimental impacts on aquatic ecosystems (Larsen *et al.*, 2004). Several studies on PWW treatments have been studied in the last two decades, including treatments to reduce/eliminate as well as recover many compounds from pharmaceutical wastewater. In view of the environmental impact and STP load as discussed earlier, it is necessary to develop treatment methodologies at source, i.e. pre-treatment for pharmaceutical wastewater. So, pre-treatment at source could reduce the environmental impact, the burden on STP and also provide the potential recovery of compounds (Lin *et al.*, 2013; Mutiyar & Mittal, 2014; Nguyen Dang Giang *et al.*, 2015), along with bio-energy

Table 11.1 Comparison of various wastewater treatment methodologies for particular drug category and its compounds with their removal efficiencies.

Drug Category	Compound Name	Treatment Combination Methodology	Pilot/Full Scale (P/F)	Removal (%)	References
Beta-blockers	Atenolol	UVA/TiO_2	P	67	Hapeshi *et al.* (2010)
		WWTP+Hybrid SF-CW	P	>70	Anderson *et al.* (2013)
		WWTP+CAS	F	78	Radjenović *et al.* (2009)
		HWWTPs+ASP		>86	Al Qarni *et al.* (2016a)
		WWTP+MBR	P	99	Kovalova *et al.* (2012a)
		WWTP+ PAC		92	Mailler *et al.* (2014)
Sedatives/Psychiatric drugs	Carbamazepine (CAR)	WWTP+CAS+ MBRs	F	n.e.	Radjenović *et al.* (2009)
		WWTP+Hybrid SF-CW	P	20–50	Anderson *et al.* (2013)
		WWTP+Membrane SBR+PAC	P	>99	Alvarino *et al.* (2016a)
		WWTP ASP + TF	F	n.e.	Kasprzyk-Hordern *et al.* (2009)
		WWTP+ PAC		97	Mailler *et al.* (2014)
		HWWTPs+ASP		>86	Al Qarni *et al.* (2016a)
	Diazepam (DIA)	WWTP+Membrane SBR+PAC	P	>99	Alvarino *et al.* (2016a)
Analgesics/anti-inflammatories	Acetaminophen (ACT)	WWTP+CAS+ MBRs	F	>99	Radjenović *et al.* (2009)
	Diclofenac (DIC)	WWTP+HSSF-CW	P	20–50	Ávila *et al.* (2013)
		WWTP+ PAC	F	85	Mailler *et al.* (2014)
		WWTP+CAS+ MBRs	F	>65	Radjenović *et al.* (2009)

(*Continued*)

Table 11.1 Comparison of various wastewater treatment methodologies for particular drug category and its compounds with their removal efficiencies (*Continued*).

Drug Category	Compound Name	Treatment Combination Methodology	Pilot/Full Scale (P/F)	Removal (%)	References
		WWTP+MBR+Post treatment	F	>90	Langenhoff *et al.* (2013)
	Ibuprofen (IBU)	WWTP+HSSF-CW	P	50–70	Ávila *et al.* (2013)
		WWTP+CAS+ MBRs	F	>99	Radjenović *et al.* (2009)
		WWTP+MBR+Post treatment		100	Langenhoff *et al.* (2013)
	Ketoprofen	STP+MBR		97	Park *et al.* (2020)
		STP+A_2O		75	
		STP+ SBR		51	
		STP+ MBBR		40	
	Naproxen (NAP)	WWTP+CAS+ MBRs		>91	Radjenović *et al.* (2009)
	Paracetamol (PA)	WWTP+ASP+TF		94	Kasprzyk *et al.* (2009)
		HWWTPs+ASP		>95	Al Qarni *et al.* (2016a)
		HWWTP+MBR	P	>99	Kovalova *et al.* (2012a)
Antibiotics	Ciprofloxacin (CFL)	WWTP+ PAC	F	91	Mailler *et al.* (2014)
		HWWTPs+ASP		>95	Al Qarni *et al.* (2016a)
		WWTP+ PAC+O_3+UV	F	>97	Kovalova *et al.* (2013)
	Clarithromycin (CLA)	WWTP+HSSF-CW	P	20–50	Ávila *et al.* (2013)
		HWWTPs+ASP	F	>86	Al Qarni *et al.* (2016a)
		WWTP+ PAC+O_3+UV		>97	Kovalova *et al.* (2013)
	Erythromycin (ERY)	WWTP+CAS+ MBRs	F	41	Radjenović *et al.* (2009)

	Ofloxacin (OFL)	WWTP+CAS+ MBRs		95	Radjenović *et al.* (2009)
		UVA/TiO_2 photocatalysis	P	85	Hapeshi *et al.* (2010)
		WWTP+ PAC		89	Mailler *et al.* (2014)
	Sulfamethoxazole (SUL)	WWTP+CAS+ MBRs	F	81	Radjenović *et al.* (2009)
		HWWTPs+ASP		>95	Al Qarni *et al.* (2016a)
		WWTP+ PAC		72	Mailler *et al.* (2014)
		WWTP+ PAC+O_3+UV		>97	Kovalova *et al.* (2013)
		WWTP+ Hybrid lagoon+SF-CW		>70	Anderson *et al.* (2013)
	Tetracyline (TET)	WWTP+ VSSF-CW		>70	Ávila *et al.* (2013)
Diuretics	Furosemide (FUR)	WWTP+ PAC+O_3+UV	P	>97	Kovalova *et al.* (2013)
Cytotoxic	Sorafenib (SOR)				
	Mitomycin (MIT)	Oxidation with KMnO/NaOCl		100	Benvenuto *et al.* (1993)
		Electrolytic treatment		99	Hirose *et al.* (2005)
Lipid regulators	Bezafibrate (BEZ)	WWTP+CAS+ MBRs	F	90	Radjenović *et al.* (2009)
		WWTP+ PAC	F	99	Mailler *et al.* (2014)
	Gemfibrozil (GEM)	WWTP+HSSF-CW	P	50–70	Ávila *et al.* (2013)
		WWTP+CAS+ MBRs	F	42	Radjenović *et al.* (2009)
	Simvastatin (SIM)	O_3, O_3+ H_2O_2, PAC or ClO_2	P	>10	Nielsen *et al.* (2013)
		WWTP +UF-HF	P	73.6	Sulaiman *et al.* (2015)
		WWTP +UF-SW		90.7	

Note: WWTP, wastewater treatment plants; HWWTPs, hospital wastewater treatment plants; CAS or ASP, conventional or activated sludge process; UVA/TiO_2, ultraviolet/titanium oxide; UF-HF, hollow fiber ultrafiltration membrane; UF-SW, spiral wound membrane; UF, ultrafiltration; TF, trickling filter; O_3, ozone; PAC, powdered activated carbon; H_2O_2, hydrogen peroxide; ClO_2, chlorine dioxide; A_2O, anaerobic–anoxic–oxic process; n.e., not eliminated; SF-CW, surface free water constructed wetlands; VSSF-CW, vertical subsurface flow constructed wetlands; HSSF CWs, horizontal subsurface flow constructed wetlands.

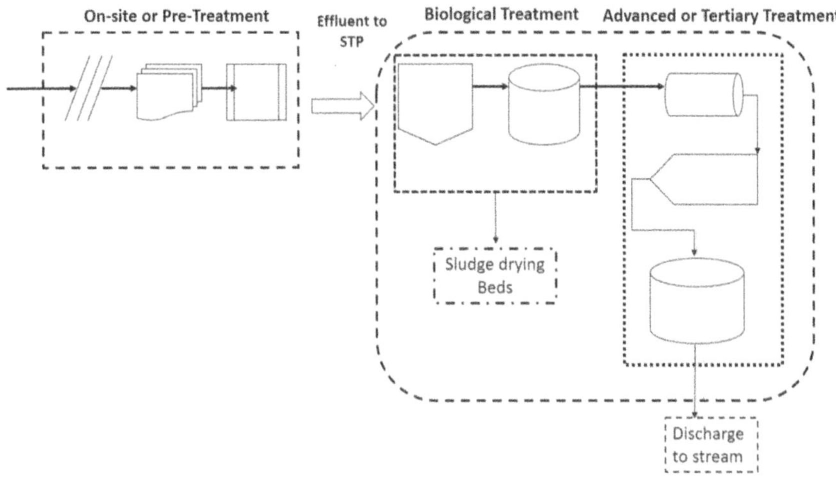

Figure 11.1 A process scheme for pharmaceutical wastewater treatment (Khan *et al.*, 2020a, b, c).

recovery from anaerobic treatments (Lin *et al.*, 2013). However, much attention is required to present a complete scenario of the pharmaceutical wastewater treatment technologies. In this section, attempts have been made to express a dedicated treatment for a particular type of drug, though it is arduous and uneconomical. A process scheme for pharmaceutical wastewater treatment, including pre-treatment, is shown in Figure 11.1.

11.2.1 Primary treatment

The principal objective of wastewater treatment, such as sedimentation, coagulation-flocculation and flotation, are generally to temporarily hold the wastewater in a quiescent basin where heavy solids can settle to the bottom while oil, grease and lighter solids float to the surface. This treatment for PWWs has been found incapable because such treatment is primarily useful for removing conventional physico-chemical parameters such as total suspended solids (TSS), biochemical oxygen demand (BOD), chemical oxygen demand (COD), pH, dissolved and total organic carbon (DC, TOC), nitrate, phosphate and, to some extent, *Escherichia coli* and fecal coliforms. However, several researchers have demonstrated that the majority of pharmaceuticals are not significantly removed by primary treatment such as sedimentation, coagulation–flocculation and flotation (Carballa *et al.*, 2005; Suarez *et al.*, 2009). Similarly, combined micro-electrolysis, Fenton oxidation, and coagulation sedimentation methods were studied by considering the removal efficiencies and operation costs. The reaction conditions were optimized and the results show that the proposed method is a promising treatment for wastewater because it is efficient and easy to operate (Qin *et al.*, 2016). A coagulation-flocculation treatment was carried out by Carballa *et al.* (2005), on eight pharmaceuticals from lipophilic, neutral and acidic compound groups. The reported removal rates of lipophilic compounds such as galaxolide and tonalide, were 35–60% and neutral compounds like diazepam and carbamazepine were 40–50 and 20–35%, respectively. Minimum removal was reported for diclofenac (20–45%), carbamazepine (20–35%) and naproxen (10–30%). Similar results were observed by Suarez *et al.* (2009), for the removal of pharmaceuticals from hospital wastewater by physico-chemical treatment (coagulation-flocculation) in combination with flotation. The maximum removal rates reported were musks (77–83%), followed by diclofenac (46%), naproxen (42%), and ibuprofen (23%). The rest of the PPCPs remained unaffected by coagulation-flocculation treatment (Suarez *et al.*, 2009). However, such treatment in combination with a secondary (biological) level shows higher removal rates compared to primary treatment (Al Qarni *et al.*, 2016a; Clara *et al.*, 2005; Jelic *et al.*, 2011; Peng *et al.*, 2006; Pal *et al.*, 2010; Rogers, 1996; Verlicchi *et al.*, 2010).

11.2.2 Secondary or biological treatment

Usually, biological wastewater treatment systems will be unsuccessful or perform poorly if prior treatment such as primary treatment, equalization and conditioning upstream is not addressed properly, particularly for hospital effluents containing various residual pharmaceuticals. In addition, the biological system requires state-of-the-art design and realization with efficient equipment (aeration, mixing), reliable on-line instruments and modulating automatic process controls. Some factors that should be taken into account when designing hospital wastewater treatment plants include land requirement, construction cost, operation cost, maintenance and landscape. Various treatment options used for efficient reduction, recovery and disposal of various types of pharmaceuticals originating from hospitals or the pharmaceutical industry can be divided into three categories, as shown in Figure 11.2.

11.2.2.1 Treatment scheme integrated with activated sludge process

The activated sludge process (ASP) is usually adopted due to its simple operation, and lower maintenance and energy consumption, whereas modified activated sludge requires trained staff, a high construction cost and relatively high operation and maintenance costs (Al Qarni *et al.*, 2016b; Polesel *et al.*, 2016; Verlicchi *et al.*, 2012b). The pharmaceuticals removal through ASP is explicitly due to microorganisms' potential to degrade organic compounds in the wastewater. However, the ability to break down the pharmaceuticals varies among compound groups. Some drugs degrade up to 98% within 2–5 days, such as sulfamethoxazole (antibiotic), ibuprofen (analgesic) and acetylsalicylic acid (a breakdown product of aspirin). On the other hand, roxithromycin (antibiotic)

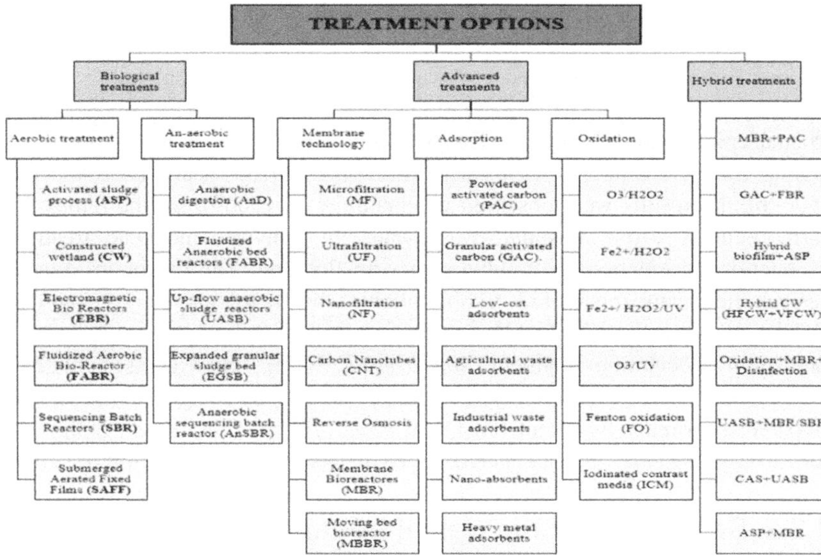

Note:	O3	Ozone/
	H2O2	Hydrogen peroxide
	Fe2+	Ferrous ion
	UV	Ultra-Violet
	HFCW	Horizontal flow constructed wetland
	VFCW	Vertical flow constructed wetland
	ABR	Anaerobic Baffled reactor
	BASR	

Figure 11.2 Recognized treatment option for PWW.

requires 5–10 days to degrade, while drugs like carbamazepine (antiepileptic drug) and diazepam (psychoactive drug) have shown no significant degradation even after 20 days. Therefore, alternative methods are required to destroy trace amounts of toxic drugs in the wastewater. A typical activated sludge process is shown in Figure 11.3.

Researchers investigated two hospital wastewater treatment plants integrated with ASP in Saudi Arabia. Their findings showed high removal efficiencies (more than 99% on average) for pharmaceutical compounds such as paracetamol, sulfamethoxazole, ciprofloxacin, and caffeine, followed by atenolol, carbamazepine, and clarithromycin (more than 86% on average). It was found that elevated temperatures and high levels of sunlight were considered as the main factors that enhanced the removal of these compounds (Al Qarni *et al.*, 2016b).

Table 11.1 depicts the removal efficiency for different pharmaceuticals in WWTPs integrated with secondary and tertiary treatment with full or pilot scale units. It can be noticed that the percentages removal through ASP integrated unit for atenolol, carbamazepine, paracetamol, clarithromycin and sulfamethoxazole were more than 78%, whereas the removal for other pharmaceuticals was not

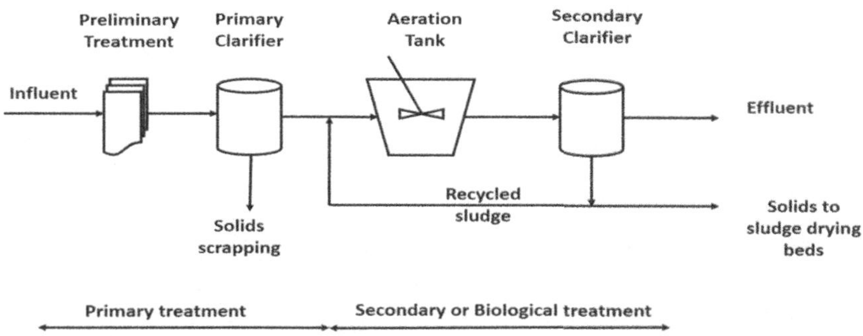

Figure 11.3 A typical activated sludge process.

significant. Moreover, the removal percentages were greatly enhanced by integrating units such as MBR, TS and ASP to the system. In fact, sometimes removal varies even for the same compound in different WWTPs.

11.2.2.2 Treatment scheme integrated with anaerobic digestion

Anaerobic digestion (AD) of pharmaceutical concentrates is an effective disposal technique in conjunction with biogas production. The pharmaceutical residues treatment by six sanitization technologies (pasteurization, thermal hydrolysis, advanced oxidation processes using Fenton's reaction, ammonia treatment, thermophilic dry digestion, and thermophilic anaerobic digestion) were investigated (Malmborg & Magnér, 2015). This approach has demonstrated that AD: (1) is the most effective technology in reducing a wide spectrum of organic substances; (2) AD temperature shows no major effect on the reduction of pharmaceutical residues; (3) AD exhibits significant correlation between lipophilicity and the persistence of pharmaceutical residues; (4) thermal hydrolysis efficiently removed steroid estrogens from sludge; (5) pasteurization, ammonia treatment, and thermophilic dry digestion achieved relatively modest reductions; (6) carbamazepine, citalopram, estrone, 17β-estradiol, 17α-ethinylestradiol, fluoxetine, ibuprofen, propranolol, and sertraline showed persistence to most treatments. Another investigated result showed that AD feasibly removes pharmaceuticals by generating high specific biogas production values and efficiently eliminates colistin sulphate. The highest biogas production from pharmaceutical sludge is 499.46 mL/g TS at an inoculum to substrate ratio (ISR) of 10.32 (Yin *et al.*, 2015).

11.2.2.3 Treatment scheme integrated with constructed wetlands (CWs)

Constructed wetlands have been extensively investigated to be efficient for PWW treatment for municipal wastewater, hospital effluents, pharmaceutical industry effluent, landfill leachate, contaminated ground water, agricultural wastewater and urban runoff. The popularity of this system is due to its low maintenance and it depicts similar functioning as an alternative treatment system to conventional biological treatment, such as the activated sludge process. The wastewater treatment in CW relies mainly on biological and physico-chemical processes to some extent, under a natural environment. Generally, constructed wetlands are land-based wastewater treatment systems which consist of shallow ponds or rectangular trenches, whereas in horizontal and vertical subsurface flow constructed wetlands (SSFs), wastewater infiltrates through vegetated gravel beds confined by a liner. The potential of SSFs has been found significant to remove contaminants from urban wastewater and numerous researchers showed favorable results, particularly for treating wastewater from small communities (Dires *et al.*, 2019; Nadeem A Khan *et al.*, 2020a, b; Özengin & Elmaci, 2016; Vymazal *et al.*, 2017). It has been found that some pharmaceuticals are efficiently reduced as follows: (i) ibuprofen under aerobic conditions, (ii) clofibric acid and diclofenac under anaerobic conditions and (iii) halogenated pollutants under anoxic conditions (Tran *et al.*, 2016). In other investigations it was found that aerobic conditions are, in general, more efficient than anaerobic ones for removing most emerging contaminants (Zwiener & Frimmel, 2003; Matamoros *et al.*, 2008).

On the other hand, an experimental study showed significant removal efficiencies for carbamazepine, ibuprofen, and sulfadiazine for a planted reactor (with *P. australis*) and an unplanted reactor (with Leca) were 89.23 and 95.94, 89.50 and 94.73, and 67.20 and 93.68%, respectively. The observed results indicate high removal efficiencies for both planted and unplanted reactors for carbamazepine and ibuprofen, whereas for sulfadiazine, the removal efficiency was higher for unplanted reactors as compared to the planted reactors (Özengin & Elmaci, 2016). In other investigations the authors found desirable sulfonomide resistance genes (SRGs) and antibiotics resistance genes (ARGs) removal results using constructed wetlands (Chen & Zhang, 2013; Chen *et al.*, 2016b; Li *et al.*, 2020; Xu *et al.*, 2015). Li *et al.* (2014) extensively reviewed the constructed wetland systems and found CW unable to eliminate pharmaceuticals completely, except for salicylic acid, paracetamol, caffeine, tetracycline, and sulfa drugs (Li *et al.*, 2014). However, constructed wetlands could be a promising technology for the treatment of pharmaceutical contaminants such as ibuprofen, clofibric acid, diclofenac, ketroprofene, carbamazepine, sulfadiazine, depending upon the mechanism and reactor conditions. Hence, the feasibility of constructed wetlands to eliminate pharmaceuticals needs comprehensive understanding on the removal efficiencies, removal mechanisms, and the influences of design and environmental factors.

11.2.2.4 Treatment scheme integrated with electro bioreactor (EBR) technology

EBR technology, the latest entrant into the field of wastewater treatment, has some positives such as higher quality of treated water, smaller space for installation and ease of operation. All of these are not found in the conventional activated sludge processes. This system runs automatically through the auto device which operates the deionization ratio in the different stages, thus, it self-analyses, through its own intelligent auto-detection system. One important feature of the EBR process is the utilization of electromagnetic de-ionization in the biological process water tank, which can help in generating high quality permeate from the wastewater emanating from households, hospital effluents, cooling tower blow down, etc. The size of the aeration tank can help in reducing very high sludge concentrations in the aeration tank by 4–5 times. Another positive attribute of the reactor is that it does not require any further treatment using sand filters, activated carbon filters, etc.

In the case of STPs with EBR technology, sewage effluent from the equalization tank is made to pass through a bio-chamber. In a bio-chamber, the process of biological treatment takes place under controlled environmental conditions. Uniform distribution and mixing with a controlled air supply and microbial inoculants act as additives to control the process. This process is continuous and completely automated since the sludge formed is automatically carried by the cake former and cutter. The series of guard feeders are also used for the biological reaction to decompose organic substances, including in the raw sewage effluent.

Several investigations were conducted on integrated EBR with electrocoagulation/electrically enhanced reactors for the treatment of municipal wastewater. For instance, Cui *et al.* (2009) studied an integrated membrane bioreactor with electrocoagulation process on a laboratory scale to enhance total phosphorus (TP) removal from synthetic domestic wastewater to reach 0.5 mg TP L^{-1} in treated effluent. Similarly, Wei *et al.* (2012) investigated nutrient removal in an electrically-enhanced membrane bioreactor (EMBR) at room temperature (20 $\pm$ 2°C) with synthetic feed and two months at 10°C with real sewage. Their findings showed removal rates of more than 99% of ammonium, more than 95% of dissolved COD, and more than 90% ortho-phosphorus in the EMBR. However, as far as the removal of pharmaceutical residues from the EBR is concerned, it is quite evident that pharmaceutical residues of different classes are prevalent in the influents from hospitals. The sources of these can be well understood, depending on the different activities of the hospital. The variation of particular type of pharmaceutical residues in the influent is attributed to the type of treatment facility, season, type of prescription, consumption, metabolism, etc.

Dhingra *et al.* (2018) studied EBR with a post bioreactor and V-Guard polishing units which helps in the effective removal of trace pollutants. In the tube-settler, there was a removal of more than half of the residues like diclofenac, ofloxacin, diazepam and simvastatin by about 68, 62.4, 61.6 and 53.21%, respectively, in EBR treatment. Some pharmaceuticals such as ibuprofen, carbamazepine, erythromycin, frusemide were not removed significantly by EBR, i.e. around 25, 29.7, 13.5 and 41.0%, respectively (Dhingra *et al.*, 2018).

11.2.2.5 Treatment scheme integrated with aerobic or anaerobic fluidized bed reactor (FAB) technology

Fluidized aerobic bed reactor (FAB) technology is an adaptation of the attached growth aerobic biological treatment. As far as COD removal is concerned, aerobic fluidized bed reactor (FBR) technology was outstanding in COD removal efficiency at an organic loading rate (OLR) of 9.4 g COD/L.d (95%), and this gradually decreased to 76.5% in OLR of 24.2 g COD/L.d. Also with an increase in the organic loading rate the contribution of the aerobic reactor in removing COD gradually decreased. In this system, the anaerobic reactor played the most important role in the removal of COD, and the aerobic moving-bed biofilm reactor (MBBR) was actually needed to remediate the anaerobic treated wastewater (Jafari *et al.*, 2013).

Based on the state of biomass fixation, the aerobic biofilm reactors are grouped into: reactors with suspended biomass, mixed reactors, and reactors with fixed/attached biomass (Lazarova & Manem, 1994). Among the several types of aerobic biofilm reactors, the fixed/attached biomass reactors with moving beds, particularly the two- and three-phase aerobic inverse fluidized bed biofilm reactors (AIFBBRs), have gained much attention in recent years, in which the solid carrier particles are made to move continuously by hydraulic means (von Sperling, 2007).

As far as pharmaceutical removal is concerned, the removals of four pharmaceuticals present at trace levels using a staged anaerobic fluidized membrane bioreactor (SAFMBR) in municipal wastewater was investigated. Maximum removal efficiency was obtained for naproxen (63.8–72.1%), whereas the removal efficiencies of ibuprofen (25.0–32.4%), carbamazepine (16.6–18.5%), and diclofenac (non-detected) remained low. For carbamazepine with a high acid dissociation constant ($pKa = 13.9$) or diclofenac with a high octanol-water partition coefficient ($\log KOW = 4.51$), sorption onto the sludge could be one critical process affecting their concentrations in the water phase. Higher HRTs could partially improve the treatment removal efficiencies of naproxen (63.8–72.1%), at increased HRT from 4.5 to 7.5 h. HRT control was the main responsible factor for the removal of these pharmaceuticals, depicting the importance of knowing the properties of a pharmaceutical for selection of a proper treatment approach (Chen *et al.*, 2019).

In another study, ibuprofen and naproxen were reported to be biodegradable under prevailing anaerobic conditions, as sorption onto sludge could be another explanation affecting the concentration of ibuprofen that possesses a high sorption coefficient to sludge (Park *et al.*, 2020; Rodgers & Zhan, 2003). Relatively high efficiency was achieved in treating antibiotics such as metronidazole from wastewaters by modified fluidized bed reactor combined with an Fe/charcoal based micro electrolysis system. The system efficiency depends on factors such as the initial concentration of the pollutant, the initial solution pH, electrodes ratio, aeration, and contact time (Malakootian *et al.*, 2019). For microbial control, immobilization of bacteria on solid surfaces can take place in fluidized bed systems. Many species of bacteria (and also other microorganisms) have the capacity to adhere to the supporting matrices (Rodgers & Zhan, 2003). A fluidized bed batch reactor was employed for removing pharmaceuticals and endocrine disruptors from a hospital wastewater using Trametes versicolor in both sterile and nonsterile conditions. The overall removal efficiency of pharmaceutically active compounds were 83.2 and 53.3%, reported in sterile and nonsterile conditions, respectively. In addition, from 51 studied pharmaceutically active compounds, 46 compounds were eliminated successfully (Cruz-Morató *et al.*, 2014).

11.2.2.6 Treatment scheme integrated with submerged aerobic fixed film (SAFF) technology

Submerged aerobic fixed film (SAFF) is a modified form of a typical anaerobic hybrid reactor with improved efficiency and requirement of area for installation. It is quite effective where nitrification, as well as denitrification, takes place and is a rate limiting phenomenon. In this, the media is placed in an aerobic zone for improving the effectiveness even at low temperatures (Deegan *et al.*, 2011).

As far as pharmaceutical residues are concerned, SAFF is not a promising technology for some pharmaceuticals such as diclofenac, ibuprofen, carbamazepine, diazepam, erythromycin, ofloxacin, furosemide, simvastatin, and sorafenib (Dhingra *et al.*, 2018). After secondary treatment the effluent shows efficient removal of ibuprofen, ofloxacin and simvastatin (55.1, 91.5 and 53.3%, respectively), whereas pharmaceuticals such as diclofenac, carbamazepine, diazepam, erythromycin, and frusemide were reduced by only 24.7, 25.3, 15.6, 39.6 and 33.1%, respectively, after tertiary treatment. Thus, SAFF tertiary treatment was not considered remarkable. Hence, this low reduction rate can be efficiently enhanced by adding a polishing unit (Gautam *et al.*, 2017).

11.2.2.7 Treatment scheme integrated with sequencing batch reactor (SBR) technology

Sequencing batch reactor (SBR) is a sophisticated and modern activated sludge organic treatment system. In SBR, cycles are filled, react, settle, draw and idle, independently (Rogers, 1996). SBR differs from the traditional treatment activated sludge system, since all the above said processes take place in one reactor itself. There may be no decrease in sludge, or biomass concentration, when compared to the activated sludge device in this process. SBR is also helpful for elimination of organic matter and nutrients. Due to this benefit, it has been used worldwide for wastewater treatment for many years. In this process there is no sludge return, hence can be attained in small areas (Ileri *et al.*, 2003). Nowadays, SBR is determined as a useful system for the reduction of pharmaceuticals. In some studies, anaerobic sequencing batch reactor (AnSBR) is also employed for pharmaceuticals treatment in the presence of oxygen under dark conditions. Pharmaceuticals up to 40 mg/L can be treated and beyond that, it results in toxicity to the microbial community. Moreover, the effect of doses on antibiotics was visible on the microbial network, using AnSBR (Khan *et al.*, 2019).

11.2.2.8 Treatment scheme integrated with upflow anaerobic sludge blanket UASB reactors

Upflow anaerobic sludge blanket (UASB) reactors are also used as pre-treatment containing pharmaceutical drugs, particularly antibiotics under high organic loading up to 21.02 kg COD/m^3/d and with pH 8.26 to attain higher efficiency of COD (Yi *et al.*, 2017). A number of modified forms of USBR reactors such as expanded granular sludge blanket (EGSB), which has high up-flow velocity in order to separate sludge from granules and upflow anaerobic stage reactor (UASR), which has combined properties of upflow septic tank/baffled reactor (USBR) and anaerobic baffled reactor. It benefits from separating acidogenesis and methanogenesis and allows reduction of biomass washout, quick hydraulic recovery and shock loading. The feasibility of UASR was carried out for pre-treatment for HWW containing tylosin and avilamycin antibiotics and achieved quite efficient results (Khan *et al.*, 2019).

In another study in the Netherlands, significant results were shown for domestic and industrial wastewaters using a UASB reactor, due to its simple operation, economy and the possibility of energy recovery. Studies focusing on modified UASB reactors, though containing contradictory results, have been reported such as the influence of climatic variations and the treatment of wastewaters containing complex or toxic substances (Nnaji, 2014). Several researchers have investigated many kinds and combinations of the UASB reactor which have enhanced the capacity of suitability for the removal of micropollutants from WWT facilities such as micro-electrolysis (ME), UASB, anoxic/oxic (A/O) activated sludge process and biological aerated filter (BAF) (Qi *et al.*, 2015).

11.2.2.9 Advanced treatment

In the past decade, several advanced treatment technologies were considered as effective options for upgrading existing wastewater treatment plants as well as providing pre-treatment with high level treatment, particularly when wastewater is taken into account for reclamation. These advanced treatment technologies are primarily classified as membrane technology, adsorption and oxidation technology. However, these technologies are still not efficient in removing all pharmaceuticals from wastewater and demand certain arrangements of treatment unit's for dedicated pharmaceutical drug removal. For instance, Paulus *et al.* (2019), studied antibiotic and antibiotic resistance gene (ARG) concentrations in hospital, communal wastewaters and wastewater treatment plants in two Dutch cities. They selected on-site treatment using advanced technologies, including MBR, ozonation, granulated activated carbon (GAC) and UV for hospital wastewater. The on-site hospital wastewater samples exhibited the presence of 13 antibiotic resistance genes out of 711 antibiotic concentrations analyzed, thus there is a need for on-site treatment of high-risk point sources of antibiotic resistance genes. Their findings suggested MBR treatment for the efficient reduction of antibiotic resistance genes and ozonation for antibiotic reduction (Paulus *et al.*, 2019).

11.2.2.10 Membrane technology

Membrane technology was traditionally adopted in desalination and water treatment, and it is now widely used in pharmaceutical wastewater treatment. In a typical plant, polymeric membranes are used in potable water reuse in areas with water scarcity issues. These membranes carry out microfiltration and ultrafiltration and are sometimes used in conjunction with techniques such as reverse osmosis and nanofiltration. However, they also offer certain limitations such as membrane fouling, stability and material related issues (Warsinger *et al.*, 2015). Nanofiltration as tertiary treatment satisfactorily removed stimulants and antibiotics in a pilot plant and was found to be highly promising in achieving complete removal of the contaminants (Egea-Corbacho *et al.*, 2019).

Clara *et al.* (2005) investigated eight pharmaceuticals, two polycyclic musk fragrances and nine endocrine disrupting chemicals in three WWTPs installed with MBRs, conventional activated sludge plants (ASP) and ultrafiltration membranes in a pilot scale. The different solid retention times (SRTs) were compared to access the treatment efficiencies. Carbamazepine was not removed in any of the adopted treatment facilities and other compounds such as bisphenol-A, ibuprofen, and bezafibrate were almost completely removed. In the WWTP facility, the removal potential of MBRs correspond to SRT. No significant removal was observed by ultrafiltration membranes (Clara *et al.*, 2005). Therefore, the removal efficiency of WWTP can be enhanced by combining advantageous methods such as reverse osmosis and nanofiltration due to lower power inputs and costs of operation. Moreover, the overall removal efficiency varies from 70–90%, except carbamazepine which may be due to the relatively low retention of pharmaceuticals in nanofiltration (NF) (Beier *et al.*, 2010). Hence, the membrane process has several merits such as rapid installation and compact design to be accommodated in existing facilities. However, some demerits are power driven operation and pollutant residual as the treatment is based on separation instead of biodegradation.

Nowadays, MBRs are becoming more prominent technology for the treatment of hospital wastewater, due to efficient modifications and economic viability of the employed membranes. Especially, the high biomass concentrations and long SRT are favorable for the biodegradation of organic pollutants, resulting in

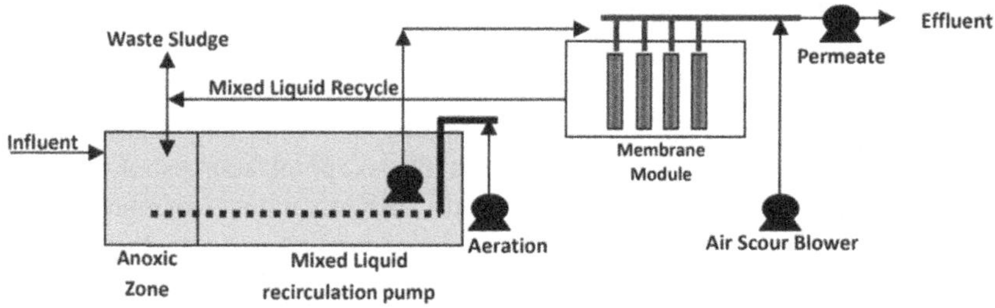

Figure 11.4 A typical membrane bioreactor (MBR) treatment.

high rate treatment systems. These characteristic parameters of MBR technology are not only favorable to organic matter removal, but are also supposed to enhance biodegradation efficiency of micro-organic pollutants. Thus, biological degradation of micropollutants such as pesticides, pharmaceuticals and personal care products are significantly removed in wastewater treatment systems (Sipma *et al.*, 2010). Several membrane methods for the removal of pharmaceuticals wastewater have been investigated by membrane bioreactors (Khan *et al.*, 2020a, b, c; Kovalova *et al.*, 2012b; Prasertkulsak *et al.*, 2016) and moving bed biofilm reactors (Andersen *et al.*, 2013), as shown in Table 11.1, with regard to the compounds analyzed in this study. Figure 11.4 shows a typical MBR treatment, including primary treatment.

Membranes can be used in external modules, or these can be submerged. Membranes can also be defined by their pore length. Hence, microfiltration membranes, having a pore size of around 0.2–0.4 μm, and ultrafiltration membranes, having a pore size of 0.03 and 0.06 μm, are employed for such purposes (Krzeminski *et al.*, 2012).

Usually, MBR is used as aerobic pre-treatment, followed by ozonation, UV/H_2O_2 and reverse osmosis for the removal of various contaminants (Langenhoff *et al.*, 2013). The MBR effectiveness depends on the hydraulic retention time and sludge retention time. However, for HWW, SRT is important (Verlicchi *et al.*, 2015). MBR treatment cannot remove micropollutants completely alone, and thus require a polishing unit such as ozonation, activated carbon, photo-degradation, etc. In addition, the MBR process is expansive, followed by either ozonation, packed activate carbon or filtration (Nielsen *et al.*, 2013).

MBBRs (Figure 11.5) are based on biofilms that vary from 1–4 cm diameter plastic carriers which are suspended and mixed in a reactor. In this way, MBBRs are usually recommended over biofilm reactors (Rodgers & Zhan, 2003). In addition, MBBRs have already been recognized as compact and robust reactors for wastewater treatment in respect of enhancing nitrification. In 2016, researchers studied the MBR process and found it as a substitute for ASP, due to its high micropollutant removal efficiency (Tran *et al.*, 2016). In addition, the reported removal efficiency was elevated by 15–42% in comparison to ASP,

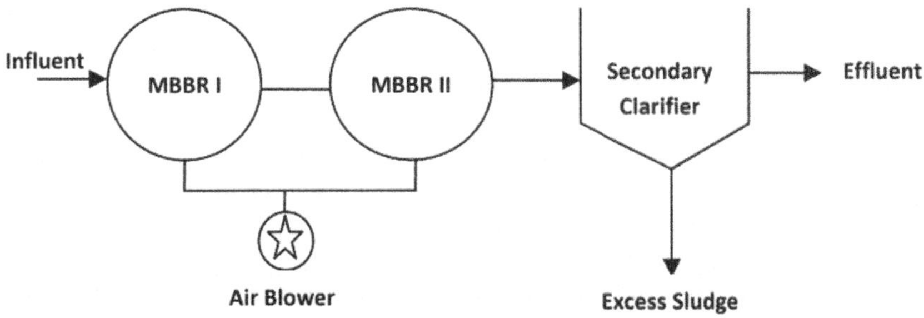

Figure 11.5 A typical moving-bed biofilm reactor (MBBR) followed by secondary clarifier.

although the removal efficiencies of some antibiotics in both the systems are almost the same.

Furthermore, MBBRs have been applied for treating industrial wastewater, particularly wastewater from pharmaceutical industries, due to their robustness and high pharmaceutical degradation capability (Andersen *et al.*, 2013). The micropollutants cannot be removed by MBR treatment alone, but need polishing units like ozonation, activated carbon, photo-degradation, etc., for achieving complete removal. A disadvantage of this method is that this will increase the overall treatment cost (Nielsen *et al.*, 2013). Recently, hybrid MBR, such as submerged sponge-MBR for COD, nitrogen, nitrification, denitrification removal and sponge-MBR coupled with ozonation, significantly removed pharmaceuticals from hospital wastewater (Nguyen *et al.*, 2016; Vo *et al.*, 2019).

11.2.2.11 Adsorption

Adsorption based commercial materials such as a granular activated carbon (García-Mateos *et al.*, 2015), powdered activated carbon (Mailler *et al.*, 2014) and low-cost adsorbents such as sawdust, clay, red mud, biochars, chitosan, agricultural and industrial wastes (De Andrade *et al.*, 2018), have been extensively used in the treatment of PWW, but these adsorbents need to be regularly regenerated using different methods, such as thermal regeneration, chemical regeneration (Nahm *et al.*, 2012) and catalytic ozonation (Alvarez *et al.*, 2009). It was found to have the capability to remove endocrine disruptors (Liu *et al.*, 2009). Many treatment studies were carried out by employing different forms of activated carbon to give efficient results (Mailler *et al.*, 2014; Yu *et al.*, 2014). In addition, the adsorption process produces a significant amount of sludge containing pharmaceutical residues, which needs to be treated further using thermal incineration, as direct disposal methods (open burning and landfilling) may lead to soil or surface water pollution (Pal, 2018). The basic advantage of using adsorption is that it does not produce by-products, which may give rise to toxic secondary compounds from the parent compound.

GAC and powdered activated carbon (PAC) adsorbents provide a higher surface area, porosity, and adsorption capacity, and have potential efficiency to remove pharmaceuticals even in trace amounts (De Andrade *et al.*, 2018). Researchers studied efficient removal (99.6%) for 28 antibiotics present in a PAC based pilot plant (Xu *et al.*, 2007). In another study, significant removal efficiencies were reported on fixed bed adsorption by activated carbon up to 95 and 98% for diclofenac and caffeine, respectively (De Andrade *et al.*, 2018).

Nowadays, single-, double- and multi-walled carbon nanotubes are commonly used for the removal of pharmaceuticals from wastewater (Liu & Wang, 2013; Liu *et al.*, 2014). Up to 70% adsorption capacity was reported at the pH range 3–7 (García-Mateos *et al.*, 2015). Some other nano-absorbents like TiO_2, FeO, aluminum etc. reported significant removal of antibiotics on external surface adsorption, followed by antiparticle diffusion (Beiginejad *et al.*, 2015; Li *et al.*, 2017). Some heavy metals such as arsenic, lead, mercury, cadmium and chromium in wastewater have shown high adsorption capacities due to their higher surface area. Another investigation reported a high removal rate of pharmaceuticals by carbon nano-tubes within the pH range of 4.0–10.0 (Wang *et al.*, 2016).

11.2.2.12 Advance oxidation

Nowadays, advanced oxidation processes are one of the most efficient methods for the removal of pharmaceutical compounds, despite their higher operational costs due to electric-driven operation and chemicals utilization (Sipma *et al.*, 2010). Kovalova *et al.*'s (2013) investigation on pilot-scale hospital wastewater treatment plant includes primary, biological and five post-treatment technologies including ozone (O_3), O_3/H_2O_2, PAC, and low pressure UV light with and without TiO_2 for the reduction of 56 micropollutants. The elimination efficiencies reported by ozonation, PAC, and UV were 90, 86, and 33%, respectively. Thus, the majority of analyzed compounds can be efficiently eliminated by ozonation and PAC, whereas some recalcitrant compounds were significantly removed by two post-treatments, such as the iodinated contrast media (ICM) and UV. The coupling treatments such as O_3/H_2O_2 and UV/TiO_2 are insignificant in the elimination of compounds as compared to O_3 and UV alone (Kovalova *et al.*, 2013).

Several investigations showed significant removal of antibiotics by UV, H_2O_2 or its combination (O_3-UV, O_3/ H_2O_2, UV- H_2O_2) (Yi *et al.*, 2017). These higher removal rates were usually dominated by higher specific surface areas and thus absorbing capacity (Yi *et al.*, 2017). The advanced oxidation process is a widely used technology for hospital and industries wastewater treatment, the greatest advantage it offers is the treatment of antibiotics. Recent situations of antibiotic/anti-microbial resistance bacteria (ARB/AMR) is a critical public health issue, hence AOP application can prevent further resistance. The biggest challenge of AOPs is its oxidation by-products and the processing cost. However, AOPs such as H_2O_2/UV, Fe^{2+}/H_2O_2, $Fe^{2+}/H_2O_2/UV$, O_3/H_2O_2, O_3/UV are effective for pharmaceuticals removal without chemical utilization.

11.2.2.13 Hybrid treatments

As of now, several hybrid models of aerobic and anaerobic methods are available which can be considered the most competent and potential methods of bioremediation as they impart excellent removal efficiencies and broad command over the treatment, thereby exhibiting multiple advantages for the treatment of micropollutants. Moreover, biological treatment in combination with advanced or polishing methods provide outstanding removal efficiencies.

Alvarino *et al.* (2016a, b) studied two hybrid MBRs using different types of membrane (flat sheet microfiltration (MF), and hollow fiber ultrafiltration (UF)) run under aerobic conditions with direct dosing of PAC for the reduction of ten organic micropollutants (OMPs). Diclofenac and roxithromycin reduction were significantly improved by adding an ultrafiltration (UF) unit to these types of membrane, which may be attributed to sorption and/or further biotransformation action (Alvarino *et al.*, 2017).

In another study, biofilm-based sand column systems exhibited higher potential to efficiently remove pharmaceuticals (95%) such as biphenylol, p-chlorocresol, chlorophene, 5-fluorouracil, gemfibrozil, ibuprofen, ketoprofen, naproxen, triclosan, and valproic acid (Onesios *et al.*, 2009). A comparative study of MBR and activated sludge systems for the removal of pharmaceuticals suggested MBRs superiority over ASP, especially for slowly degradable pharmaceuticals removal, due to the relatively long sludge ages, which leads to the development of distinct microbial communities. However, for the treatment of degradable compounds and recalcitrant compounds, the ASP and MBR treatments yield almost the same amounts of effluent. Several investigations show MBR superior to ASP treatment for most pharmaceuticals of intermediate products. This indicates that much research is still required to gain a better insight into the biodegradation potential of pharmaceuticals in biological wastewater treatment (Sipma *et al.*, 2010).

In hybrid biofilm activated sludge processes, the degradation of seven compounds which are usually recalcitrant in classical activated sludge treatment, such as diclofenac, propranolol, iopromide, iohexol, iomeprol tebuconazole and propiconazole, are removed efficiently by advanced treatment (biofilm reactor-slow sand filtration), i.e. 41, 94, 58, 57 and 85% respectively (Casas & Bester, 2015).

In many studies which are based on MBBR coupled with MBR in which physical parameters were analyzed with different retention times had shown promising trends due to their biodegradation pathways. Besides, MBR has some physical complications in operation as discussed in previous sections. UASB is also a good alternative coupled with aerobic MBR with low HRT. This system can achieve efficient micropollutant removal as compared with conventional processes. An innovative plant configuration based on a UASB reactor coupled to a hybrid aerobic membrane bioreactor was studied in terms of OMPs removal

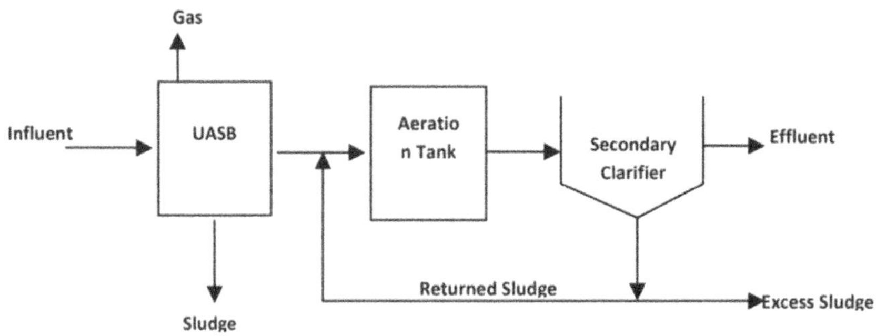

Figure 11.6 A typical CAS (aerobic) +UASB (anaerobic) hybrid treatment.

(>95%). Under anaerobic conditions, both sulfamethoxazole and trimethoprim have shown more than 84% biodegradable efficiencies (Alvarino *et al.*, 2016b; Khan *et al.*, 2019). Similarly, a combined conventional activated sludge (CAS) unit and a UASB reactor efficiently removed 16 pharmaceutical and personal care products (PPCPs) (Alvarino *et al.*, 2014). A typical treatment under aerobic and anaerobic conditions consisting of hybrid treatment CAS + UASB units are shown in Figure 11.6.

Regarding all the hybrid and combined processes, cost of technology involved, types of wastewater, temperature and, most importantly, financial conditions for running these projects play a vital role. In addition, based on the characteristics of the pharmaceuticals residue in the wastewater, multiple treatment options are efficiently designed and installed. Every treatment has its own merits and demerits, hence proper investigation of the characteristics of particular wastewater could help to design an efficient combination to effectively eliminate pollutants.

11.3 COMPARISON OF TECHNOLOGIES FOR TREATMENT OF HOSPITAL EFFLUENTS

In order to decide on the most efficient treatment technology for pharmaceuticals reduction, the removal rate at various points of primary, biological and tertiary treatments are accessed by considering various parameters. Recently, Park *et al.* (2020) analyzed 27 pharmaceuticals in the unit processes of four different sewage treatment plants (STPs) and found overall target pharmaceuticals were reduced by 88–95% (covering-stimulant caffeine, acetaminophen, naproxen, acetylsalicylic acid and cefradine. To compare the pharmaceutical removal efficiencies of the biological treatment processes consisting of a membrane bioreactor (MBR) process, sequencing batch reactor (SBR) process, anaerobic–anoxic–oxic (A_2O) process, and MBBR process, a quantitative meta-analysis was executed. The following hierarchy of removal efficiency by biological processes in decreasing order is: MBR > SBR > A_2O > MBBR. For the tertiary treatment, PAC exhibits the maximum removal efficiency (18–63%) for gemfibrozil, ibuprofen, ketoprofen, atenolol, cimetidine, and trimethoprim (Park *et al.*, 2020).

In India, Gautam *et al.* (2017) explored comparative analysis consisting of the operation and maintenance (O&M) of several aerobic treatment technologies: MBBR, MBR, SBR, extended aeration (EA), and SAFF (Gautam *et al.*, 2017). This study suggests SBR as most cost-effective treatment technology and MBR as the most expensive among various available treatment options. The results shows the sequence of treatment facilities with respect to cost wise efficiencies as follows: MBBR > SBR > EA > SAFF > MBR.

In another investigation, Guillossou *et al.* (2019) compared the effluent concentrations of a WWTP and a pilot-scale advanced treatment by activated carbon adsorption in comparison to conventional treatment (Guillossou *et al.*, 2019). They found the sequence of treatment efficiencies as follows: Primary treatment (<20%) > biological treatment (>60%). Further, advanced treatment was >30% for recalcitrant compounds whereas higher elimination (>60%) was obtained for all pharmaceuticals except azithromycin and sulfamethoxazole over the WWTP and the activated carbon pilot. A higher dose of activated carbon or the combination with another advanced treatment should be used to achieve a satisfactory removal of those compounds.

In 2017, the authors investigated a combined wastewater from domestic and industrial sources in China, that comprises various treatment units such as grit chamber, anaerobic/anoxic/oxic (A_2/O), sedimentation and UV disinfection (Ashfaq *et al.*, 2017). The overall treatment process involves screen, grit chamber, anaerobic/anoxic/oxic (A_2/O), sedimentation and UV disinfection (Ashfaq *et al.*, 2017). They found the sequence of treatment efficiencies in decreasing order as follows: Primary treatment (screen + grit chamber) > anaerobic process (>80%) > aerobic treatment > (secondary sedimentation + UV disinfection).

Furthermore, Park *et al.* (2017) studied the performance of the tertiary treatment process of the 604 STPs with a treatment capacity of more than 500 m^3/d, >60% are designed, consisting disk filters and filtration systems (53%) and sedimentation/flotation (39%) to total phosphorus reduction efficiency. On comparing the pharmaceutical removal efficiencies in various units like coagulation/disk filter (CD), coagulation/sedimentation (CS), rapid coagulation-sedimentation (RCS) processes, UV and PAC, the statistically meta-analysis showed no significant removal efficiencies of pharmaceuticals and SRE decreased in the sequence of PAC > CD $\approx$ CS > RCS > UV. This pattern indicates no significant effect on the removal of pharmaceuticals, due to the very small molecular size of the pharmaceuticals as compared to the pore size of the disc filters, thus these disk filters are inefficient at reducing pharmaceuticals (Park *et al.*, 2017). Similarly, UV treatment did not show any significant effect on the removal of pharmaceuticals as UV disinfection in the WWTP was not effective in removing most trace contaminants (Yang *et al.*, 2014). Moreover, the results depict higher removal efficiency of PAC out of five tertiary treatment processes (Park *et al.*, 2020).

11.4 CONCLUSION

The dedicated treatment of particular categories of drugs has attracted significant attention over the last two decades. After reviewing various technologies, advanced treatment such as membrane technology in combination with an oxidation process have been proven to be potent technologies for a wide array of pharmaceuticals. In addition, hybrid technologies demonstrated potential advantages compared to conventional systems, not only in higher treatment efficiency and energy savings, but also because they offer both sustainable and environmentally- friendly treatment.

However, properly sited, designed, constructed, and maintained onsite wastewater treatment systems have the potential to provide an effective and efficient means of treating pharmaceuticals and thereby reducing the load on STP and protecting the environment. Moreover, such designed units impart economic and ecological advantages to the system.

Recently, Park *et al.* (2020) investigated 27 pharmaceuticals in four STPs in South Korea and a comparative evaluation standardized removal efficiency (SREs) of biological treatments showed that the SRE decreased in the order of MBR > SBR > A_2O > MBBR. In addition, among five types of tertiary treatment processes, PAC showed the most efficient performance, with an increase of 18–63% in the removal efficiency for compounds including gemfibrozil, ibuprofen, ketoprofen, atenolol, cimetidine, and trimethoprim (Park *et al.*, 2020). Physico-chemical treatment is not significant in removing trace contaminants and thus has no influence on the operational parameters, such as size, charge, polarity, and functional group. On the other hand, numerous advanced treatments, such as ozonation-activated carbon adsorption, could be a proven combination for the removal of pharmaceuticals as stated by Guillossou *et al.* (2019). To adopt efficient treatment technologies, the role of scientists, consultants, operators and government agencies is highly important to eliminate pharmaceuticals for new, upgrading, and existing water distribution systems. Hence, this chapter presents useful information to those who design or operate in either onsite or urban STPs regarding pharmaceuticals in wastewater.

REFERENCES

Al Qarni H., Collier P., O'Keeffe J. and Akunna J. (2016a). Investigating the removal of some pharmaceutical compounds in hospital wastewater treatment plants operating in Saudi Arabia. *Environmental Science and Pollution Research International*, **23**, 13003–13014. https://doi.org/10.1007/s11356-016-6389-7

Al Qarni H., Collier P., O'Keeffe J. and Akunna J. (2016b). Investigating the removal of some pharmaceutical compounds in hospital wastewater treatment plants operating in Saudi Arabia. *Environmental Science and Pollution Research International*, **23**, 13003–13014. https://doi.org/10.1007/s11356-016-6389-7

Alvarez P., Beltrán F., Masa F. and Pocostales P. (2009). A comparison between catalytic ozonation and activated carbon adsorption/ozone-regeneration processes for wastewater treatment. Applied Catalysis B, Environmental, **92**, 393–400. https://doi.org/10.1016/j.apcatb.2009.08.019

Alvarino T., Suarez S., Lema J. M. and Omil F. (2014). Understanding the removal mechanisms of PPCPs and the influence of main technological parameters in anaerobic UASB and aerobic CAS reactors. *Journal of Hazardous Materials*, **278**, 506–513. https://doi.org/10.1016/j.jhazmat.2014.06.031

Alvarino T., Komesli O., Suarez S., Lema J. M. and Omil F. (2016a). The potential of the innovative SeMPAC process for enhancing the removal of recalcitrant organic micropollutants. *Journal of Hazardous Materials*, **308**, 29–36. https://doi.org/10.1016/j.jhazmat.2016.01.040

Alvarino T., Suárez S., Garrido M., Lema J. M. and Omil F. (2016b). A UASB reactor coupled to a hybrid aerobic MBR as innovative plant configuration to enhance the removal of organic micropollutants. *Chemosphere*, **144**, 452–458. https://doi.org/10.1016/j.chemosphere.2015.09.016

Alvarino T., Torregrosa N., Omil F., Lema J. M. and Suarez S. (2017). Assessing the feasibility of two hybrid MBR systems using PAC for removing macro and micropollutants. *Journal of Environmental Management*, **203**, 831–837. https://doi.org/10.1016/j.jenvman.2016.03.023

Andersen H. R., Chhetri R. K., Hansen K. M. S., Christensson M., Sundmark K., Sund C., Escola M., Bester K., Møller T., Litty K., Lindholst S., Thoft A. and Kragelund L. C. (2013). Staged MBBR optimised for pharmaceutical biodegradation and ozonation of hospital wastewater. Danish Technological Institute. Conference Paper.

Anderson J. C., Carlson J. C., Low J. E., Challis J. K., Wong C. S., Knapp C. W. and Hanson M. L. (2013). Performance of a constructed wetland in Grand Marais, Manitoba, Canada: removal of nutrients, pharmaceuticals, and antibiotic resistance genes from municipal wastewater. *Chemistry Central Journal*, **7**, 54. https://doi.org/10.1186/1752-153X-7-54

Ashfaq M., Li Y., Wang Y., Chen W., Wang H., Chen X., Wu W., Huang Z., Yu C.-P. and Sun Q. (2017). Occurrence, fate, and mass balance of different classes of pharmaceuticals and personal care products in an anaerobic-anoxic-oxic wastewater treatment plant in Xiamen, China. *Water Research*, **123**, 655–667. https://doi.org/10.1016/j.watres.2017.07.014

Ávila C., Reyes C., Bayona J. M. and García J. (2013). Emerging organic contaminant removal depending on primary treatment and operational strategy in horizontal subsurface flow constructed wetlands: influence of redox. *Water Research*, **47**, 315–325. https://doi.org/10.1016/j.watres.2012.10.005

Beier S., Köster S., Veltmann K., Schröder Fr H. and Pinnekamp J. (2010). Treatment of hospital wastewater effluent by nanofiltration and reverse osmosis. *Water Science and Technology*, **61**(1), 1691–1698. https://doi.org/10.2166/wst.2010.119

Beiginejad H., Amani A., Nematollahi D. and Khazalpour S. (2015). Thermodynamic study of the electrochemical oxidation of some aminophenol derivatives: experimental and theoretical investigation. *Electrochimica Acta*, **154**, 235–243. https://doi.org/10.1016/j.electacta.2014.12.042

Benvenuto J. A., Connor T. H., Monteith D. K., Laidlaw J. L., Adams S. C., Matney T. S. and Theiss J. C. (1993). Degradation and inactivation of antitumor drugs. *Journal of Pharmaceutical Sciences*, **82**, 988–991. https://doi.org/10.1002/jps.2600821007

Carballa M., Omil F. and Lema J. M. (2005). Removal of cosmetic ingredients and pharmaceuticals in sewage primary treatment. *Water Research*, **39**, 4790–4796. https://doi.org/10.1016/j.watres.2005.09.018

Casas M. E. and Bester K. (2015). Can those organic micro-pollutants that are recalcitrant in activated sludge treatment be removed from wastewater by bio film reactors (slow sand filters)? *Science of the Total Environment*, **506–507**, 315–322. https://doi.org/10.1016/j.scitotenv.2014.10.113

Chen H. and Zhang M. (2013). Effects of advanced treatment systems on the removal of antibiotic resistance genes in wastewater treatment plants from Hangzhou, China. *Environmental Science & Technology. United States*, **47**(15), 8157–8163. doi: 10.1021/es401091y.

Chen J. *et al.* (2016). Removal of antibiotics and antibiotic resistance genes from domestic sewage by constructed wetlands: Effect of flow configuration and plant species, *Science of the Total Environment*, **571**, 974–982. https://doi.org/10.1016/j.scitotenv.2016.07.085

Chen W.-H., Wong Y.-T., Huang T.-H., Chen W.-H. and Lin J.-G. (2019). Removals of pharmaceuticals in municipal wastewater using a staged anaerobic fluidized membrane bioreactor. *International Biodeterioration and Biodegradation*, **140**, 29–36. https://doi.org/10.1016/j.ibiod.2019.03.008

Clara M., Strenn B., Gans O., Martinez E., Kreuzinger N. and Kroiss H. (2005). Removal of selected pharmaceuticals, fragrances and endocrine disrupting compounds in a membrane bioreactor and conventional wastewater treatment plants. *Water Research*, **39**, 4797–4807. https://doi.org/10.1016/j.watres.2005.09.015

Cruz-Morató C., Lucas D., Llorca M., Rodriguez-Mozaz S., Gorga M., Petrovic M., Barceló D., Vicent T., Sarrà M. and Marco-Urrea E. (2014). Hospital wastewater treatment by fungal bioreactor: removal efficiency for pharmaceuticals and endocrine disruptor compounds. *Science of the Total Environment*, **493**, 365–376. https://doi.org/10.1016/j.scitotenv.2014.05.117

Cui B. *et al.* (2009). Evaluating the ecological performance of wetland restoration in the Yellow River Delta, China, *Ecological Engineering*, **35**(7), 1090–1103. https://doi.org/10.1016/j.ecoleng.2009.03.022

De Andrade J. R., Oliveira M. F., Da Silva M. G. C. and Vieira M. G. A. (2018). Adsorption of pharmaceuticals from water and wastewater using nonconventional low-cost materials: a review. *Industrial and Engineering Chemistry Research*, **57**, 3103–3127. https://doi.org/10.1021/acs.iecr.7b05137

Deegan A. M., Shaik B., Nolan K., Urell K., Oelgemöller M., Tobin J. and Morrissey A. (2011). Treatment options for wastewater effluents from pharmaceutical companies. *International Journal of Environmental Science and Technology*, **8**, 649–666. https://doi.org/10.1007/BF03326250

Dhingra A., Ahmed W., Gautam S., Khan N. A. and Ahmed S. (2018). Identification of priority pharmaceutical residues in hospital effluent. *Global Journal of Engineering Science and Research*, **5**, 420–435.

Dires S., Birhanu T. and Ambelu A. (2019). Use of broken brick to enhance the removal of nutrients in subsurface flow constructed wetlands receiving hospital wastewater. *Water Science and Technology*, **79**(1), 156–164. https://doi.org/10.2166/wst.2019.037

Egea-Corbacho A., Gutiérrez Ruiz S. and Quiroga Alonso J. M. (2019). Removal of emerging contaminants from wastewater using nanofiltration for its subsequent reuse: full–scale pilot plant. *Journal of Cleaner Production*, **214**, 514–523. https://doi.org/10.1016/j.jclepro.2018.12.297

García-Mateos F. J., Ruiz-Rosas R., Marqués M. D., Cotoruelo L. M., Rodríguez-Mirasol J. and Cordero T. (2015). Removal of paracetamol on biomass-derived activated carbon: modeling the fixed bed breakthrough curves using batch adsorption experiments. *Chemical Engineering Journal* **279**, 18–30. https://doi.org/10.1016/j.cej.2015.04.144

Gautam S., Ahmed S., Dhingra A. and Fatima Z. (2017). Cost-effective treatment technology for small size sewage treatment plants in India. *Journal of Scientific and Industrial Research*, **76**, 249–254.

Guillossou R., Le Roux J., Mailler R., Vulliet E., Morlay C., Nauleau F., Gasperi J. and Rocher V. (2019). Organic micropollutants in a large wastewater treatment plant: what are the benefits of an advanced treatment by activated carbon adsorption in comparison to conventional treatment? *Chemosphere*, **218**, 1050–1060. https://doi.org/10.1016/j.chemosphere.2018.11.182

Hapeshi E., Achilleos A., Vasquez M. I., Michael C., Xekoukoulotakis N. P., Mantzavinos D. and Kassinos D. (2010). Drugs degrading photocatalytically: kinetics and mechanisms of ofloxacin and atenolol removal on titania suspensions. *Water Research*, **44**, 1737–1746. https://doi.org/10.1016/j.watres.2009.11.044

He X., Pelaez M., Westrick J. A., O'Shea K. E., Hiskia A., Triantis T., Kaloudis T., Stefan M. I., de la Cruz A. A. and Dionysiou D. D. (2012). Efficient removal of microcystin-LR by UV-C/H_2O_2 in synthetic and natural water samples. *Water Research*, **46**, 1501–1510. https://doi.org/10.1016/j.watres.2011.11.009

Hirose J., Kondo F., Nakano T., Kobayashi T., Hiro N., Ando Y., Takenaka H. and Sano K. (2005). Inactivation of antineoplastics in clinical wastewater by electrolysis. *Chemosphere*, **60**, 1018–1024. https://doi.org/10.1016/j.chemosphere.2005.01.024

Ileri R., Sengil I. A., Kulac S. and Damar Y. (2003). Treatment of mixed pharmaceutical industry and domestic wastewater by sequencing batch reactor. Journal of Environmental Science and Health. Part A, ToxicHazardous Substances and Environmental Engineering, **38**, 2101–2111. https://doi.org/10.1081/Ese-120023336

Jafari J., Mesdaghinia A., Nabizadeh R., Farrokhi M. and Mahvi A. H. (2013). Investigation of anaerobic fluidized bed reactor/aerobic moving bed bioreactor (AFBR/MMBR) system for treatment of currant wastewater. *Iranian Journal of Public Health*, **42**, 860–867.

Jelic A., Gros M., Ginebreda A., Cespedes-Sánchez R., Ventura F., Petrovic M. and Barcelo D. (2011). Occurrence, partition and removal of pharmaceuticals in sewage water and sludge during wastewater treatment. *Water Research*, **45**, 1165–1176. https://doi.org/10.1016/j.watres.2010.11.010

Jung Y. J., Kim W. G., Yoon Y., Kang J. W., Hong Y. M. and Kim H. W. (2012). Removal of amoxicillin by UV and UV/H2O2processes. *Science of the Total Environment*, **420**, 160–167. https://doi.org/10.1016/j.scitotenv.2011.12.011

Kasprzyk-Hordern B., Dinsdale R. M. and Guwy A. J. (2009). The removal of pharmaceuticals, personal care products, endocrine disruptors and illicit drugs during wastewater treatment and its impact on the quality of receiving waters. *Water Research*, **43**, 363–380. https://doi.org/10.1016/j.watres.2008.10.047

Khan N. A., Khan S. U., Ahmed S., Farooqi I. H., Yousefi M., Mohammadi A. A. and Changani F. (2019). Recent trends in disposal and treatment technologies of emerging-pollutants – A critical review. *Trends in Analytic Chemistry*, **122**, 1–15. https://doi.org/10.1016/j.trac.2019.115744

Khan A. H., Khan N. A., Ahmed S., Dhingra A., Singh C. P., Khan S. U., Mohammadi A. A., Changani F., Yousefi M., Shamshad A., Vambol S., Vambol V., Khursheed A. and Ali I. (2020a). Application of advanced oxidation processes followed by different treatment technologies for hospital wastewater treatment. *Journal of Cleaner Production*, **269**, 1–14. https://doi.org/10.1016/j.jclepro.2020.122411

Khan N. A., Ahmed S., Farooqi I. H., Ali I., Vambol V., Changani F., Yousefi M., Vambol S., Khan S. U. and Khan A. H. (2020b). Occurrence, sources and conventional treatment techniques for various antibiotics present in hospital wastewaters: a critical review. *Trends in Analytic Chemistry*, **129**, 1–15. https://doi.org/10.1016/j.trac.2020.115921

Khan N. A., El Morabet R., Khan R. A., Ahmed S., Dhingra A., Alsubih M. and Khan A. R. (2020c). Horizontal sub surface flow constructed wetlands coupled with tubesettler for hospital wastewater treatment. *Journal of Environmental Management*, **267**, 110627. https://doi.org/10.1016/j.jenvman.2020.110627

Kolpin D. W., Furlong E. T., Meyer M. T., Thurman E. M., Zaugg S. D., Barber L. B. and Buxton H. T. (2002). Pharmaceuticals, hormones, and other organic wastewater contaminants in U.S. streams, 1999–2000: a national reconnaissance. *Environmental Science and Technology*, **36**, 1202–1211. https://doi.org/10.1021/es011055j

Kovalova L., Siegrist H., Singer H., Wittmer A. and McArdell C. S. (2012a). Hospital wastewater treatment by membrane bioreactor: performance and efficiency for organic micropollutant elimination. *Environmental Science and Technology*, **46**, 1536–1545. https://doi.org/10.1021/es203495d

Kovalova L., Siegrist H., Singer H., Wittmer A. and McArdell C. S. (2012b). Hospital wastewater treatment by membrane bioreactor: performance and efficiency for organic micropollutant elimination. *Environmental Science and Technology*, **46**, 1536–1545. https://doi.org/10.1021/es203495d

Kovalova L., Siegrist H., Gunten U., Von, Eugster J., Hagenbuch M., Wittmer A., Moser R. and Mcardell C. S. (2013). Elimination of micropollutants during post-treatment of hospital wastewater with powdered activated carbon, ozone and UV. *Environmental Science and Technology*, **47**, 7899–7908. https://doi.org/10.1021/es400708w

Krzeminski P., Gil J. A., van Nieuwenhuijzen A. F., van der Graaf J. H. J. M. and van Lier J. B. (2012). Flat sheet or hollow fibre – comparison of full-scale membrane bio-reactor configurations. *Desalination and Water Treatment*, **42**, 100–106. https://doi.org/10.1080/19443994.2012.682963

Langenhoff A., Inderfurth N., Veuskens T., Schraa G., Blokland M., Kujawa-Roeleveld K. and Rijnaarts H. (2013). Microbial removal of the pharmaceutical compounds ibuprofen and diclofenac from wastewater. *BioMed Research International*, **2013**, 1–9. https://doi.org/10.1155/2013/325806

Larsen T. A., Lienert J., Joss A. and Siegrist H. (2004). How to avoid pharmaceuticals in the aquatic environment. *Journal of Biotechnology*, **113**, 295–304. https://doi.org/10.1016/j.jbiotec.2004.03.033

Lazarova V. and Manem J. (1994). Advances in biofilm aerobic reactors ensuring effective biofilm activity control. *Water Science and Technology*, **29**(10–11), 319–327. https://doi.org/10.2166/wst.1994.0775

Li Y., Zhu G., Ng W. J. and Tan S. K. (2014). A review on removing pharmaceutical contaminants from wastewater by constructed wetlands: design, performance and mechanism. *Science of the Total Environment*, **468–469**, 908–932. https://doi.org/10.1016/j.scitotenv.2013.09.018

Li W., Wang J., He G., Yu L., Noor N., Sun Y., Zhou X., Hu J. and Parkin I. P. (2017). Enhanced adsorption capacity of ultralong hydrogen titanate nanobelts for antibiotics. *Journal of Materials Chemistry*, **5**, 4352–4358. https://doi.org/10.1039/C6TA09116D

Li H., Cai Y., Gu Z., Yang Y. L., Zhang S., Yang X. L. and Song H. L. (2020). Accumulation of sulfonamide resistance genes and bacterial community function prediction in microbial fuel cell-constructed wetland treating pharmaceutical wastewater. *Chemosphere*, **248**, 126014. https://doi.org/10.1016/j.chemosphere.2020.126014

Lin H., Peng W., Zhang M., Chen J., Hong H. and Zhang Y. (2013). A review on anaerobic membrane bioreactors: applications, membrane fouling and future perspectives. *Desalination*, **314**, 169–188. https://doi.org/10.1016/j.desal.2013.01.019

Liu Y. and Wang J. (2013). Degradation of sulfamethazine by gamma irradiation in the presence of hydrogen peroxide. *Journal of Hazardous Materials*, **250–251**, 99–105. https://doi.org/10.1016/j.jhazmat.2013.01.050

Liu Z. H., Kanjo Y. and Mizutani S. (2009). Removal mechanisms for endocrine disrupting compounds (EDCs) in wastewater treatment – physical means, biodegradation, and chemical advanced oxidation: a review. *Science of the Total Environment*, **407**, 731–748. https://doi.org/10.1016/j.scitotenv.2008.08.039

Liu F. F., Zhao J., Wang S., Du P. and Xing B. (2014). Effects of solution chemistry on adsorption of selected pharmaceuticals and personal care products (PPCPs) by graphenes and carbon nanotubes. *Environmental Science and Technology*, **48**, 13197–13206. https://doi.org/10.1021/es5034684

Luo Y., Guo W., Ngo H. H., Nghiem L. D., Hai F. I., Zhang J., Liang S. and Wang X. C. (2014). A review on the occurrence of micropollutants in the aquatic environment and their fate and removal during wastewater treatment. *Science of the Total Environment*, **473–474**, 619–641. https://doi.org/10.1016/j.scitotenv.2013.12.065

Mailler R., Gasperi J., Coquet Y., Deshayes S., Zedek S., Cren-Oliv C., Cartiser N., Eudes V., Bressy A., Caupos E., Moilleron R., Chebbo G. and Rocher V. (2014). Study of a large scale powdered activated carbon pilot: removals of a wide range of emerging and priority micropollutants from wastewater treatment plant effluents. *Water Research*, **72**, 315–330. https://doi.org/10.1016/j.watres.2014.10.047

Malakootian M., Kannan K., Gharaghani M. A., Dehdarirad A., Nasiri A., Shahamat Y. D. and Mahdizadeh H. (2019). Removal of metronidazole from wastewater by Fe/charcoal micro electrolysis fluidized bed reactor. *Journal of Environmental Chemical Engineering*, **7**, 1–10. https://doi.org/10.1016/j.jece.2019.103457

Malmborg J. and Magnér J. (2015). Pharmaceutical residues in sewage sludge: effect of sanitization and anaerobic digestion. *Journal of Environmental Management*, **153**, 1–10. https://doi.org/10.1016/j.jenvman.2015.01.041

Matamoros V., Caselles-Osorio A., García J. and Bayona J. M. (2008). Behaviour of pharmaceutical products and biodegradation intermediates in horizontal subsurface flow constructed wetland. A microcosm experiment. *Science of the Total Environment*. **394**, 171–176. https://doi.org/10.1016/j.scitotenv.2008.01.029

Mutiyar P. K. and Mittal A. K. (2014). Occurrences and fate of selected human antibiotics in influents and effluents of sewage treatment plant and effluent-receiving river Yamuna in Delhi (India). *Environmental Monitoring and Assessment*, **186**, 541–557. https://doi.org/10.1007/s10661-013-3398-6

Nahm S. W., Shim W. G., Park Y.-K. and Kim S. C. (2012). Thermal and chemical regeneration of spent activated carbon and its adsorption property for toluene. *Chemical Engineering Journal*, **210**, 500–509. https://doi.org/10.1016/j.cej.2012.09.023

Nguyen Dang Giang C., Sebesvari Z., Renaud F., Rosendahl I., Hoang Minh Q. and Amelung W. (2015). Occurrence and dissipation of the antibiotics sulfamethoxazole, sulfadiazine, trimethoprim, and enrofloxacin in the Mekong Delta, Vietnam. *PLoS One*, **10**, 1–24. https://doi.org/10.1371/journal.pone.0131855

Nguyen T. T., Bui X. T., Vo T. D. H., Nguyen D. D., Nguyen P. D., Do H. L. C., Ngo H. H. and Guo W. (2016). Performance and membrane fouling of two types of laboratory-scale submerged membrane bioreactors for hospital wastewater treatment at low flux condition. *Separation and Purification Technology*, **165**, 123–129. https://doi.org/10.1016/j.seppur.2016.03.051

Nielsen U., Hastrup C., Klausen M. M., Pedersen B. M., Kristensen G. H., Jansen J. L. C., Bak S. N. and Tuerk J. (2013). Removal of APIs and bacteria from hospital wastewater by MBR plus O_3, $O_3 + H_2O_2$, PAC or ClO_2. *Water Science and Technology*, **67**(4), 854–862. https://doi.org/10.2166/wst.2012.645

Nnaji C. C. (2014). A review of the upflow anaerobic sludge blanket reactor. *Desalination and Water Treatment*, **52**, 4122–4143. https://doi.org/10.1080/19443994.2013.800809

Onesios K. M., Yu J. T. and Bouwer E. J. (2009). Biodegradation and removal of pharmaceuticals and personal care products in treatment systems: a review. *Biodegradation*, **20**, 441–466. https://doi.org/10.1007/s10532-008-9237-8

Özengin N. and Elmaci A. (2016). Removal of pharmaceutical products in a constructed wetland. *Iranian Journal of Biotechnology*, **14**, 221–229. https://doi.org/10.15171/ijb.1223

Pal P. (2018). Treatment and disposal of pharmaceutical wastewater: toward the sustainable strategy. *Separation and Purification Reviews*, **47**, 179–198. https://doi.org/10.1080/15422119.2017.1354888

Pal A., Gin K. Y. H., Lin A. Y. C. and Reinhard M. (2010). Impacts of emerging organic contaminants on freshwater resources: review of recent occurrences, sources, fate and effects. *Science of the Total Environment*, **408**, 6062–6069. https://doi.org/10.1016/j.scitotenv.2010.09.026

Park J., Yamashita N., Park C., Shimono T., Takeuchi D. M. and Tanaka H. (2017). Removal characteristics of pharmaceuticals and personal care products: comparison between membrane bioreactor and various biological treatment processes. *Chemosphere*, **179**, 347–358. https://doi.org/10.1016/j.chemosphere.2017.03.135

Park J., Kim C., Hong Y., Lee W., Chung H., Jeong D. H. and Kim H. (2020). Distribution and removal of pharmaceuticals in liquid and solid phases in the unit processes of sewage treatment plants. *International Journal of Environmental Research and Public Health*, **17**, 687. https://doi.org/10.3390/ijerph17030687

Paulus G. K., Hornstra L. M., Alygizakis N., Slobodnik J., Thomaidis N. and Medema G. (2019). The impact of on-site hospital wastewater treatment on the downstream communal wastewater system in terms of antibiotics and antibiotic resistance genes. *International Journal of Hygiene and Environmental Health*, **222**, 635–644. https://doi.org/10.1016/j.ijheh.2019.01.004

Peng X., Wang Z., Kuang W., Tan J. and Li K. (2006). A preliminary study on the occurrence and behavior of sulfonamides, ofloxacin and chloramphenicol antimicrobials in wastewaters of two sewage treatment plants in Guangzhou, China. *Science of the Total Environment*, **371**, 314–322. https://doi.org/10.1016/j.scitotenv.2006.07.001

Polesel F., Andersen H. R., Trapp S. and Plósz B. G. (2016). Removal of antibiotics in biological wastewater treatment systems – A critical assessment using the activated sludge modeling framework for xenobiotics (ASM-X). *Environmental Science and Technology*, **50**, 10316–10334. https://doi.org/10.1021/acs.est.6b01899

Prasertkulsak S., Chiemchaisri C., Chiemchaisri W., Itonaga T. and Yamamoto K. (2016). Removals of pharmaceutical compounds from hospital wastewater in membrane bioreactor operated under short hydraulic retention time. *Chemosphere*, **150**, 624–631. https://doi.org/10.1016/j.chemosphere.2016.01.031

Qi Y., He S., Wu S., Dai B. and Hu C. (2015). Utilization of micro-electrolysis, up-flow anaerobic sludge bed, anoxic/oxic-activated sludge process, and biological aerated filter in penicillin G wastewater treatment. *Desalination and Water Treatment*, **55**, 1480–1487. https://doi.org/10.1080/19443994.2014.926259

Qin Z., Liu S., Liang S., Kang Q., Wang J. and Zhao C. (2016). Advanced treatment of pharmaceutical wastewater with combined micro-electrolysis, Fenton oxidation, and coagulation sedimentation method. *Desalination and Water Treatment*, **57**, 25, 369–25,378. https://doi.org/10.1080/19443994.2016.1155174

Quinn B., Gagné F. and Blaise C. (2008). An investigation into the acute and chronic toxicity of eleven pharmaceuticals (and their solvents) found in wastewater effluent on the cnidarian, Hydra attenuata. *Science of the Total Environment*, **389**, 306–314. https://doi.org/10.1016/j.scitotenv.2007.08.038

Radjenović J., Petrović M. and Barceló D. (2009). Fate and distribution of pharmaceuticals in wastewater and sewage sludge of the conventional activated sludge (CAS) and advanced membrane bioreactor (MBR) treatment. *Water Research*, **43**, 831–841. https://doi.org/10.1016/j.watres.2008.11.043

Rodgers M. and Zhan X.-M. (2003). Moving-medium biofilm reactors. *Reviews in Environmental Science and Biotechnology*, **2**, 213–224. https://doi.org/10.1023/B:RESB.0000040467.78748.1e

Rogers H. R. (1996). Sources, behaviour and fate of organic contaminants during sewage treatment and in sewage sludges. *Science of the Total Environment*, **185**, 3–26. https://doi.org/10.1016/0048-9697(96)05039-5

Sipma J., Osuna B., Collado N., Monclús H., Ferrero G., Comas J. and Rodriguez-Roda I. (2010). Comparison of removal of pharmaceuticals in MBR and activated sludge systems. *Desalination*, **250**, 653–659. https://doi.org/10.1016/j.desal.2009.06.073

Suarez S., Lema J. and Omil F. (2009). Pre-treatment of hospital wastewater by coagulation-flocculation and flotation. *Bioresource Technology*, **100**, 2138–2146. https://doi.org/10.1016/j.biortech.2008.11.015

Sui Q., Cao X., Lu S., Zhao W., Qiu Z. and Yu G. (2015). Occurrence, sources and fate of pharmaceuticals and personal care products in the groundwater: a review. *Emerging Contaminants*, **1**, 14–24. https://doi.org/10.1016/j.emcon.2015.07.001

Sulaiman S., Khamis M., Nir S., Lelario F., Scrano L., Bufo S. A., Mecca G. and Karaman R. (2015). Stability and removal of atorvastatin, rosuvastatin and simvastatin from wastewater. *Environmental Technology*, **36**, 3232–3242. https://doi.org/10.1080/09593330.2015.1058422

Tran N. H., Chen H., Reinhard M., Mao F. and Gin K. Y. H. (2016). Occurrence and removal of multiple classes of antibiotics and antimicrobial agents in biological wastewater treatment processes. *Water Research*, **104**, 461–472. https://doi.org/10.1016/j.watres.2016.08.040

Verlicchi P., Galletti A., Petrovic M. and Barceló D. (2010). Hospital effluents as a source of emerging pollutants: an overview of micropollutants and sustainable treatment options. *Journal of Hydrology*, **389**, 416–428. https://doi.org/10.1016/j.jhydrol.2010.06.005

Verlicchi P., Al Aukidy M., Galletti A., Petrovic M. and Barceló D. (2012a). Hospital effluent: investigation of the concentrations and distribution of pharmaceuticals and environmental risk assessment. *Science of the Total Environment*, **430**, 109–118. https://doi.org/10.1016/j.scitotenv.2012.04.055

Verlicchi P., Al Aukidy M. and Zambello E. (2012b). Occurrence of pharmaceutical compounds in urban wastewater: removal, mass load and environmental risk after a secondary treatment – A review. *Science of the Total Environment*, **429**, 123–155. https://doi.org/10.1016/j.scitotenv.2012.04.028

Verlicchi P., Al Aukidy M. and Zambello E. (2015). What have we learned from worldwide experiences on the management and treatment of hospital effluent? – An overview and a discussion on perspectives. *Science of the Total Environment*, **514**, 467–491. https://doi.org/10.1016/j.scitotenv.2015.02.020

Vo T. K. Q., Bui X. T., Chen S. S., Nguyen P. D., Cao N. D. T., Vo T. D. H., Nguyen T. T. and Nguyen T. B. (2019). Hospital wastewater treatment by sponge membrane bioreactor coupled with ozonation process. *Chemosphere*, **230**, 377–383. https://doi.org/10.1016/j.chemosphere.2019.05.009

von Sperling M. (2007). Biological Wastewater Treatment – Activated Sludge and Aerobic Biofilm Reactors. IWA Publishing, UK.

Vymazal J., Dvořáková Březinová T., Koželuh M. and Kule L. (2017). Occurrence and removal of pharmaceuticals in four full-scale constructed wetlands in the Czech Republic – the first year of monitoring. *Ecological Engineering*, **98**, 354–364. https://doi.org/10.1016/j.ecoleng.2016.08.010

Wang Y., Ma J., Zhu J., Ye N., Zhang X. and Huang H. (2016). Multi-walled carbon nanotubes with selected properties for dynamic filtration of pharmaceuticals and personal care products. *Water Research*, **92**, 104–112. https://doi.org/10.1016/j.watres.2016.01.038

Warsinger D. M., Swaminathan J., Guillen-Burrieza E., Arafat H. A. and Lienhard V. J. H. (2015). Scaling and fouling in membrane distillation for desalination applications: a review. *Desalination*, **356**, 294–313. https://doi.org/10.1016/j.desal.2014.06.031

Wei *et al.* (2012). Functional cloning and characterization of antibiotic resistance genes from the chicken gut microbiome, Applied and environmental microbiology. 2012/01/27. *American Society for Microbiology*, **78**(8), 3028–3032. doi: 10.1128/AEM.06920-11.

WHO (2014). Safe Management of Wastes from Health-care Activities. World Health Organization, Geneva Switzerland.

Xu W., Zhang G., Li X., Zou S., Li P., Hu Z. and Li J. (2007). Occurrence and elimination of antibiotics at four sewage treatment plants in the Pearl River Delta (PRD), South China. *Water Research*, **41**, 4526–4534. https://doi.org/10.1016/j.watres.2007.06.023

Xu J. *et al.* (2015). Occurrence of antibiotics and antibiotic resistance genes in a sewage treatment plant and its effluent-receiving river, *Chemosphere*, **119**, 1379–1385. https://doi.org/10.1016/j.chemosphere.2014.02.040

Yang W., Zhou H. and Cicek N. (2014). Treatment of organic micropollutants in water and wastewater by UV-based processes: a literature review. *Critical Reviews in Environmental Science and Technology*, **44**, 1443–1476. https://doi.org/10.1080/10643389.2013.790745

Yi Q., Zhang Y., Gao Y., Tian Z. and Yang M. (2017). Anaerobic treatment of antibiotic production wastewater pretreated with enhanced hydrolysis: simultaneous reduction of COD and ARGs. *Water Research*, **110**, 211–217. https://doi.org/10.1016/j.watres.2016.12.020

Yin F., Wang D., Li Z., Ohlsen T., Hartwig P. and Czekalla S., 2015. Study on anaerobic digestion treatment of hazardous colistin sulphate contained pharmaceutical sludge. *Bioresource Technology*, **177**, 188–193. https://doi.org/10.1016/j.biortech.2014.11.091

Yu W., Xu L., Qu J. and Graham N., 2014. Investigation of pre-coagulation and powder activate carbon adsorption on ultrafiltration membrane fouling. *Journal of Membrane Science and Technology*, **459**, 157–168. https://doi.org/10.1016/j.memsci.2014.02.005

Zwiener C. and Frimmel F. H. 2003. Short-term tests with a pilot sewage plant and biofilm reactors for the biological degradation of the pharmaceutical compounds clofibric acid, ibuprofen, and diclofenac. *Science of the Total Environment*, **309**, 201–211. https://doi.org/10.1016/S0048-9697(03)00002-0

Chapter 12

Agricultural water reuse

Ahmed Kettab[1] *and Hanane Bouanani*[2]

[1]*International Consultant-Expert, Associate Professor University of Bouira – Algeria; Professor/Research Director of the National Polytechnic School – Algiers, Member of French Water Academy*

[2]*Engineer, PhD student at the National Polytechnic School of Algiers*

ABSTRACT

The selection of the ideal destination for the disposal or reuse of a wastewater treatment by-product requires careful evaluation of the water or agricultural amendments and of the health and environmental risks. Consequently, the characterization of the treated water and its by-products constitutes a pillar for the selection of the optimal management option and for the estimation of the tolerance of the receiving environment is supported by legislation established by the standardization of national and international institutions. Conventional release or reuse limits cover a broad spectrum of parameters that are subject to continual updating due to the continued emergence of micropollutants and other contaminants with cumulative effect.

Keywords: Contaminants, Conventional, Health and environmental risks, Ideal destination, Legislation.

Contents

12.1 Introduction 306
12.2 Water Resources in Algeria and Worldwide 307
12.3 Treated Wastewater as a Complementary Resource 310

 Pharmaceutical Wastewater Treatment Technologies: Concepts and Implementation Strategies
Editor(s): Nadeem A Khan, Sirajuddin Ahmed, Viola Vambol and Sergij Vambol
doi: 10.2166/9781789061338_0305

12.4 Reuse Destinations of Treated Wastewater 311
12.4.1 Uses in agriculture 311
12.4.2 Urban and peri-urban use 311
12.4.3 Industrial use 312
12.4.4 Statistics on TWWR in Algeria 312
12.4.5 Agricultural reuse of treated wastewater 312
12.4.5.1 Advantages 314
12.4.5.2 Algerian situation 315
12.4.5.3 Risks and impacts 316
12.5 Agricultural Manuring Using Dehydrated Sludge 320
12.5.1 Human contamination 321
12.5.2 Environmental risks 321
12.5.3 Agricultural development 321
12.6 Conclusion 322
References 322

12.1 INTRODUCTION

In Algeria, the constraints of the climate, demographic growth, economic and social transformations are at the origin of an ever-increasing demand for water. At the same time, irrigation has become a necessity, given the importance of the climatic water deficit and the intensification of agriculture. However, faced with these demands, water resources are scarce and insufficient. Thus, in order to meet all of the country's water needs and to reserve good quality water for drinking water supply, one solution would be to use marginal quality water in agriculture. Water resources in Algeria are limited, vulnerable and unevenly distributed (Bixio *et al.*, 2005).

For a population of 38 million inhabitants, renewable water resources are 550 m^3/year per inhabitant. This average is very low compared to the Middle East and North Africa, which is 1250 m^3 or the world's average individual potential, which is 7500 m^3. The limit for water scarcity is 1000 m^3/year/inhabitant, therefore, Algeria is a country where water is scarce. In these water deficit conditions, the agriculture sector is the largest demand for water. In this context, the use of treated wastewater has become a necessity and it must be an integral part of the strategy for mobilizing all available resources. In Algeria, water consumption has increased steadily over the past decade (Kettab, 2001).

It also leads to the development of water pollution, which sequentially can infest the environment and alter groundwater. The polluted water discharged is around 900 million cubic meters, or around 20% of the surface water stored in dams. Algeria suffers from a chronic lack of water and all solutions are being considered to limit the consequences. Thus, a policy for the mobilization and use of conventional water (underground and surface) as well as unconventional water (desalination of seawater, saline water, wastewater) has recently been implemented (Tose, 2006). As part of treated wastewater management, a legal

Table 12.1 Treated wastewater reuse perspectives in Algeria in 2006 (Karef *et al.*, 2013; Kettab *et al.*, 2017).

	2010	2015	2020	2025	2030
Intercepted wastewater volumes (Hm^3/year)	501	708	972	1208	1271
Treated wastewater volumes (Hm^3/year)	257	399	561	705	762
Potential irrigable surface without storage (ha)	21,773	34,485	48,279	59,876	64,431
Potential irrigable surface without 100% storage (ha)	36,288	57,476	80,466	99,794	107,385

frame to protect users and managers has been put in place for better protection of the aquatic environment. The treated water must meet certain standards; which are given in Table 12.1 with regard to compliance with the standards for the use of treated water and the risks of contamination by these treated waters. Similarly, the conditions for using this water have been clearly established (Tables 12.2 and 12.3) (Tose, 2006).

12.2 WATER RESOURCES IN ALGERIA AND WORLDWIDE

Whether used for human consumption, agriculture, recreation, transportation or energy production, water has an undeniable value. The water cycle that used to be natural and continuous has since been altered by human activity. Increased population growth, accelerated urbanization and industrialization, as well as the

Table 12.2 Algerian irrigation norms (Kettab *et al.*, 2017).

Parameter	Unit	Regulatory Limits
Temperature	°C	<30
pH	—	6.5–8.5
Dissolved oxygen	mg O_2/L	>5
Total suspended solids	mg/L	<30
Biochemical oxygen demand	mg/L	<40
Chemical oxygen demand	mg/L	<90
Total nitrogen	mg/L	<50
Phosphorous	(PO_4) mg/L	<02
Grease	mg/L	<20
Fecal coliforms	CFU/100 mL	<1000 CFU/100 mL

Source: ANRH, Algiers.

Table 12.3 Algerian treated wastewater reuse standards (Karef *et al.*, 2013; Kettab, 2001; Kettab *et al.*, 2017).

Parameters	Unit	Standard Limits		
		FAO (1985)	**WHO (1989)**	**JORA (2012)**
		6.5–8.4		6.5–8.5
CE	ds/m	<0.7 No restriction 0.7–3.0 medium restriction >3.0 High restriction		3
Total suspended solids	mg/L	<30		30
Chemical oxygen demand	mg O_2/L	<40		90
Biochemical oxygen demand	mg O_2/L	<10		30
NO_3^-	mg/L	50		30
NO_2^-	mg/L	<1		Not available
NH_4^+	mg/L	<2		Not available
PO_4^{3-}	mg/L	<0.94		Not available
SAR	meq/L	<3 No restriction 3–9 Medium >9 High restriction		Not available
Total coliforms	CFU/100 mL	Not available		Not available
Streptococci	CFU/100 mL	1000		Not available
Salmonella	CFU/1 L	Absence		Not available

growth of large-scale crops, are constantly increasing the demand for freshwater (Kettab, 2001); 96.5% of the earth's water is salty and only 2.5% of this resource consists of fresh water in addition to the fact that this portion is hardly accessible.

In order to assess water availability properly, we must consider the amount of water available for all possible uses (economic, social and natural). The degradation of water quality also engenders its scarcity, making it unavailable, since polluted water can become partially or entirely unfit for use. Water scarcity around the world has become one of the most important concerns for water organization managers. According to the Kyoto conference in 2003, two billion people would not have access to drinking water in 2015 (Biswas, 1987). Mediterranean countries are among the most affected regions by this problem (Bdour *et al.*, 2009).

According to Figure 12.1, 844 million inhabitants, which equals 11% of the world's population, still have no access to potable water. It is true that this figure is less alarming than the 19% noted in 2000 but there is still a long way to go because at the same time, 50% of the global water supply comes from overexploitation of underground aquifers causing landslides, salt intrusion and rising sea levels. India, alongside the countries of north-eastern Africa, remains the best example by having about twenty million wells (Kettab *et al.*, 2017).

In order to meet the increasing needs linked to urban demography, it is possible to:

- fight against individual or collective waste;
- accelerate the natural cycle of renewal;
- create new resources through desalination and wastewater recycling.

Algeria has limited, irregular and very unevenly distributed natural resources. With the exception of the fossil waters of the Sahara (Terminal Complex and Intermediate Continental), natural hydraulic resources are mainly located in the north of the country. Ninety per cent of all surface flows are found on the littoral region (7% of the area of the territory), the remaining 10% being shared between the high plateaux and Saharan basins (Kettab, 2001).

Algeria's water resources are estimated at 19.2 billion m^3, of which 12.2 billion m^3 are surface water, i.e. 12 billion m^3 are located in the northern regions and 0.2

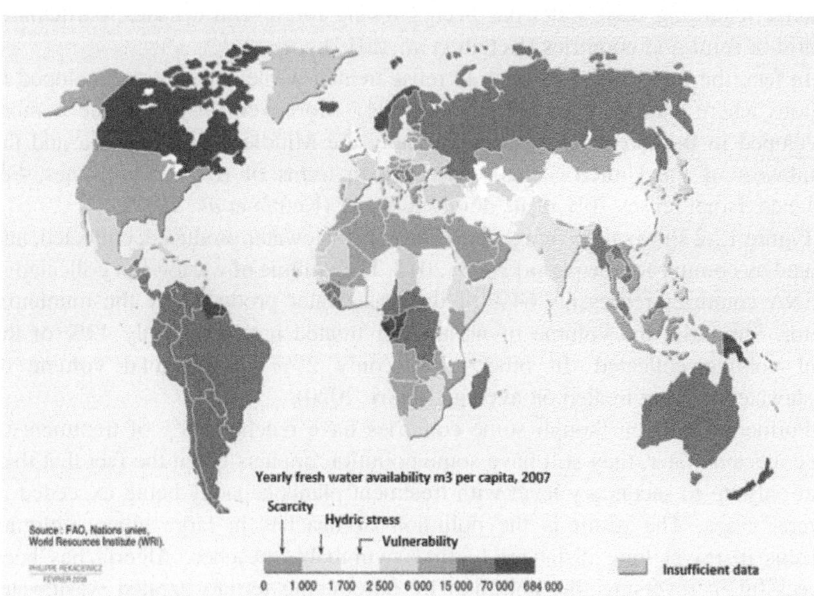

Figure 12.1 Yearly fresh water availability (Kettab, 2001).

billion m^3 for the entire territory of the Algerian Sahara and the country's underground resources are estimated at 7 billion m^3 where the north only represents 2 billion m^3. Actually, all northern underground resources are being mobilized, in addition to 70% of the surface resources (Kettab, 2001; Tose, 2006).

Despite the considerable efforts made to mobilize the country's water resources, Algeria is experiencing a shortage situation in the short term. According to the World Bank, Algeria is ranked among the poorest countries in water potential, below the theoretical threshold of 1000 m^3/inhabitant/year which will evolve to 400 m^3/inhabitant/year by 2020. Water requirements by 2020 will amount to more than 8.3 billion m^3 per year. This currently represents almost double of the volume available (Biswas, 1987).

The pollution of water resources poses a threat to public health, the environment and biological balance. To this end, the purification of wastewater is therefore an imperative necessity if we want to protect our water resources and also to increase them by recycling treated water for use for industrial or agricultural purposes (Kettab, 2001).

12.3 TREATED WASTEWATER AS A COMPLEMENTARY RESOURCE

Water reuse is defined as the use of treated wastewater that meets certain quality criteria according to the desired use. The international advancement of this practice regarding wastewater has been growing for several decades, particularly in arid or semi-arid countries (Kettab *et al.*, 2017).

In fact, the first projects aiming to reuse treated wastewater were developed in regions where water is deficient (Exall, 2004). Moreover, this technique is more developed in the threatened regions, notably the Middle East, Australia and the Southwest of the United States, especially in terms of recycled volumes. For instance, Israel reuses 70% of its domestic water (Kettab *et al.*, 2017).

Figure 12.2 shows the average quantities of wastewater produced, collected, and treated by country for the period 2008–2012. The volume of wastewater collected in MENA countries represents 64% of the wastewater produced by the municipal sector. Similarly, the volume of wastewater treated represents only 43% of the total volume collected. In other terms, only 27% of the total volume of wastewater is being treated on average (Zekri, 2020).

Furthermore, even though some countries have reached 80% of treatment of the collected water, they still have some pollution impacts given the fact that they treat only up to secondary level with treatment plants capacity being exceeded in several cases. The result is the pollution of beaches in large cities, obliging citizens to travel long distances to find swimmable beaches. Algeria has been successful in reversing this pollution by introducing tertiary treated wastewater (TTWW) treatment and improving the quality of seawater for swimming in the capital's beaches (Kettab *et al.*, 2017; Page *et al.*, 2010).

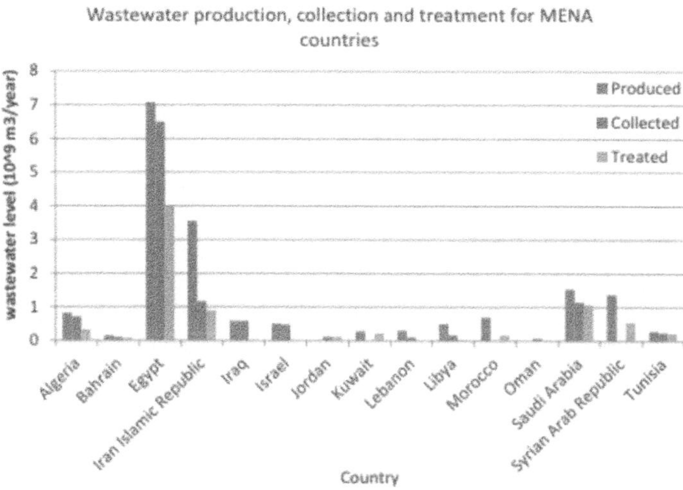

Figure 12.2 Wastewater production, collection, and treatment in MENA. (Source: FAO-AQUASTAT 2018, data for 2008–2012) (Zekri, 2020).

12.4 REUSE DESTINATIONS OF TREATED WASTEWATER

The different types of reuse according to four categories: agricultural, urban, peri-urban uses and groundwater recharge, industrial use and mixed use. Globally, treated wastewater in agriculture, industry and domestic use covers 70% respectively; 20 and 10% of their water demand (Hamoda, 2004).

12.4.1 Uses in agriculture

The majority of wastewater reuse projects are related to agricultural use, irrigation reuse is mainly present in countries known to be agricultural, with low water potentialities. Indeed, this practice is widely used where agriculture encounters quantitative water problems. Irrigation or watering can be done either on crops that are intended for human consumption or on crops that are assigned to the transformation or the use of other sectors. Irrigation using treated wastewater is a cost-effective way to reduce environmental waste and to benefit from supply even during drought events (Anctil, 2008).

12.4.2 Urban and peri-urban use

In urban and peri-urban areas, there can be several uses: irrigation, fire-fighting, air conditioning of buildings, flushing, street washing and supply of decorative fountains. Treated water use rapidly develops and becomes a fundamental element of the integrated water management policy in large cities. To this end, urban reuse is divided according to the applications allowing public open access or restricted access. It is therefore possible to use lower quality water for places where access is controlled (golf courses, restriction of wetlands) (Kettab, 2014). However, uncontrolled areas (parks, schoolyards, etc.) must benefit from better quality due to the increased contacts. For such uses, the standards governing the quality of wastewater are very strict and close to drinking water quality regulations of the WHO in 2006 (Blinda, 2011).

12.4.3 Industrial use

Many industries re-use their own wastewater using a closed cyclic system in order to save resources and decrease their waste. This practice is now a technical and economic reality. According to the WHO, for some countries, recycled water provides 85% of the global industrial requirement. The largest water consuming sectors are thermal and nuclear power plants (cooling water). The quality of reused water is regulated and depends on the type of industrial application or production, the characteristics of the treated wastewater, the treatment system and the required quality criteria (Kettab *et al.*, 2017).

12.4.4 Statistics on TWWR in Algeria

According to the Algerian water resources authorities in 2003, wastewater represents one of the components of the global water potential beside surface and groundwater. A target plan for the production of additional quantities of water for different uses has been planned (Kettab, 2014; Kettab *et al.*, 2017).

In Algeria, the coverage rate of the sewerage network is around 95%. The overall volume of discharged water at the national level was estimated to be 750 million m^3 in 2010, this volume will presumably exceed 1.5 billion m^3 by 2020 (Kettab, 2014).

Upstream discharges from dams indirectly contribute to the recharge of aquifers, treated wastewater that is discharged into seas represented 77% of the total treated wastewater volume in 2004, this resource is considered lost, therefore its recovery must be prioritized (Kettab *et al.*, 2008).

12.4.5 Agricultural reuse of treated wastewater

The agricultural sector is the largest consumer of water resources given the decrease in water supplies observed for several decades. Farmers, particularly those from mainland regions, are interested in the use of wastewater. According to the FAO, the treatment of wastewater and its use in irrigation is an attractive option, especially in arid and semi-arid areas, as it represents a source of additional renewable and low-cost fertilizers and water (Andreadakis, 2003).

According to Figure 12.3, the largest water use occurs in the agriculture sector, primarily for irrigation. It accounts for an estimated 70% of total freshwater withdrawals. In many drier countries, agricultural water use accounts for more than 90% of total withdrawals. At a time when there is a need to increase

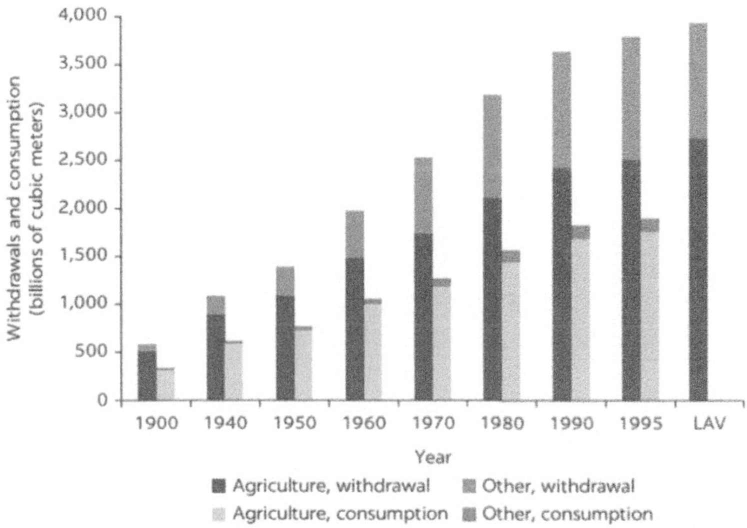

Source: Scheierling and Tréguer 2016a, based on FAO 2016a; Shiklomanov and Rodda 2003.
Note: LAV = latest available value.

Figure 12.3 Global irrigation withdrawals and consumption (Zekri, 2020).

agricultural production with depleting water resources, the focus is directed towards improving agricultural water productivity and efficiency. However, the report finds issues with the methods used for measuring agricultural water productivity and efficiency, and the application of these concepts and methods. Hence, it recommends designing interventions with the local hydrologic, policy, and institutional contexts in mind (Andreadakis, 2003).

The reuse of treated wastewater is a widespread practice in regions of the world affected by shortages of water resources. It is very developed in the United States, where 34 states often developed very demanding regulations or recommendations relating to the agricultural use of wastewater. Mention should also be made of Mexico and several countries in South America, Australia, South Africa, Japan, China and the Persian Gulf countries. Some countries and states (Australia, California, Florida, Israel, Jordan, etc.) aim to satisfy 10–30% of their water demand with this alternative resource over the next 5–10 years (Ecosse, 2001; Kettab *et al.*, 2017).

During the past 10 years, the yearly reuse of wastewater has been oscillating from 10 to 29%, in Europe, USA and China and has reached 41% in Australia. However, only 5% of wastewater treated on a global scale is currently reused, which represents a global volume of around 7.1 billion m^3 per year, or 0.18% of the world water demand. It is obvious that wastewater reuse (with or without treatment) has been promoted worldwide, based on real experiences (Kettab *et al.*, 2017).

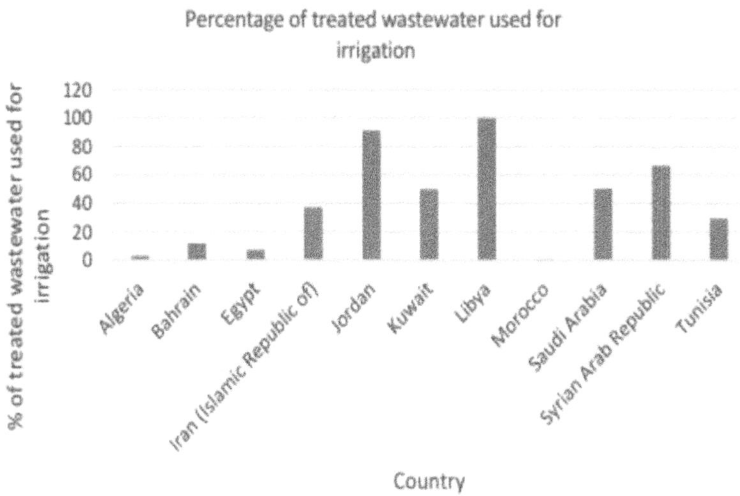

Figure 12.4 Percentage of treated wastewater reused for irrigation in selected MENA countries (FAO and AQUASTAT data for the period 2008–2012) (Zekri, 2020).

Figure 12.4 shows the percentages of the reused volumes of treated wastewater for irrigation purposes in MENA regions. On average, 37% of the treated wastewater is being unevenly explored among countries. Some countries such as Jordan and Libya use 90 and 100%, respectively, of the treated wastewater. However, overall, most of the treated wastewater (63%) is directly released on surface waters (Zekri, 2020).

12.4.5.1 Advantages

Environmental benefits. Reuse makes it possible to mobilize an additional water resource. We should therefore see a decrease in withdrawals (Toze, 2006). Indeed, this practice makes it possible to reduce the use of resources such as surface water and groundwater, and also to avoid their overexploitation. Reuse makes it possible to avoid, or at least to reduce, the quality of direct discharges of wastewater into receiving environments. This practice is generally also accompanied by better control of sanitation, and therefore, less pollutants (Couture, 2006; Kettab *et al.*, 2017).

Social advantages. The treated wastewater can be used at all times and therefore is always available for crop irrigation, even in the event of drought. In addition, this type of reuse yields higher than conventional irrigation with fresh water, even when using artificial fertilizer (Catherine *et al.*, 2008). Thus, the WHO confirms the increase in yields of crops intended for human consumption leads to greater availability of foodstuffs (Kettab, 2014; Kettab *et al.*, 2008).

Economic benefits: Treated wastewater reuse is more cost-effective than other developed techniques to provide fresh water. Treated wastewater production costs less than deep groundwater supply, water importation and desalination (Wu *et al.*, 2018).

12.4.5.2 Algerian situation

According to the Algerian water authorities, Algeria struggles with the scarcity of conventional water resources and can no longer afford to decline the possibility of reusing the enormous quantities of produced wastewater (Kettab *et al.*, 2017; Zekri, 2020). Pluvial waters, dam waters and boreholes are no longer enough to meet water needs. This explains Algeria's current ambition to treat one billion m^3 of wastewater for the irrigation of 100,000 hectares. Algeria, which has a treated water volume of 650,000 m^3, devotes 65% of its water resources to the agricultural sectors (Anctil, 2008; Kettab *et al.*, 2017). In the deficit conditions in water resources, the agricultural sector is one that requires the largest water volumes. In 2006, 900,000 hectares, or 10.5%, of the useful agricultural area was irrigated, and 78% of this area was irrigated with groundwater and 13% with surface water (Kettab *et al.*, 2017).

The extension of irrigated areas in small and medium hydraulic sectors, despite encouragement for the development of agriculture, has caused a considerable rise in individual boreholes and dangerous over-exploitation of certain underground aquifers (Kettab *et al.*, 2017). Large irrigated perimeters (GPI) are monitored under the jurisdiction of the National Office of Irrigation and Drainage (ONID). These perimeters are irrigated by dams and boreholes in the north of the country while in the south, irrigations are ensured from deep boreholes in the Albian aquifers. Their area is around 200,000 hectares, crops grown in GPIs in 2008, arboriculture 64.6%, market gardening 28.5%, industrial crops 6.1% and the rest in cereals and fodder from small perimeters and areas of irrigation (private production) (Kettab *et al.*, 2017).

Besides, the reuse of purified wastewater for irrigation must primarily concern areas lacking conventional water. The reuse of treated wastewater in agriculture has become a reality, 8% of irrigated land, especially in small and medium hydraulic sectors, receives untreated wastewater (Catherine *et al.*, 2008). This practice continues to develop downstream of urban centres, large and small agglomerations.

As a result, great efforts have to be deployed at all levels, technical, institutional and regulatory, to improve the level of use with the minimum of risk. The confrontation of water resources needs by 2013 showed a significant deficit which will be filled by the introduction of treated wastewater from agricultural areas (Kettab *et al.*, 2017). Among the treatment plants operated by the National Board for Sanitation (ONA) across the country, some are concerned with the reuse of treated wastewater in agriculture. At the end of 2011, the volume reused to irrigate more than 10,000 hectares of agricultural area was estimated at 17 million m^3/year (Anctil, 2008; Kettab *et al.*, 2017).

The potential for reusing treated wastewater for agricultural purposes will increase significantly from around 17 million m^3 in 2011 to around 250 million m^3 in 2014. Projects that have been led by the Algerian Ministry of Water Resources show that a volume of 850 million m^3 will be reused in agricultural irrigation towards the end of 2015 (Kettab *et al.*, 2017). The reuse of treated wastewater, particularly for agricultural purposes, has become one of the main axes of the strategy for the water resources sector in Algeria (Table 12.1) (Kettab, 2001; Kettab *et al.*, 2017). These include the El-Karma station in the wilaya of Oran, some of the treated wastewater is used in agricultural irrigation in the Macta plain, covering an area of 8100 hectares, while the perimeter of Hannaya, extending over 912 hectares, is irrigated from the purified water of the treatment plant of Ain Hout, Chétouane in Tlemcen. In Constantine, the irrigation perimeter from the WWTP of Hamma Bouziane covers an area of 327 hectares, while the perimeter of Dahmouni is 1214 hectares in the wilaya of Tiaret. The irrigation perimeter, from the WWTP of Bordj Bou Areridj, is around 350 hectares (Kettab *et al.*, 2008; Ramya *et al.*, 2019).

There are two pilot research projects undertaken in collaboration with UNESCO and the European Union on the reuse of treated water in agriculture. There is also the draft decree regulating the reuse of treated wastewater for agricultural purposes, the general study on the reuse of treated wastewater at the national level which will establish a master plan for the reuse of treated wastewater for agricultural and industrial purposes, and to define the volumes and parameters of each reuse. This is established through a feasibility study based on three pilot projects and the development of a draft standards for the reuse of treated wastewater (Kettab, 2001; Kettab *et al.*, 2017).

The reuse of treated wastewater mainly for agriculture can save nearly 0.9 billion m^3 of water (30% of the volume intended for DWS and industry). This option will give the possibility of treating wastewater before it is discharged, hence the reduction of pollution and water-borne diseases (Kettab *et al.*, 2008, 2017).

12.4.5.3 Risks and impacts

Human contamination. The risks associated with reusing wastewater for agricultural purposes are (Bassompierre, 2007; Kettab *et al.*, 2008):

- microbiological risk;
- chemical risk;
- environmental risk;
- microbiological risk.

In the case of agriculture, it has been proven for a long time that the pathogenic microorganisms originating from animals cannot penetrate nor survive inside plants. Microorganisms are therefore found on the surface of plants and on the soil. The plants and their leaves create a cool, humid environment (evaporation). There may therefore be contamination during plant growth or harvesting. Pathogens survive longer on soil than on plants (Amy *et al.*, 1996; Kettab *et al.*, 2017).

Consumer protection involves regulating the sanitary quality of water used for irrigation. On the other hand, the bacteriological quality depends on the cultures practiced. For some, a level of 10^4 and 10^5 coliform bacteria will be acceptable. For others, this level may drop to less than 100/100 mL, thus requiring disinfection of the water (Karef *et al.*, 2013).

The method of irrigation has a direct influence on the risk: thus, underground or gravity irrigation can adversely affect the quality of ground and surface water, direct contamination can occur during the maintenance of the irrigation system. Sprinkler irrigation creates contaminating aerosols that can be transported over long distances, while gravity line and flood irrigation expose workers to high health risks, especially when working the land without protection. The WHO recommendations in 2006 foresaw some risk levels according to irrigation techniques and types of crops (Kettab, 2014; Nakib *et al.*, 2015b).

Chemical risk. Beyond the overall effect of certain constituents of wastewater on irrigated crops such as salinity, wastewater can potentially create toxicity due to a high concentration of certain elements such as boron and some heavy metals. The most usual contamination mode for trace elements is the consumption of cultivated plants, in which they accumulate (Kettab *et al.*, 2017; Nakib *et al.*, 2015b).

Not all trace elements are toxic, many are essential in small quantities for plant growth (Fe, Mn, Mo, Zn). However, in excessive amounts they can cause undesirable accumulations in plant tissues and a reduction in their growth. For this purpose, metals have been categorized according to their dangerousness and their agronomic interest (Nakib *et al.*, 2015b).

Chlorosis and necrosis, most often due to iron deficiency induced by excess metal, accompanied by a drop in yield, are the major symptoms of toxicity due to metals. Metabolic disorders can be manifested before the very appearance of visual symptoms (photosynthesis reduction, activation of stress enzymes). Cadmium often induces an inhibition of photosynthesis and imbalance of other elements (Kettab *et al.*, 2017; Leone *et al.*, 2007).

The phytotoxicity of lead and mercury is rarely monitored despite the fact that they are very toxic to plants, leading to the destruction of the root meristems and stopping the growth of the plant. Selenium is easily absorbed by plants or it can be transformed into organic Se which can affect the metabolism of sulphur and generate chlorosis (Kettab *et al.*, 2017; Nakib *et al.*, 2015b).

It is highly important to note that the risk posed by heavy metals depends on their potential toxicity and the level of exposure (concentration, duration), moreover, some are good for plant growth. They are easily eliminated by physical treatment (decantation) and are generally found in sludge (Karef *et al.*, 2013; Nakib *et al.*, 2015b).

Environmental and agricultural alterations. Soil salinization – the salinization of the soil from irrigated waters results from the combined effects of several factors (climate, soil characteristics, topography of the land, cultivation techniques, irrigation pipes, etc.). Indeed, each factor will contribute, according to its state, to the accentuation or the attenuation of the salinization of the soil (Leone *et al.*, 2007).

Irrigation water. The choice of the area of wastewater reuse depends essentially on the quality of the effluent, the types of crops, the irrigation system and the soil conditions. The knowledge of the physicochemical quality of this wastewater is therefore fundamental to predict possible impacts on the receiving environment (Kettab *et al.*, 2017; Leone *et al.*, 2007).

In fact, when reused for agricultural irrigation, their quality is mainly determined by the salinity and especially the sodium ion content. A study in 2006 found that irrigation with treated wastewater loaded with Na++ leads to an increase in the contents of this element throughout the profile of a "Typical Haplustax" type soil. in addition, they reported that the calcium provided by the treated wastewater leads to the release of sodium from the soil absorbing complex in exchangeable form, which facilitates its migration in depth as well as its bioavailability (Kettab *et al.*, 2017; Toze, 2006).

In the study of another case of irrigation with treated wastewater, the salt load of which was 1.8 g/L showed, after two years of monitoring, an increase in the electrical conductivity of the soil from 0.92 to 1.93 dS/m over the first 40 cm of the soil (Toze, 2006). This corresponds to a salt supply in the order of 6.25 T/ha. This rise in soil salinity resulting from irrigation by wastewater is also reported by other authors (Kettab *et al.*, 2017; Yadav *et al.*, 2002).

Irrigation mode. The mode, dose and frequency of irrigation have a direct influence on the process of salinization of the soil. Under these conditions, a dose higher than the needs of the soil is favourable to the phenomenon of salt leaching. This keeps the soil salinity at a reasonable level, especially if the internal and external drainage is adequate. There has been a study on the effect of treated wastewater irrigation on the chemical properties of the soil by testing two different irrigation systems; the first system is surface, while the second is subsurface (Heidarpour *et al.*, 2007).

Consequently, it has been demonstrated that the subsurface irrigation system increased the electrical conductivity of the soil in the surface horizons. On the other hand, the opposite result is observed with the surface irrigation system where the electrical conductivity is higher in the lower levels. According to the authors, the increase in EC in the first case is due to a capillary rise in salts on the surface following evapotranspiration. In the second case, it is rather the leaching of salts by irrigation water, which is at the origin of the rise in the conductivity level in the deep layers (Kettab *et al.*, 2017).

As a result, surface irrigation better allows the leaching of salts to the depths. Close irrigation periods lead to the same result. Frequent irrigation leads to the leaching of the sodium provided by the treated wastewater to the soil depth, which is manifested by an increase in the rate of exchangeable sodium. The conventional method of flood irrigation is more suitable for desalinating the soil profile using irrigation than any other localized irrigation method (Heidarpour *et al.*, 2007).

The soil cover. The characteristics of the soil, which play a key role in accentuating or attenuating the salinization process, are the texture, permeability and lime content. An experiment has been carried out using saltwater irrigation on two soils of different textures, the first sandy and the second rather clayey. They found that the salinity reached high levels in the clay soil. At the end of the experiment, the salinity was between 15.8 and 19.1 mS/cm in the clay-textured soil, while it was between 1.2 and 6 mS/cm in the sandy soil (Dominguez-Mariani *et al.*, 2004).

Miyamoto and Chacon (2006) conducted a statistical study on the influence of certain soil parameters on the salinity of the soil when it is irrigated by high salinity water. They found that soils with a clay texture or those with a calcareous, impermeable or moderately permeable level are the most vulnerable to salinization (Leone *et al.*, 2007; Villagra & Cavagnaro, 2006).

Wastewater irrigation affects certain soil parameters over time. Thus, a slight decrease in pH is observed in some basic soils (Yadav *et al.*, 2002). This decrease is explained by a leaching from irrigation water of active limestones which are responsible for the alkalinity of the soil. Wastewater, through its fertilizing power, also causes an increase in the rate of organic matter and soil nutrients. However, these nutrients stimulate the microbiological activity of the soil, which promotes the mineralization of organic matter leading to a decrease in soil conductivity (Kettab *et al.*, 2017; Toze, 2006).

Besides, irrigation by wastewater, intensifies the microbiological activity of the soil and decreases the hydraulic conductivity of the soil due to the formation of bacterial biofilms, which clog the porosity of the soil (Dominguez-Mariani *et al.*, 2004; Kettab *et al.*, 2017).

Accumulation of metals in the soil. Although the concentration of metals in mainly treated wastewater is low, irrigation can ultimately lead to the accumulation of these elements in the soil. Indeed, the retention of metallic elements by the soil is governed by various mechanical, physico-chemical and even biological phenomena (Karef *et al.*, 2013).

Incontestably, the metals transported to the ground by wastewater are not all in an assimilable or biodegradable form. In fact, the chemical form of metals will depend on intrinsic soil conditions such as pH, clay content, etc. (Kettab *et al.*, 2017).

The accumulation of metallic elements after irrigation with raw sewage is often observed. Thus, Mapanda *et al.* (Miyamoto & Chacon, 2006) found an increase in the contents of metallic elements in the surface horizons of soils irrigated by raw wastewater, lasting more or less significant periods. The levels found in the surface horizons are much higher than the ones found in the subsurface horizons and in the control soil (Nakib *et al.*, 2015b). It has been found that using more than 40 years of practicing irrigation with raw sewage or mixed with river water led to an increase in the levels of metals in the soil (Kettab *et al.*, 2017; Nakib *et al.*, 2015b).

The role of certain physico-chemical parameters of the soil, such as pH and total organic carbon (TOC), on the retention or release of metals from treated wastewater has also been reported. Indeed, in the plots irrigated for more than 100 years, the authors noted a decrease in the pH of the soil. Following this effect, the metals absorbed on the soil aggregates were released in a more exchangeable form, which consequently facilitates their migration into the soil as well as their bioavailability (Heidarpour *et al.*, 2007).

Impact of wastewater irrigation on the water table. Irrigation or the spreading of wastewater on agricultural land is one of the techniques for artificial recharge of groundwater commonly used in arid or semi-arid regions.

Effect of sewage irrigation on crop plants. Due to their nutrient content and their richness in trace elements, when it is reused for irrigation wastewater leads to an improvement in the yields of cultivated plants. Rusan *et al.* (Mapanda *et al.*, 2007) found an increase in the biomass of a fodder plant when it is irrigated with either raw or treated wastewater. Thus, the recorded yields exceeded 110% for plants irrigated by both types of water. Similarly, irrigation with wastewater leads to significant enrichment of the tissue of plants cultivated with trace elements (Yadav *et al.*, 2002).

The trace elements, which are generally immobilized in the upper layers of the soil, can cause long-term risks for plant development. Indeed, only a few trace elements (boron, iron, manganese, zinc, copper and molybdenum) are recognized as necessary for the development of plants in very small quantities (Nakib *et al.*, 2015b).

Irrigation using wastewater can provide these elements, but also other trace elements that not essential to the plant such as lead, mercury, cadmium, boron, fluor, aluminium, nickel, chromium, selenium and tin. The bioavailability of these elements in the soil can cause their accumulation in plant tissues and in some cases; the contents of these elements can reach phytotoxicity thresholds (Nakib *et al.*, 2015b).

12.5 AGRICULTURAL MANURING USING DEHYDRATED SLUDGE

There are three main ways of sewage sludge recycling: agricultural recovery, thermal oxidation and landfill. In our case, we are interested in agricultural development. Agricultural development encounters some struggles from agricultural by-products: animal droppings and by-products from the agro industries from the animal and vegetable processing sectors (Dominguez-Mariani *et al.*, 2004; Karef *et al.* 2014; Kettab *et al.*, 2017).

No scientific or ecological data makes it possible to compare the biomass from agricultural and urban origins. However, the sludge in urban areas poses immense problems as for their use in agriculture and their outlets (Kettab, 2001). Waste sludge, the ultimate product of water purification, is a source of bad odours. This inconvenience can only be overcome by deep burring. Composting is a way to control odours and to change the slime into conventional potting soil. It should be noted that compared to the spreading of animal excrement, the smelly odours due to sludge are minimal. The States of the European Union prohibit the use of sludge on grassland, 30 days before grazing (Villagra & Cavagnaro, 2006).

12.5.1 Human contamination

For vegetable crops, it is forbidden to spread during the entire growing period and on soils that are intended for vegetable cultivation as well as fruits that are intended for raw consumption. It is also prohibited to use manure 18 months before cultivation (French standards) (Djillali *et al.*, 2020; Kettab *et al.*, 2017; Miyamoto & Chacon, 2006).

The main manuring methods are (Djillali *et al.*, 2020; Kettab *et al.*, 2017; Mapanda *et al.*, 2007):

- ingestion of animal and plant products in which pollutants have bio-accumulated;
- inhalation is a way of occupational exposure, the health risks of which are placed under the control of occupational medicine;
- skin contact with sludge. The toxic product can penetrate human skin and the risk mainly depends on the contact time and especially the lipo-solubility of the pollutant. This last factor highly affects the transcutaneous diffusion which is much more important for organic matter than metals;
- ingestion of ground and surface water contaminated by percolation.

12.5.2 Environmental risks

So far, most studies that have been led on the trace elements dynamics in the soil after sludge manuring had the aim of nutritive transfer from the soil to the plant but the results showed that the contamination has mostly been transferred to the experimental product. In addition, the cumulative effect imposes the eradication of these elements before manuring to break the contamination cycle (Rusan *et al.*, 2016).

12.5.3 Agricultural development

The beneficial effects of sludge use in agriculture have been proven by many researchers. Indeed, their application improves the physical, chemical and biological properties of the soil. The biomass and the yields are increased: apple trees; winter wheat sorghum and millet (Boumalek *et al.*, 2019).

A study has been led on a digested urban sludge in the national agronomic school of Algiers and has confirmed the sludge fertilizing action and also showed that the results were similar to the adjacent potato tubers farmyards (Nakib *et al.*, 2015a).

Another study in 2005 (Ait Hamou & Hermouche, 2007; Debba *et al.*, 2005) used lagoon sludge and sludge from dairy products and showed that the yields, which had undergone sludge treatments, can reach twice the control plots. The two major constraints that must be managed when opting for agricultural recovery of sludge are trace metals and pathogens. Health risks linked to the presence of pathogenic germs must also be taken into account, even if these remain low, especially for products that have been previously composted or dehydrated. In addition, the purifying power of the soil has its limits. Whatever the immediate danger that can be caused by pathogens, they remain relatively more controllable (Kettab *et al.*, 2017).

12.6 CONCLUSION

Treated wastewater constitutes a non-negligible water resource, whether in the municipal, industrial or agricultural field, where sewage sludge also constitutes a rich and recoverable fertilizer. As the agricultural components are quite delicate and in order to take maximum advantage of the by-products of water purification on an agricultural level, regular and meticulous monitoring of all the matrices linked to this technique is essential. It is imperative to plan the characterization of the by-products, the soil, the tolerance and the sensitivity of the crop as well as the climatic conditions.

It is imperative to have good management and evaluation strategies regarding the economic and environmental profitability of any urban or agricultural project that is likely to explore wastewater treatment by-products in order to approach successful results without also compromising public health or the receiving environment.

REFERENCES

Ait Hamou R. and Hermouche M. (2007). Etude de l`effet des boues résiduaires sur la production de la pomme de terre en comparaison avec le fumier de ferme et la litière de volaille (Study of the effect of waste sludge on potato production compared to farm manure and poultry litter). Recherche Agnomique (Agronomic Research Journal), National Agronomic School, Algeria.

Amy G., Debroux J. F., Arnold R. and Wilson L. G. (1996). Preozonation for enhancing the biodegradability of wastewater effluent in a potable-recovery soil aquifer treatment (SAT) system. *Revue des Sciences de l'eau*, **9**(3), 365–380.

Anctil F. (2008). L'eau et ses enjeux (Water Encounters). University of Laval, Quebec, p. 228.

Andreadakis A. (2003). Wastewater reuse criteria in Grèce. *Global Nest, the International Journal*, **5**(1), 9–14.

Audic J. M. (2002). Présentation des divers éléments des filières de traitement des eaux usées, Traitement des eaux usées urbaines – Urban Wastewater treatment stages. Lyonnaise des eaux, Lyon, France, 523 pp.

Bassompierre C. (2007). Procédé à boues activées pour le traitement d'effluents papetiers: de la conception d'un pilote à la validation de modèles (Activated sludge process for wastewater originating from paper industries: Pilot station). Thèse Doctorat, Institut National Polytechnique De Grenoble, pp. 25–42.

Bdour A., Hamidi M. R. and Tarawneh L. (2009). Perspective on sustainable wastewater treatment technologies and reuse option in the urban areas of the Mediterranean region. *Desalinisation*, **237**(1–3), 339.

Biswas A. K. (1987). Role of wastewater reuse in planning and management. In: Treatment and Reuse of Sewage Effluent for Irrigation. Butterworth Scientific, Guilford, UK, p. 286.

Bixio D., De Heyder B., Cikurel H., Miska V., Joksimovic D. and Ravazzini A. (2005). Municipale wastewater reclamation, where do we stand? An overview of treatment technology and management practice. *Water Science and Technology Water Supply*, **5**(1), 77–85.

Blinda M. (2011). Stratégie méditerranéenne pour le développement durable, efficience d'utilisation de l'eau (Mediterranean strategy for sustainable development, efficient use). Blue plan. UNEP report. Athena, Greece, 58 p.

Boumalek W., Kettab A., Bensacia N., Bruzzoniti M. C., Othman D. B., Mandi L., Nacer Chabaca M. and Benziada S. (2019). Specification of sewage sludge arising from a domestic wastewater treatment plant for agricultural uses. *Desalination and Water Treatment*, **143**, 178–183.

Catherine B., Alain H. and Jean M. H. (2008). Technologies d'épuration en vue d'une réutilisation des eaux usées traitées. Rapport final, Convention de partenariat ONEMA (treatment technologies for reuse, final report, ONEMA partnership). Cemagref, National Centre of Management of Fauna Reserves, Beijing, 100 pp.

Couture I. (2006). Principaux critères pour évaluer la qualité de l'eau en micro-irrigation, Colloque sur l'irrigation l'eau, source de qualité et de rendement (Main Criteria for water quality assessment towards micro irrigation). Symposium on Water as a Qualitative and Quantitative Resource. CRAAQ, Quebec, Canada.

Debba P., Van Ruitenbeek F., Van Der Meer F., Carranza E. and Stein A. (2005). Optimal field sampling for targeting minerals using hyperspectral data. *Remote Sensing of Environment*, **99**, 373–386.

Djillali Y., Nacer Chabaca M., Benziada S., Bouanani H., Mandi L., Concetta Bruzzoniti M., Boujelbene N. and Kettab A. (2020). Effect of treated wastewater on strawberry. *Desalination and Water Treatment*, **181**, 338–345.

Dominguez-Mariani E., Carrillo-Chaver A., Ortega A. and Orozco- Esquivel M. T. (2004). Wastewater reuse in Valsequillo agricultural area, Mexico, environmental impact. Grand water. *Water, Air and Soil Pollution*, **155**, 251–267.

Ecosse D. (2001). Techniques alternatives en vue de subvenir à la pénurie d'eau dans le mond (Alternative techniques to fight against water scarcity). Graduation Thesis. Faculty of Sciences, University of Amiens, France, p. 15, 17, 21, 26.

Exall K. (2004). A review a water reuse and recycling with reference to Canadian practice and potential, 2. Application-Review article. *Water Quality Research Journal*, **39**(1), 13–28.

Hamoda M. F. (2004). Water strategies and potential of water reuse in the south Mediterranean countries. *Desalination*, **165**, 31–41.

Heidarpour M., Mostafazadeh-Fard B., Abedi Koupai J. and Malekian R. (2007). The effects of treated wastewater on soil chemical properties using subsurface and surface irrigation methods. *Agricultural Water Management*, **90**, 87–94.

Karef S., Nakib M., Kettab A., Benmaamar S., Benziada S. and Bouguerra K. (2013). Study of performance indicators of a sewage treatment plant (Algeria). 3rd International Water Conference, Algiers.

Karef S., Kettab A. and Nakib M. (2014). Characterization of byproducts from wastewater treatment of Medea (Algeria) with a view to agricultural reuse. *Desalination and Water Treatment*, **52**, 2201–2207.

Kettab A. (2001). Water resources in Algeria: strategies, investments, and vision. *Desalination*, **136**(1–3), 25–33.

Kettab A. (2014). Water for all with quality and quantity: it is the concern of all. *Desalination and Water Treatment*, **52**, 1965–1966.

Kettab A., Metiche R. and Bennacar N. (2008). De l'eau pour un développement durable: enjeux et stratégie (Water for sustainable development: encounters and strategies). *Revue des Science de l'Eau*, **21**(2), 247–256.

Kettab A., Nacer M., Chabaca K. S., Nakib M., Boumalek W., Bouanani H., Djillali Y., Benziada S. and Benmamar S. (2017). Traitement des eaux usées urbaines et leurs réutilisations en Agriculture (Urban wastewater treatment and agricultural reuse). EUE (European University Editions).

Leone A. P., Menenti M., Buondonno A., Letizia A., Maffei C. and Sorrentino G. (2007). A field experiment on spectrometry of crop response to soil salinity. *Agricultural Water Management*, **89**(1–2), 39–48. doi: 10.1016/j.agwat.2006.12.004.

Mapanda F., Mangwayana E. N., Nyamangara J. and Giller K. E. (2007). Uptake of heavy metals by vegetables irrigated using wastewater and the subsequent risks in Harare, Zimbabwe. *Physics and Chemistry of the Earth A/B/C*, **32**, 1399–1405.

Miyamoto S. and Chacon A. (2006). Soil salinity of urban turf areas irrigated with saline water. II. Soil factors. *Landscape & Urban Planning*, **77**, 28–38.

Nakib M., Kettab A., Berreksi A. and Mandi L. (2015a). Study of the prospects for agricultural utilization of sludge produced from WWTPs in North Central Algeria. *Desalination and Water Treatment*, **55**, 1152–1166.

Nakib M., Kettab A., Bouanani H., Tebbal S. and Berreksi A. (2015b). Study of the fertilizing potential of the treated wastewater of the Kolea wastewater treatment plant (Algeria). *Desalination and Water Treatment*, **57**, 1–5.

Page D., Dillon P., Toze S., Bixio D., Genthe B., Jiménez Cisneros B. E. and Wintgens T. (2010). Valuing the subsurface pathogen treatment barrier in water recycling via aquifers for drinking supplies. *Water Research*, **44**(6), 1841–1852. doi: 10.1016/j.watres.2009.12.008.

Ramya T., Premkumar P., Thanarasu A., Velayutham K., Dhanasekaran A. and Sivanesan S. (2019). Degradation of pesticide-contaminated wastewater (coragen) using electrocoagulation process with iron electrodes. *Desalination and Water Treatment*, **165**, 4–11.

Rusan M. J., Albalasmeh A. A. and Malkawi H. I. (2016). Treated olive mill wastewater effects on soil properties and plant growth. *Water, Air, and Soil Pollution*, **227**, 135.

Tose S. (2006). Reuse of effluent water benefits and risks. *Agricultural Water Management*, **86**, 147–159.

Toze S. (2006) Reuse of effluent water: Benefits and risks. *Agricultural Water Management*, **80**, 146–159. doi: 10.1016/j.agwat.2005.07.010

Villagra P. E. and Cavagnaro J. B. (2006). Water stress effects on the seedling growth of Prosopis argentina and Prosopis alpataco. *Journal of Arid Environments*, **64**(3), 390–400. doi: 10.1016/j.jaridenv.2005.06.008

Wu G., He K., Miao J., Yin Q. and Zhao Y. (2018). Strategies for sustainable wastewater treatment based on energy recovery and emerging compounds control: A mini review. *Desalination and Water Treatment*, **127**.

Yadav R. N. S., Sharm I. M. P., Kamthe S. D., Tajuddin A., Yadav S. and Tejra R. K. (2002). Performance evaluation of sugarcane chopper harvester. *Sugar Tech*, **4**(3–4), 117–122.

Zekri S. (2020). Water Policies in MENA Countries. Springer, Cham.

Chapter 13

Advanced technologies for removing triclosan from wastewater: a state-of-the-art review

Mariem Chtourou[1,2], Albert Galizia[2], Victoria Salvadó[3] and Hèctor Monclús[2]

[1]*Laboratory of Materials Sciences and Environment, Faculty of Sciences of Sfax, University of Sfax, Tunisia*
[2]*LEQUIA, Institute of the Environment, University of Girona, M. Aurèlia Capmany, 69, 17003 Girona, Spain*
[3]*Department of Chemistry, University of Girona, M. Aurèlia Capmany, 69, 17003 Girona, Spain*

ABSTRACT

Triclosan is a broad spectrum antimicrobial agent used in pharmaceutical and personal care products and there is increasing concern about its presence in surface waters and the risks it poses to the environment. Reports from around the world indicate that detectable levels in surface waters range from ng/L to low µg/L and, because triclosan interacts with aquatic organisms and ecosystems, it can now be found in the human food chain. In this work, the efficacy of triclosan removal from aqueous solutions and raw wastewater has been reviewed and summarised. While advanced treatments such as (i) adsorption, (ii) advanced oxidation processes, (iii) membrane processes, and (iv) membrane bioreactors are the most efficient triclosan removal technologies, some combined treatment processes such as membrane bioreactor/activated carbon, advanced oxidation processes/activated carbon or activated carbon/membrane have also been used here and the paper reviews how effective they are. Due to the diversity in nature of TCS, one single treatment technology is incapable of controlling it because each treatment method has their own drawbacks that need to be resolved. As a result, combining different

Editor(s): Nadeem A Khan, Sirajuddin Ahmed, Viola Vambol and Sergij Vambol
doi: 10.2166/9781789061338_0327

advanced methods is an interesting research topic. Reviewing the different methods for TCS removal has revealed that future research should focus on combining ozonation and AC adsorption for TCS removal. In addition to this, widely implemented and extensively accepted technologies, such as MBR, could be upgraded with AC, thus facilitating TCS biodegradation while also avoiding some of the limiting factors that occur when the MBR technology is used separately.

Keywords: Adsorption, Advanced oxidation processes, Advanced treatment technology, Membrane, Membrane bioreactor, Triclosan, Wastewater.

Contents

13.1 Introduction ... 328
13.2 TCS Removal Technologies ... 329
13.2.1 Adsorption ... 329
13.2.1.1 Activated carbon, GAC and PAC ... 329
13.2.1.2 Carbon nanotubes (CNTs) ... 331
13.2.1.3 Other sorbents ... 333
13.3 Advanced Oxidation Process (AOP) ... 333
13.3.1 Ozonation (O_3) ... 336
13.3.2 Fenton and photo-Fenton oxidation ... 336
13.3.3 Photolysis ... 337
13.4 Physical Separation by Membrane Process ... 338
13.4.1 The MBR process ... 340
13.5 Discussion ... 341
13.6 TCS Removal by Hybrid Combined Treatments ... 341
13.7 Conclusions ... 343
References ... 344

13.1 INTRODUCTION

Triclosan (TCS) or "5-chloro-2-(2,4-dichlorophenoxy)-phenol" is a common synthetic antimicrobial agent used in more than 700 different industrial, personal care and pharmaceutical products (PPCPs). Today, TCS can be found in personal hygiene products (e.g. toothpaste, mouthwash and soaps), in household cleaners and textiles (e.g. sportswear and bedclothes) as well as in shoes and carpets, where it acts as an antifungal and antibacterial agent. The amount of TCS found in such products ranges between 0.1 and 0.3% (Sabaliunas *et al.*, 2003; Schweizer, 2001) and, being an external-use product, it is almost always released directly into the environment as it is washed down the drain to wastewater treatment plants (WWTPs) to be partially removed by conventional treatment techniques (Benotti *et al.*, 2009). Nevertheless, because it is not completely removed by these treatment processes, residual TCS is continuously being released into receiving waters via wastewater discharge.

TCS, which may have endocrine disruptor effects, is a chloroform precursor if it is combined with Cl^- and exhibits a tendency to accumulate and persist in biosolids (Sioufi *et al.*, 1977). It has been estimated that up to 50% of the TCS in WWTP influent will remain in the biosolids of the WWTPs themselves. Consequently, applying biosolids containing residues of TCS as agro-fertilizers for the land poses a potential risk to ecosystems and eventually to human health (Chalew & Halden, 2009). TCS is a hydrophobic compound with a high $logK_{OW}$ of 4.76 and it is likely to be adsorbed into organic particles in the sewage sludge treatment process and then be transported to the land via sludge applications (Ogunyoku & Young, 2014)

This highlights the importance wastewater and sludge management strategies have and the need for advanced methods like: (i) adsorption; (ii) oxidation and advanced oxidation processes (AOP); (iii) membrane processes; (iv) membrane bioreactors MBR; and (v) combined processes: MBR/adsorption, AOP/ adsorption. In particular, the AOP/membrane combination for removing TCS has been studied extensively in recent decades. However, no published reviews have focussed on comparing the overall effectiveness of major sewage treatment technologies in eliminating a range of TCS, or on identifying the potential of increasing TCS elimination by enhancing existing treatment processes.

13.2 TCS REMOVAL TECHNOLOGIES

13.2.1 Adsorption

To enhance the adsorption capacity for TCS, various adsorbents have been researched and developed to adsorb TCSs in aqueous solutions. This section will focus specifically on carbon-based adsorptive materials i.e. granular activated carbon (GAC), powdered activated carbon (PAC), carbon nanotubes (CNTs), dried sludge, organo-zeolites (OZs) and biochar. The different adsorbents' TCS removal efficiencies are summarized in Table 13.1.

13.2.1.1 Activated carbon, GAC and PAC

Activated carbon (AC) is an adsorbent synthesised from petroleum residue, coal, or wood. AC is an important treatment for removing residual micropollutants as it can remove 80% of the micropollutants found in municipal WWTP effluents (Bui *et al.*, 2016; Theydan & Ahmed, 2012). While AC removal performance is based on the physico-chemical properties of the sorbent (AC), such as surface area, pore size distribution, surface charge or oxygen content, it is also affected by the properties (shape, size, charge, and hydrophobicity) of the solute (Jung *et al.*, 2015). AC comes in two forms: granular (GAC) and powdered activated carbon (PAC); both of which can be used to remove TCS from wastewaters through the adsorption process.

GAC, on the one hand, tends to adsorb hydrophobic organic compounds such as TCS and results in high (up to 100% in some cases) removal rates (Hernández-Leal

Table 13.1 TCS adsorbents and removals efficiencies.

Adsorbent	Water Sample	TCS Concentration (ng L^{-1})	Conditions				Removal Efficiency (%)	References
			Q (mg L^{-1})	pH	Temp (°C)	Time (min)		
GAC	Secondary effluent	108	20	—	—	10 and 20	19	Mailler *et al.* (2016)
	Secondary effluent*	1–40 × 10^3	—	7.7–7.8	25	35	100	Paredes *et al.* (2016)
	Grey water	2032	84	8.4	25	—	95	Hernández-Leal *et al.* (2011)
PAC	Secondary effluent	135	10	—	—	—	23	Mailler *et al.* (2015)
	Wastewater	2400	100	7.5	20.1	—	100	Sheng *et al.* (2016)
	Surface water	10–250	5	8.2	—	240	98	Westerhoff *et al.* (2005)
CNTs	Water	50–2000 × 10^3	—	7	—	—	—	Cho *et al.* (2011)
Dried sludge	Secondary sedimentation*	20 × 10^3	—	6.5	25	360	100	Tohidi and Cai (2016)
OZs	Wastewater*	40 × 10^6	—	6–10	—	—	100	Lei *et al.* (2013)
Biochar	Secondary effluent	300 × 10^3	—	—	—	—	32	Tong *et al.* (2016)

*Synthetic. Q = Adsorbent concentration.

et al., 2011; Katsigiannis *et al.*, 2015; Paredes *et al.*, 2016; Patiño *et al.*, 2015). The residual TCS in secondary effluent are barely removed by the fixed-bed of GAC (Ma *et al.*, 2017). PAC, on the other hand, has (depending on the water matrix and AC concentration), demonstrated similar efficacy. For example, when using 100, 50 or 10 mg L^{-1} of PAC, the removal percentage can be 100, 95 or 75%, respectively (Mailler *et al.*, 2015; Sheng *et al.*, 2016)

In the case of GAC, screening, with respect to hydraulic capacity and carbon usage rate, is a necessity if the specific product being used has not previously been tested on similar wastewater. If the explicit task is to remove as much pharmaceutical residue as possible from new wastewater, then using PAC would be recommended. However, the high performing GAC process is also able of attaining the same performance levels as the PAC process (Kårelid *et al.*, 2017).

For both GAC and PAC, the external surface area of the adsorbent materials and the availability of the mesoporous volume have been shown to be decisive in the amount of adsorbate retained (Patiño *et al.*, 2015). For AC adsorbents, TCS removal is influenced by the combination of $\log(K_{OW})$ solute and pKa (TCS >7), which means it is highly adsorbed by AC (Shanmuganathan *et al.*, 2015). Other factors, like the oxygen-containing functional groups or pore blockage, can also affect the TCS adsorption capabilities of AC.

Increasing the amount of oxygen-containing functional groups on the surface of the AC decreases the adsorption rates of organic solutes, indicating that there is a preferential adsorption for water molecules over organic solutes at these sites (Li *et al.*, 2002). Pore blockage and direct site competition are considered to be the most common mechanisms that would affect AC sorption in the presence of natural organic matter (NOM) (Behera *et al.*, 2010). The inorganic constituents of water, such as Ca, form complexes with the NOM in water or precipitate as $CaCO_3$, thus reducing the adsorption capacity of the AC filters (Pastrana-Martínez *et al.*, 2010).

13.2.1.2 Carbon nanotubes (CNTs)

Carbon nanotubes (CNTs) are graphene cylinder sheets with one or multiple layers. The standard industrial types are as follows: single-walled carbon nanotubes (SWCNT), which are composed of a single cylinder; double-walled carbon nanotubes (DWCNT), which are made up of two layers, and multi-walled carbon nanotubes (MWCNT), which are composed of multiple layers rolled up into a seamless cylinder with a nanoscale outer diameter of between 1 and 30 nm (Jackson *et al.*, 2013) (Figure 13.1(a)).

The adsorption properties of CNTs depend on various factors, including adsorption site, which can be external, interstitial, internal or grooves (Figure 13.1(b)), surface area, purity and/or surface chemistry (Cho *et al.*, 2011). The surface chemistry of CNTs with the chemical properties of PPCPs, and aqueous solution chemistry (pH, ionic strength, dissolved organic matter, etc.) all play an important role in CNTs' PPCP sorption.

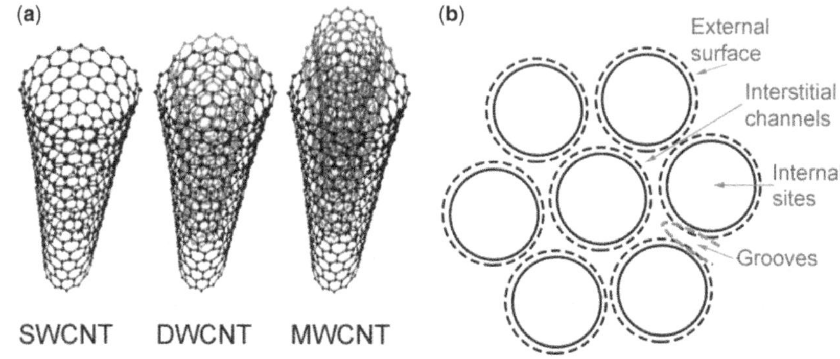

Figure 13.1 CNTs: (**a**) industrial types; (**b**) adsorption sites. Adapted from Jackson *et al.* (2013) and Cho *et al.* (2011).

It has also been reported that larger molecules, such as TCS, have a much greater adsorption rate than smaller molecules because smaller molecules need to be diffused to the inner sites of MWNTs, which is time-consuming and results in extremely low diffusivities (Jung *et al.*, 2015).

By comparing the SWCNT and SWCNT, Cho *et al.* (2011) demonstrated the effect that a specific area has on TCS removal. Higher TCS removal was observed using SWCNT versus MWCNT, while Wang *et al.* (2016) reported that MWCNT efficiency levels increase as TCS feeding concentrations decrease. Also, the surface of a CNT can be modified by attaching functional groups to make it more selective. Li *et al.* (2017) compared TCS adsorption using a non-functionalized CNT (NF-CNT) with two functionalized CNTs (CNTs-OH, CNTs-COOH), and demonstrated that TCS sorption decreased in the following order: NF-CNTs, CNTs-OH, CNTs-COOH because of the surface charge density. This is because functionalized CNT has more negative charge surface density than NF-CNTs which affects the electrostatic repulsion between TCS and CNTs at low and neutral pH values.

Different studies have demonstrated that pH plays a vital role and that the effect is greater for NF-CNTs than for functionalized CNTs. For all CNTs, TCS sorption decreases when pH increases due to the electrostatic repulsion between TCS and CNTs (Li *et al.*, 2017). Specifically, when using SWCNT and MWCNT, the electrostatic interactions motivate a greater TCS removal at pH 4 than at pH 10 or 7 (Wang *et al.*, 2015).

The competitive sorption effects of natural organic material are affected by the hydrophobicity and concentration of TCS compounds and are pronounced except in high equilibrium concentrations for hydrophobic TCS compounds with a strong affinity to CNTs (Cho *et al.*, 2011). Special functional groups that are capable of forming specific bonding with the targeted PPCP attach to the surface of the MWCNT, thus eliminating the NOM competition or the pre-coagulation phase that is commonly employed to enhance NOM removal (Wang *et al.*, 2016).

Carbon nanotubes are highly promising for TCS adsorption because they are sensitive to the presence of TCS molecules and, therefore, could be used as a filter system for wastewater treatment (Castro *et al.*, 2017).

13.2.1.3 Other sorbents

Despite the abundance of studies in published research on the removal of pollutants with various adsorbents, there are very few about using adsorption as a treatment process for pharmaceutical compounds (Kyzas *et al.*, 2015).

Dried sludge, which is abundant, cheap and easy to use, can be considered an effective potential sorbent for TCS removal from wastewater as it tends to adsorb considerable amounts of this compound. However, as in the AC removal methods, pH, temperature and adsorbent concentration affects the efficiency of the TCS removal. For instance, acidic conditions or lowering the temperature enhances TCS sorption capacity. Although the uptake of compounds improved with an increase in sludge concentration, the adsorption capacity dropped significantly (Tohidi & Cai, 2016).

Organo-zeolites (OZs) were prepared by loading cetylpyridinium bromide onto natural zeolite to remove TCS from an aqueous solution. Lei *et al.* (2013) found that the TCS adsorption capacities are strongly dependent on the pH solution and the nature of the surface charge of the OZs; which are a little higher in acidic and neutral pH conditions. As such, OZs could be used as effective adsorbents for TCS removal from wastewater.

Biochar adsorption capacity is related to high surface area, hydrophobicity, and potential interaction with TCS functional groups including hydrogen bonding and π-stacking. In wastewater, acid-treated biochar also effectively adsorbs TCS, albeit at a decreased capacity and removal rate because of the competition from other organic constituents (Tong *et al.*, 2016).

Also, recent studies shows natural cork can also be used to remove micro plutons by adsorption. Mallek *et al.* (2018) used powder cork particles (of <2 mm) and 10 mL of different individual phenolic compounds solutions.

While adsorption of TCS onto AC, CNT and other adsorbents is a promising technique, several problems must be solved to pave the way for any large-scale application. For instance, TCS with a high molecular weight is more sensitive to competition of organic matter. Furthermore, high pH levels could inhibit TCS adsorption onto all the standard adsorbents.

13.3 ADVANCED OXIDATION PROCESS (AOP)

The most commonly used AOP techniques are ozonation, heterogeneous photocatalysis with semiconductors, Fenton and photo-Fenton (Klavarioti *et al.*, 2009). Table 13.2 outlines the different TCS removal processes using AOPs.

Table 13.2 TCS oxidants and removals efficiencies.

AOP Method	Water Sample	TCS Concentration (L^{-1})	Conditions		Removal Efficiency (%)	References
			Oxidant Concentration	Other Conditions		
Ozonation	Wastewater effluent	5 mg	O_3 4 mg L^{-1}	DOC 7.5 mg L^{-1}	100	Suarez *et al.* (2007)
	STP effluent	246 ng	O_3 0.34 mmol L^{-1}	15 min – pH 7.63	78	Rosal *et al.* (2010)
	Surface water	102 ng	O_3 2.5 mg L^{-1}	pH 7	99	Snyder *et al.* (2006)
	Grey water	48 ng	O_3 10 mg L^{-1}	20 min	>87	Hernández-Leal *et al.* (2011
Fenton	Wastewater	10 mg	Fe^{3+} 1 mg L^{-1} + H_2O_2 20 mg L^{-1}	60 min – 25°C	100	Munoz *et al.* (2012)
	Wastewater	34.5 μmol	$BiFeO_3$ 0.5 mg L^{-1} + H_2O_2 10 mmol L^{-1}	180 min – 25°C	82.7	Song *et al.* (2012)
	Secondary effluent	100 μg	Fe^{2+} 5 mg L^{-1} + H_2O_2 50 mg L^{-1}	280 min – pH 3	88	Klamerth *et al.* (2010a)
	Secondary effluent	100 μg	Fe^{2+} 5 mg L^{-1} + H_2O_2 50 mg L^{-1}	280 min	87	Klamerth *et al.* (2010b)

Photo Fenton	Secondary effluent	135 ng	Fe^{2+} 5 mg L^{-1} + H_2O_2 50 mg L^{-1} + 550 W m^{-2}	17°C – pH 2.5–10 min	100	De la Cruz *et al.* (2012)
	Wastewater*	5 mg	Fe^{2+} 2 nM L^{-1} + H_2O_2 5 nM L^{-1} + 4.2 mW cm^{-2}	60 min	100	Son *et al.* (2010)
Photolysis UV	Secondary effluent	150 μg	UV 290–400 nm	pH 7.85	62	Wang *et al.* (2017a)
	Wastewater effluent	135 ng	550 W m^{-2} + H^2O^2 50 mg L^{-1}	30 min – pH 2.5–17°C	100	De la Cruz *et al.* (2012)
Photocatalyst UV/TiO_2	Wastewater effluent	100 μg	Solar UV lamp 30 W, b400 nm	90 min	100	Miranda-García *et al.* (2011)
	Wastewater	31.8 mg	TiO_2 200 mg L^{-1} + UV lamp 300–400 nm	240 min – pH 6.5	99.7	Constantin *et al.* (2018)
	Wastewater	9 mg	TiO_2 100 mg L^{-1} + UV lamp 365 nm	240 min–6 h	95	Yu *et al.* (2006)
UV/ZnO	Wastewater	10 mg	Mercury vapor lamp 125 W, 200–300 nm	90 min	>90	Kosera *et al.* (2017)

*Synthetic.

13.3.1 Ozonation (O_3)

Ozonation (O_3) appears to provide an effective technique for enhancing TCS removal because it reacts with the phenol ring, rapidly oxidizing and eliminating the TCS. In a direct reaction, 90% of the TCS can be removed in 5 min (Rozas *et al.* 2017). That said, the O_3-TCS reaction can be affected by the water matrix and the dissolved organic carbon, pH and contact time can also affect TCS elimination.

Suarez *et al.* (2007) demonstrated that DOC can also compete with TCS as a target compound for the O3. When they increased the DOC concentration from 7.5 to 12.4 mg L^{-1} the TCS removal rate was reduced from 100 to 58%, even when the O_3 dosage was raised to 50%.

This treatment process can convert TCS into different products: 2,4-dichlorophenol, chlorocatechol, chlororesorcinol, monohydroxy-TCS or dihydroxy-TCS (Chen *et al.*, 2012). It is important to point out that the EU surface water directive 76/464/EC (The Council of the European Communities, 1976) considers 2,4-dichlorophenol to be a dangerous substance that "may cause long-term adverse effects in the aquatic environment".

To eliminate TCS from surface water with 1–5 mg/L TCS via continuous ozonation at 5 mg/L, 20–30 min ozonation time (depending on the pH) is required. For the complete oxidation of triclosan and its by-products, the ozone demand was calculated to be 13.04 mg ozone per mg of triclosan (Orhon *et al.*, 2017).

13.3.2 Fenton and photo-Fenton oxidation

Homogeneous oxidation with Fenton's reagent occurs in the presence of ferrous (Fe^{2+}) or ferric (Fe^{3+}) ions and hydrogen peroxide (H_2O_2) via a free radical chain reaction which produces hydroxyl radicals. It is a metal-catalysed oxidation reaction in which iron acts as the catalyst. The oxidation efficiency may be enhanced in the presence of UV irradiation as more hydroxyl radicals are produced in the so-called photo-Fenton reaction (Klavarioti *et al.*, 2009). Fenton oxidation in acidic conditions is an important oxidation treatment method for removing pollutants (Wang & Wang, 2016) and consequently photo-Fenton oxidation processes have been developed in recent years as a tertiary treatment to remove micropollutants at relatively low concentrations (ng L^{-1}–g L^{-1}) (Prieto-Rodríguez *et al.*, 2013).

When applying Fenton only 4.1% of the TCS was removed while that figure was 17.6% when only UV-C (254 nm light) was applied. That said, removal efficiencies up to 68% can be reached after 20 min of photo-Fenton treatment (Son *et al.*, 2010).

The high removal of total organic carbon (TOC) together with the high generation of -OH radicals through photo-Fenton improves TCS degradation because competition from organic matter has been reduced (De la Cruz *et al.*, 2012; Son *et al.*, 2010). However, even higher efficiency removals were observed when using real rather than synthetic wastewaters. This is because the presence of humic acids in real wastewater contributes more by giving electrons and hydroxyl radicals to the radicals produced by photo-Fenton (Klamerth *et al.*, 2010b). In real wastewater, Klamerth *et al.* (2010a) managed to completely remove the TCS (spiking 5 $\mu g\ L^{-1}$ of TCS), whereas removal efficiency was reduced to 87% when using synthetic wastewater.

13.3.3 Photolysis

Photolysis is when artificial or natural light interacts with target molecules and causes photochemical reactions that can lead to the degradation of target molecules to intermediate products and, when complemented with extensive light, could eventually decompose to obtain mineral end-products (Klavarioti *et al.*, 2009).

Photolysis, both direct and indirect, has been shown to cause the breakdown of many emerging contaminants. In direct photolysis, the contaminant itself absorbs photons and is degraded (Zepp & Cline, 1977), while in indirect photolysis, degradation occurs when the compound reacts to a reactive species generated by photosensitizers, which can absorb radiation, to reach an excited state (Andreozzi *et al.*, 2003).

The efficiency of photolytic degradation depends on several factors such as the absorbance spectrum of the pharmaceutical in question, the quantum yield of photolysis, the concentration of H_2O_2 employed, and the water matrix (Klavarioti *et al.*, 2009).

Many organic substances absorb radiation primarily below 250 nm, which is a small percentage of solar radiation. Therefore, photolysis occasionally occurs in the environment, often supported by natural photosensitizers called humic acids. Unfortunately, this process is inefficient and is associated with a long reaction time and low quantum yields (Gmurek *et al.*, 2017). However, this natural process can be accelerated by combining different radiation wavelengths, which is more effective rather than just one wavelength. (Tixier *et al.*, 2002; Yuval *et al.*, 2017).

TCS photodegradation rates in secondary effluent are reduced by 62% compared with Milli-Q water. The TCS degradation rate decreased with indirect photolysis which plays a negative role in TCS removal because of the presence of photosensitizers that can absorb the available light, thus shielding the target compounds from photodegradation. In the presence of light, nitrate could act as a photosensitizer producing hydroxyl radicals. This seems to facilitate and improve TCS photodegradation but TCS removal pathways are higher with direct, rather than indirect, photolysis with nitrates and DOM (Wang *et al.*, 2017a).

A limiting factor has been detected when solar irradiation is present. For instance, Agüera *et al.* (2003) observed that the presence of TCS in surface water can form highly toxic dioxin-type derivatives because TCS is a halogenated compound and could be involved in dealkylation mechanisms.

Photolysis can be enhanced to a photocatalysis process by adding metallic oxides. Photocatalysis is a process based on the excitation of a semiconductor through light. This excitation generates electron-hole pairs that, in the presence of oxygen and water, photo-generate hydroxyl radicals. The most important photocatalysts are titanium dioxide (TiO_2) and zinc oxide (ZnO) (Yu *et al.*, 2006).

The photocatalyst using TiO_2 showed a very efficient degradation and the total disappearance of TCS is obtained with sufficient irradiation time (i.e. 60 min). Moreover, no dioxin derivatives were detected. Son *et al.* (2009) showed that TCS oxidation by OH radicals can greatly reduce the production of dioxin-group intermediates as well as generate the effective mineralization of TCS.

Immobilized TiO_2 can be applied under solar irradiation to successfully degrade TCS from real waste water on a pilot plan scale, i.e. real spiked effluent (Constantin *et al.*, 2018; Miranda-García *et al.*, 2011), as well as in real secondary wastewater combined with simulated UV light (Murgolo *et al.*, 2015).

ZnO may be an interesting alternative to TiO_2 to remove TCS because of its higher efficiency compared to photolysis. Kosera *et al.* (2017) used ZnO spheres for 18 h to achieve 50% TCS removal and a 90% removal after 24 h of photolysis. A further advantage to ZnO (instead of TiO_2) is that it reduces the TCS by-product formation and thus achieves good degradation levels. As such it offers another alternative to lowering the costs of photocatalysis and using renewable energy; something which is very attractive from an environmental perspective.

The main limitation of the photocatalysis is the difficulty in separating the catalysts in the final treatment stage, especially when the catalyst is used in powdered form (Lee *et al.*, 2016). Although chemical technologies can achieve high TCS removal efficiency, they are usually costly and frequently generate toxic intermediate products (Wang & Wang, 2016). Therefore, the environmental risks they pose need to be assessed to confirm whether the photocatalysis process represents a safe alternative to conventional wastewater treatment processes or not.

13.4 PHYSICAL SEPARATION BY MEMBRANE PROCESS

Membrane separation process is a physical separation process based on filtration mechanisms. There are different kind of membranes depending on their pore size and distribution. These can be differentiated by the membrane material used, i.e. ceramic membranes or polymeric membranes. The pyemic membranes can be classified depending on their weight cut-off pore size: micro-filtration (MF), ultra-filtration (UF), nano-filtration (NF), and reverse osmosis (RO) membranes.

Membrane processes are becoming prominent in wastewater reclamation/reuse and drinking water treatments because they are effective in removing macro and microorganic pollutants. However, only NF or RO has been reported as being efficient in removing PPCPs and other dissolved contaminants from wastewater because of their size and molecular weight (MW). In other words, most pharmaceuticals are smaller than the molecular weight cut-off of most MF and UF membranes. In addition NF and RO membranes have very similar removal efficiencies (Ganiyu *et al.*, 2015; Luo *et al.*, 2014).

Sudhakaran *et al.* (2013) reported 90% TCS removal efficiency using NF polymeric membranes and Fujioka *et al.* (2014) reported similar removal efficiency using NF ceramic membranes. However, TCS removal by RO is sometimes slightly higher than NF (Table 13.3) because TCS can be diffused through the NF membrane.

The membrane process can be affected by pH and water matrix composition, and membrane fouling is also a factor to consider because the formation of a hydrophobic fouling layer on the membrane surface could interfere with the solute–membrane interaction and thus reduce the diffusive transport of TCS across the membrane (Chang *et al.*, 2012). NOM had a positive effect on TCS removal due to the TCS–humic acid complex formation that limits TCS permeation through the membrane (Ogutverici *et al.*, 2016). Consequently, significant enhancement in TCS rejection was observed when the membranes were pre-fouled with the three-model organic foulants: bovine serum albumin (BSA), alginate and humic acid (Nghiem & Coleman, 2008).

The high energy consumption of RO with high waste production, low water recovery and high maintenance tasks due to, for example, membrane RO fouling, have dismissed RO as a promising technology for TCS treatment (Lee *et al.*, 2012a). Nevertheless, membrane processes can be combined with biological processes to enhance TCS elimination as well as the performance of the membrane bioreactor (MBR).

Table 13.3 TCS membrane separation (NF/RO) and removal efficiencies.

Process	Water Sample	TCS Concentration (ng L^{-1})	Removal Efficiency (%)	References
NF	Wastewater	74	>95	Kim *et al.* (2007)
Membrane	Wastewater*	300	90	Fujioka *et al.* (2014)
RO	Wastewater	74	>95	Kim *et al.* (2007)
Membrane	Wastewater*	7.5×10^5	100	Nghiem and Coleman (2008)

*Synthetic.

13.4.1 The MBR process

MBR technology combines biodegradation using activated sludge with a solid–liquid separation using membrane filtration (usually MF or UF). A full-scale MBR sustained high and stable nutrient removal up to >95% for total nitrogen (TN) and phosphate (PO_4^{3-}P) (Phan *et al.*, 2015).

It was found that UF-MBR with initial concentrations of TSS between 4.2 and 5.5 mg L^{-1} had >90% TCS removal (Banerjee *et al.*, 2016; Phan *et al.*, 2014, 2016; Wijekoon *et al.*, 2015). Depending on the MBR conditions (aerobic, anaerobic, anoxic), TCS removal can oscillate between 70 and >90% (Table 13.4). In addition, TCS removal increases at high temperatures, and has long hydraulic retention (HRT) and sludge retention times (SRT) (Armstrong *et al.*, 2018; Trinh *et al.*, 2016).

The TCS removal by an MBR is more related to the biological rather than the filtration process. Several studies reported effective spaces to remove TCS from wastewater, such as *Trametes versicolor* (Hundt *et al.*, 2000), *Sphingomonas* sp. Rd1 (Hay *et al.*, 2001), *Nitrosomonas europaea* (Roh *et al.*, 2009) or *Sphingopyxis* sp., which can achieve 100% removal and complete dichlorination (Lee *et al.*, 2012a, b; Wang *et al.*, 2017b).

The limiting factor of the MBR is the membrane fouling because of pore blocking due to the suspended solids in the bioreactor. Operational membranes require high amounts of energy to reduce membrane fouling (typically with air-scour on membrane surface) which adds to their running costs (Grandclément *et al.*, 2017). Furthermore, high concentrations of TCS may have a toxic effect on nitrification microorganisms. TCS can inactivate *Nitrosomonas europaea* and stops nitrification activity (Qiu *et al.*, 2015). On the other hand, lower TCS concentrations and increasing the ammonia oxidation activity enhance TCS removal in nitrifying activated sludge (Chtourou *et al.*, 2018; Huang *et al.*, 2016).

Table 13.4 The removal of TCS by MBR process.

MBR	Membrane	TCS Concentration (μg L^{-1})	Removal Efficiency (%)	References
Aerobic	HF-UF	2	>91.8	Tadkaew *et al.* (2011)
Anaerobic	HF-UF	0.594 ± 0.361	90.2	Monsalvo *et al.* (2014)
Anaerobic	Ceramic	5	70	Wijekoon *et al.* (2015)
Anoxic-Aerobic	HF-UF	5	>90	Phan *et al.* (2014)

13.5 DISCUSSION

On the one hand, adsorption presents a wide range of TCS removal efficiencies, together with the removal of residuals, disinfection and oxidation products, but its greatest advantages are that it does not generate toxins. On the other hand, there are some limitations which, in a nutshell, are its high maintenance, re-generation and AC product costs (Table 13.5).

AOP processes present high removal efficiency as well as adsorption, and they also improve reaction time. However, despite this, they consume very high amounts of energy. Furthermore, AOPs generate some non-desirable toxics due to the oxidation capacity of the process (Table 13.6).

As the membrane separation process depends on the pore size of the membrane, this means that NF and RO the most promising configurations. RO presents slightly better TCS removal and lower operating costs than NF does, so it would seem to be the most appropriate membrane technology.

Thanks to the combination of biological activated sludge and membrane separation, MBR is highly effective. Several microorganisms are capable of TCS removal, but membrane pore size also plays a key role in completing TCS removal. Furthermore, MBR demands high energy consumption to cope with air-scouring and backwashing, and, as TCS releases some toxicity into the biomass that could result in being a nitrification inhibition, the MBR may require some pre-treatment to avoid the activated sludge not working as it should.

13.6 TCS REMOVAL BY HYBRID COMBINED TREATMENTS

In order to solve the technologies' limitations, some of them can be combined to create hybrid treatments. These are designed to reduce long-term treatments and energy costs and achieve high TCS removal. Filtration processes can be combined with AC (MBR-PAC/GAC, MF + GAC), AC can also work with PAOs technologies (PAC + O_3), CNT can be improved with coagulation (coagulation + MWCNT) and conventional activated sludge could use a filtration and oxidation process to remove TCS (sludge + sand filter + O_3) (Table 13.6).

While MBR technology presents high TCS removal rates, it can also be combined with AC treatments (like PAC or GAC) with high TCS concentrations in order to avoid the nitrification issues. The addition of PAC into an MBR may also reduce membrane cleaning and fouling issues, thus increasing membrane lifetime. PAC is also more efficient than GAC (Ng *et al.*, 2006). Nevertheless, AC adsorption capacity could be exhausted, thus triggering a gradual decrease in the TCS removal efficiency (Nguyen *et al.*, 2013; Park *et al.*, 1999). As well as a filtration technology, Shanmuganathan *et al.* (2015) applied NF as MF-GAC post-treatment, demonstrating no TMP increment and no flux reduction in the NF membrane and achieving 54.5 and 90% TCS removal, respectively.

Table 13.5 TCS removal process comparison.

Treatment Process	Advantages	Challenges	References
Adsorption	• A wide range of TCS removal from wastewater. • Removal of residual disinfection/oxidation products • Not generating toxic active products of TCS	• Relatively high cost in operation and maintenance • Regeneration and disposal issues of high sludge • Depend of quality • With high molecular weight is more sensitive to the competition of organic matter	Behera *et al.* (2010); Nguyen *et al.* (2013); Mailler *et al.* (2015); Sheng *et al.* (2016)
AOP	• High removal efficiency • Short degradation rate	• Energy consumption issues, operational and miniatous cost • Formation of toxic disinfection by-products	Nakada *et al.* (2007); Chen *et al.* (2012); Gomes *et al.* (2017)
Membranes (NF/RO)	• Full effective in removing TCS as pore sizes less large than TCS	• High cost of operation • High energy demand, membrane fouling and disposal issue	Kim *et al.* 2007; C. O. Lee *et al.* (2012b)
MBR	• Effective for the removal of TCS • Several microorganisms capable of depredating TCS	• High energy consumption and fouling, control heat and mass transfer • High aeration cost and robustness of membrane • High concentration of TCS inhibit the nitrification, TCS pre-treatment would be necessary	Hundt *et al.* (2000); Hay *et al.* (2001); Ng *et al.* (2006); Roh *et al.* (2009); Lee *et al.* (2012a, b); Nguyen *et al.* (2013); Phan *et al.* (2014); Qiu *et al.* (2015); Huang *et al.* (2016)

Table 13.6 TCS removal combined treatments

Process	Water Sample	TCS Concentration (L^{-1})	Removal Efficiency (%)	References
PAC + MBR/MBR + GAC	Synthetic wastewater	5 μg	>95	Nguyen *et al.* (2013)
MF + GAC	Sewage effluent	11 ng	54.5	Shanmuganathan *et al.* (2015)
PAC + O_3	Synthetic water	2.3 mg	100	Rozas *et al.* (2017)
Coagulation + MWCTN	Wastewater effluent	1 mg	93	Wang *et al.* (2018)
Activated sludge + Sand filtration + O_3	Wastewater	316 ng	>97.4	Nakada *et al.* (2007)

Regarding AOP, Rozas *et al.* (2017) showed that applying O_3 to the PAC process contributes to accelerating the oxidation of the transformed compounds that cause toxicity at lower ozone doses. Thus, this carbon based-AOP is a promising candidate to be integrated into water treatment to effectively remove emerging organic pollutants.

With respect to other AC technologies such as CNT, Wang *et al.* (2018) discussed how pre-coagulation of wastewater effluent can improve the TCS removal efficiency of MWCNT. Adding a coagulation pre-treatment can reduce the competitive adsorption of effluent organic matter (EfOM) and increase the removal efficiencies of an MWCNT membrane by 11–34%. The pre-coagulation primarily removes biopolymers and humic substances in EfOM which, in turn, enhances TCS removal when employing an MWCNT-membrane. Biopolymer removal, on the other hand, results in reduced membrane fouling during the filtration of pre-coagulated wastewater.

The combination of conventional activated sludge treatments, sand filtration, and ozonation is effective in reducing the risks associated with TCS in sewage effluent and TCS removal was highly efficient (>95%).

13.7 CONCLUSIONS

Due to the diversity of the nature of TCS, one single treatment technology is incapable of controlling it because each treatment method has its own drawbacks that need to be resolved. As a result, an interesting research topic would be to investigate taking different advanced methods and combining them. Reviewing the different methods for TCS removal here has revealed that future research should focus on combining ozonation and AC adsorption for TCS removal. In addition to this, widely-implemented and extensively-accepted technologies such as MBR could be upgraded with AC, thus facilitating TCS biodegradation while also avoiding some of the limiting factors that occur when the MBR technology is used alone.

REFERENCES

Agüera A., Fernández-Alba A. R., Piedra L., Mézcua M. and Gómez M. J. (2003). Evaluation of triclosan and biphenylol in marine sediments and urban wastewaters by pressurized liquid extraction and solid phase extraction followed by gas chromatography mass spectrometry and liquid chromatography mass spectrometry. *Analytica Chimica Acta*, **480**, 193–205. https://doi.org/10.1016/S0003-2670(03)00040-0

Andreozzi R., Marotta R. and Paxéus N. (2003). Pharmaceuticals in STP effluents and their solar photodegradation in aquatic environment. *Chemosphere*, **50**, 1319–1330. https://doi.org/10.1016/S0045-6535(02)00769-5

Armstrong D. L., Lozano N., Rice C. P., Ramirez M. and Torrents A. (2018). Degradation of triclosan and triclocarban and formation of transformation products in activated sludge using benchtop bioreactors. *Environmental Research*, **161**, 17–25. https://doi.org/10.1016/j.envres.2017.10.048

Banerjee P., Dey T. K., Sarkar S., Swarnakar S., Mukhopadhyay A. and Ghosh S. (2016). Treatment of cosmetic effluent in different configurations of ceramic UF membrane based bioreactor: toxicity evaluation of the untreated and treated wastewater using catfish (Heteropneustes fossilis). *Chemosphere*, **146**, 133–144. https://doi.org/10.1016/j.chemosphere.2015.12.004

Behera S. K., Oh S. Y. and Park H. S. (2010). Sorption of triclosan onto activated carbon, kaolinite and montmorillonite: effects of pH, ionic strength, and humic acid. *Journal of Hazardous Materials*, **179**, 684–691. https://doi.org/10.1016/j.jhazmat.2010.03.056

Benotti M. J., Trenholm R. A., Vanderford B. J., Holady J. C., Stanford B. D. and Snyder S. A. (2009). Pharmaceuticals and endocrine disrupting compounds in U.S. drinking water. *Environmental Science and Technology*, **43**, 597–603. https://doi.org/10.1021/es801845a

Bui X. T., Vo T. P. T., Ngo H. H., Guo W. S. and Nguyen T. T. (2016). Multicriteria assessment of advanced treatment technologies for micropollutants removal at large-scale applications. *Science of the Total Environment*, **563–564**, 1050–1067. https://doi.org/10.1016/j.scitotenv.2016.04.191

Castro S. M., Araújo A. B., Nogueira R. F. P. and Guerini S. (2017). Adsorption of triclosan on single wall carbon nanotubes: a first principle approach. *Applied Surface Science*, **403**, 519–524. https://doi.org/10.1016/j.apsusc.2017.01.215

Chalew T. E. A. and Halden R. U. (2009). Environmental exposure of aquatic and terrestrial biota to triclosan and triclocarban. *Journal of the American Water Resources Association*, **45**, 4–13. https://doi.org/10.1111/j.1752-1688.2008.00284.x

Chang E. E., Chang Y. C., Liang C. H., Huang C. P. and Chiang P. C. (2012). Identifying the rejection mechanism for nanofiltration membranes fouled by humic acid and calcium ions exemplified by acetaminophen, sulfamethoxazole, and triclosan. *Journal of Hazardous Materials*, **221–222**, 19–27. https://doi.org/10.1016/j.jhazmat.2012.03.051

Chen X., Richard J., Liu Y., Dopp E., Tuerk J. and Bester K. (2012). Ozonation products of triclosan in advanced wastewater treatment. *Water Research*, **46**, 2247–2256. https://doi.org/10.1016/j.watres.2012.01.039

Cho H. H., Huang H. and Schwab K. (2011). Effects of solution chemistry on the adsorption of ibuprofen and triclosan onto carbon nanotubes. *Langmuir*, **27**, 12,960–12,967. https://doi.org/10.1021/la202459g

Chtourou M., Mallek M., Dalmau M., Mamo J., Santos-Clotas E., Salah A. B., Walha K., Salvadó V. and Monclús H. (2018). Triclosan, carbamazepine and caffeine removal by activated sludge system focusing on membrane bioreactor. *Process Safety and Environal Protection*, **118**, 1–9. https://doi.org/10.1016/j.psep.2018.06.019

Constantin L. A., Nitoi I., Cristea N. I. and Constantin M. A. (2018). Possible degradation pathways of triclosan from aqueous systems via TiO_2 assisted photocatalyis. *Journal of Industrial and Engineering Chemistry*, **58**, 155–162. https://doi.org/10.1016/j.jiec.2017.09.020

De la Cruz N., Giménez J., Esplugas S., Grandjean D., De Alencastro L. F. and Pulgarín C. (2012). Degradation of 32 emergent contaminants by UV and neutral photo-fenton in domestic wastewater effluent previously treated by activated sludge. *Water Research*, **46**, 1947–1957. https://doi.org/10.1016/j.watres.2012.01.014

Fujioka T., Khan S. J., McDonald J. A. and Nghiem L. D. (2014). Nanofiltration of trace organic chemicals: a comparison between ceramic and polymeric membranes. *Separation and Purification Technology*, **136**, 258–264. https://doi.org/10.1016/j.seppur.2014.08.039

Ganiyu S. O., Van Hullebusch E. D., Cretin M., Esposito G. and Oturan M. A. (2015). Coupling of membrane filtration and advanced oxidation processes for removal of pharmaceutical residues: a critical review. *Separation and Purification Technology*, **156**, 891–914. https://doi.org/10.1016/j.seppur.2015.09.059

Gmurek M., Olak-Kucharczyk M. and Ledakowicz S. (2017). Photochemical decomposition of endocrine disrupting compounds – A review. *Chemical Engineering Journal*, **310** (Part 2), 437–456. https://doi.org/10.1016/j.cej.2016.05.014

Gomes J., Costa R., Quinta-Ferreira R. M. and Martins R. C. (2017). Application of ozonation for pharmaceuticals and personal care products removal from water. *Science of the Total Environment*, **586**, 265–283. https://doi.org/10.1016/j.scitotenv.2017.01.216

Grandclément C., Seyssiecq I., Piram A., Wong-Wah-Chung P., Vanot G., Tiliacos N., Roche N. and Doumenq P. (2017). From the conventional biological wastewater treatment to hybrid processes, the evaluation of organic micropollutant removal: a review. *Water Research*, **111**, 297–317. https://doi.org/10.1016/j.watres.2017.01.005

Hay A. G., Dees P. M. and Sayler G. S. (2001). Growth of a bacterial consortium on triclosan. *FEMS* Microbiology Ecology, **36**, 105–112. https://doi.org/10.1111/j.1574-6941.2001.tb00830.x

Hernández-Leal L., Temmink H., Zeeman G. and Buisman C. J. N. (2011). Removal of micropollutants from aerobically treated grey water via ozone and activated carbon. *Water Research*, **45**, 2887–2896. https://doi.org/10.1016/j.watres.2011.03.009

Huang C. L., Abass O. K. and Yu C. P. (2016). Triclosan: a review on systematic risk assessment and control from the perspective of substance flow analysis. *Science of the Total Environment*, **566–567**, 771–785. https://doi.org/10.1016/j.scitotenv.2016.05.002

Hundt K., Martin D., Hammer E., Jonas U., Kindermann M. K. and Schauer F. (2000). Transformation of triclosan by Trametes versicolor and Pycnoporus cinnabarinus. *Applied and Environmental Microbiology*, **66**, 4157–4160. https://doi.org/10.1128/AEM.66.9.4157-4160.2000

Jackson P., Jacobsen N. R., Baun A., Birkedal R., Kühnel D., Jensen K. A., Vogel U. and Wallin H. (2013). Bioaccumulation and ecotoxicity of carbon nanotubes. *Chemistry Central Journal*, **7**, 1–21. https://doi.org/10.1186/1752-153X-7-154

Jung C., Son A., Her N., Zoh K. D., Cho J. and Yoon Y. (2015). Removal of endocrine disrupting compounds, pharmaceuticals, and personal care products in water using carbon nanotubes: a review. *Journal of Industrial and Engineering Chemistry*, **27**, 1–11. https://doi.org/10.1016/j.jiec.2014.12.035

Kårelid V., Larsson G. and Björlenius B. (2017). Pilot-scale removal of pharmaceuticals in municipal wastewater: comparison of granular and powdered activated carbon treatment at three wastewater treatment plants. *Journal of Environmental Management*, **193**, 491–502. https://doi.org/10.1016/j.jenvman.2017.02.042

Katsigiannis A., Noutsopoulos C., Mantziaras J. and Gioldasi M. (2015). Removal of emerging pollutants through Granular Activated Carbon. *Chemical Engineering Journal*, **280**, 49–57. https://doi.org/10.1016/j.cej.2015.05.109

Kim S. D., Cho J., Kim I. S., Vanderford B. J. and Snyder S. A. (2007). Occurrence and removal of pharmaceuticals and endocrine disruptors in South Korean surface, drinking, and waste waters. *Water Research*, **41**, 1013–1021. https://doi.org/10.1016/j.watres.2006.06.034

Klamerth N., Malato S., Maldonado M. I., Agüera A. and Fernández-Alba A. R. (2010a). Application of Photo-Fenton as a tertiary treatment of emerging contaminants in municipal wastewater. *Environmental Science and Technology*, **44**, 1792–1798. https://doi.org/10.1021/es903455p

Klamerth N., Rizzo L., Malato S., Maldonado M. I., Agüera A. and Fernández-Alba A. R. (2010b). Degradation of fifteen emerging contaminants at μg L-1 initial concentrations by mild solar photo-Fenton in MWTP effluents. *Water Research*, **44**, 545–554. https://doi.org/10.1016/j.watres.2009.09.059

Klavarioti M., Mantzavinos D. and Kassinos D. (2009). Removal of residual pharmaceuticals from aqueous systems by advanced oxidation processes. *Environment International*, **35**(2), 402–417. https://doi.org/10.1016/j.envint.2008.07.009

Kosera V. S., Cruz T. M., Chaves E. S. and Tiburtius E. R. L. (2017). Triclosan degradation by heterogeneous photocatalysis using ZnO immobilized in biopolymer as catalyst. Journal of Photochemisry and Photobiology. A, Chemistry, **344**, 184–191. https://doi.org/10.1016/j.jphotochem.2017.05.014

Kyzas G. Z., Fu J., Lazaridis N. K., Bikiaris D. N. and Matis K. A. (2015). New approaches on the removal of pharmaceuticals from wastewaters with adsorbent materials. *Journal of Molecular Liquids*, **209**, 87–93. https://doi.org/10.1016/j.molliq.2015.05.025

Lee C. O., Howe K. J. and Thomson B. M. (2012a). Ozone and biofiltration as an alternative to reverse osmosis for removing PPCPs and micropollutants from treated wastewater. *Water Research*, **46**, 1005–1014. https://doi.org/10.1016/j.watres.2011.11.069

Lee D. G., Zhao F., Rezenom Y. H., Russell D. H. and Chu K. H. (2012b). Biodegradation of triclosan by a wastewater microorganism. *Water Research*, **46**, 4226–4234. https://doi.org/10.1016/j.watres.2012.05.025

Lee K. M., Lai C. W., Ngai K. S. and Juan J. C. (2016). Recent developments of zinc oxide based photocatalyst in water treatment technology: a review. *Water Research*, **88**, 428–448. https://doi.org/10.1016/j.watres.2015.09.045

Lei C., Hu Y. Y. and He M. Z. (2013). Adsorption characteristics of triclosan from aqueous solution onto cetylpyridinium bromide (CPB) modified zeolites. *Chemical Engineering Journal*, **219**, 361–370. https://doi.org/10.1016/j.cej.2012.12.099

Li L., Quinlivan P. A. and Knappe D. R. U. (2002). Effects of activated carbon surface chemistry and pore structure on the adsorption of organic contaminants from aqueous solution. *Carbon*, **40**, 2085–2100. https://doi.org/10.1016/S0008-6223(02)00069-6

Li H., Zhang W., Zhang Z. and Zhang X. (2017). Sorption of triclosan to carbon nanotubes: the combined effects of sonication, functionalization and solution chemistry. *Science of the Total Environment*, **580**, 1318–1326. https://doi.org/10.1016/j.scitotenv.2016.12.095

Luo Y., Guo W., Ngo H. H., Nghiem L. D., Hai F. I., Zhang J., Liang S. and Wang X. C. (2014). A review on the occurrence of micropollutants in the aquatic environment and their fate and removal during wastewater treatment. *Science of the Total Environment*, **473–474**, 619–641. https://doi.org/10.1016/j.scitotenv.2013.12.065

Ma D., Chen L. and Liu R. (2017). Removal of novel antiandrogens identified in biological effluents of domestic wastewater by activated carbon. *Science of the Total Environment*, **595**, 702–710. https://doi.org/10.1016/j.scitotenv.2017.03.272

Mailler R., Gasperi J., Coquet Y., Deshayes S., Zedek S., Cren-Olivé C., Cartiser N., Eudes V., Bressy A., Caupos E., Moilleron R., Chebbo G. and Rocher V. (2015). Study of a large scale powdered activated carbon pilot: removals of a wide range of emerging and priority micropollutants from wastewater treatment plant effluents. *Water Research*, **72**, 315–330. https://doi.org/10.1016/j.watres.2014.10.047

Mailler R., Gasperi J., Coquet Y., Buleté A., Vulliet E., Deshayes S., Zedek S., Mirande-Bret C., Eudes V., Bressy A., Caupos E., Moilleron R., Chebbo G. and Rocher V. (2016). Removal of a wide range of emerging pollutants from wastewater treatment plant discharges by micro-grain activated carbon in fluidized bed as tertiary treatment at large pilot scale. *Science of the Total Environment*, **542**, 983–996. https://doi.org/10.1016/j.scitotenv.2015.10.153

Mallek M., Chtourou M., Portillo M., Monclús H., Walha K., Salah A. B. and Salvadó V. (2018). Granulated cork as biosorbent for the removal of phenol derivatives and emerging contaminants. *Journal of Environmental Management*, **223**, 576–585. https://doi.org/10.1016/j.jenvman.2018.06.069

Miranda-García N., Suárez S., Sánchez B., Coronado J. M., Malato S. and Maldonado M. I. (2011). Photocatalytic degradation of emerging contaminants in municipal wastewater treatment plant effluents using immobilized TiO2 in a solar pilot plant. *Applied Catalysis B: Environmental*, **103**, 294–301. https://doi.org/10.1016/j.apcatb.2011.01.030

Monsalvo V. M., McDonald J. A., Khan S. J. and Le-Clech P. (2014). Removal of trace organics by anaerobic membrane bioreactors. *Water Research*, **49**, 103–112. https://doi.org/10.1016/j.watres.2013.11.026

Munoz M., de Pedro Z. M., Casas J. A. and Rodriguez J. J. (2012). Triclosan breakdown by Fenton-like oxidation. *Chemical Engineering Journal*, **198–199**, 275–281. https://doi.org/10.1016/j.cej.2012.05.097

Murgolo S., Petronella F., Ciannarella R., Comparelli R., Agostiano A., Curri M. L. and Mascolo G. (2015). UV and solar-based photocatalytic degradation of organic pollutants by nano-sized TiO_2 grown on carbon nanotubes. *Catalysis Today*, **240**, 114–124. https://doi.org/10.1016/j.cattod.2014.04.021

Nakada N., Shinohara H., Murata A., Kiri K., Managaki S., Sato N. and Takada H. (2007). Removal of selected pharmaceuticals and personal care products (PPCPs) and endocrine-disrupting chemicals (EDCs) during sand filtration and ozonation at a municipal sewage treatment plant. *Water Research*, **41**, 4373–4382. https://doi.org/10.1016/j.watres.2007.06.038

Ng C. A., Sun D. and Fane A. G. (2006). Operation of membrane bioreactor with powdered activated carbon addition. *Separation Science and Technology*, **41**, 1447–1466. https://doi.org/10.1080/01496390600634632

Nghiem L. D. and Coleman P. J. (2008). NF/RO filtration of the hydrophobic ionogenic compound triclosan: transport mechanisms and the influence of membrane fouling. *Separation and Purification Technology*, **62**, 709–716. https://doi.org/10.1016/j.seppur.2008.03.027

Nguyen L. N., Hai F. I., Kang J., Nghiem L. D., Price W. E., Guo W., Ngo H. H. and Tung K. L. (2013). Comparison between sequential and simultaneous application of activated carbon with membrane bioreactor for trace organic contaminant removal. *Bioresource Technology*, **130**, 412–417. https://doi.org/10.1016/j.biortech.2012.11.131

Ogunyoku T. A. and Young T. M. (2014). Removal of triclocarban and triclosan during municipal biosolid production. *Water Environmental Research*, **86**, 197–203. https://doi.org/10.2175/106143013x13807328849378

Ogutverici A., Yilmaz L., Yetis U. and Dilek F. B. (2016). Triclosan removal by NF from a real drinking water source – Effect of natural organic matter. *Chemical Engineering Journal*, **283**, 330–337. https://doi.org/10.1016/j.cej.2015.07.065

Orhon K. B., Orhon A. K., Dilek F. B. and Yetis U. (2017). Triclosan removal from surface water by ozonation – Kinetics and by-products formation. *Journal of Environmental Management*, **204**, 327–336. https://doi.org/10.1016/j.jenvman.2017.09.025

Paredes L., Fernandez-Fontaina E., Lema J. M., Omil F. and Carballa M. (2016). Understanding the fate of organic micropollutants in sand and granular activated carbon biofiltration systems. *Science of the Total Environment*, **551–552**, 640–648. https://doi.org/10.1016/j.scitotenv.2016.02.008

Park H., Choo K. H. and Lee C. H. (1999). Flux enhancement with powdered activated carbon addition in the membrane anaerobic bioreactor. *Separation Science and Technology*, **34**, 2781–2792. https://doi.org/10.1081/SS-100100804

Pastrana-Martínez L. M., López-Ramón M. V., Fontecha-Cámara M. A. and Moreno-Castilla C. (2010). Batch and column adsorption of herbicide fluroxypyr on different types of activated carbons from water with varied degrees of hardness and alkalinity. *Water Research*, **44**, 879–885. https://doi.org/10.1016/j.watres.2009.09.053

Patiño Y., Díaz E. and Ordóñez S. (2015). Performance of different carbonaceous materials for emerging pollutants adsorption. *Chemosphere*, **119**, S124–S130. https://doi.org/10.1016/j.chemosphere.2014.05.025

Phan H. V., Hai F. I., Kang J., Dam H. K., Zhang R., Price W. E., Broeckmann A. and Nghiem L. D. (2014). Simultaneous nitrification/denitrification and trace organic contaminant (TrOC) removal by an anoxic-aerobic membrane bioreactor (MBR). *Bioresource Technology*, **165**, 96–104. https://doi.org/10.1016/j.biortech.2014.03.094

Phan H. V., Hai F. I., McDonald J. A., Khan S. J., Zhang R., Price W. E., Broeckmann A. and Nghiem L. D. (2015). Nutrient and trace organic contaminant removal from wastewater of a resort town: comparison between a pilot and a full scale membrane bioreactor. *International Biodeterioration and Biodegradation*, **102**, 40–48. https://doi.org/10.1016/j.ibiod.2015.02.010

Phan H. V., Hai F. I., Zhang R., Kang J., Price W. E. and Nghiem L. D. (2016). Bacterial community dynamics in an anoxic-aerobic membrane bioreactor - Impact on nutrient and trace organic contaminant removal. *International Biodeterioration and Biodegradation*, **109**, 61–72. https://doi.org/10.1016/j.ibiod.2016.01.002

Prieto-Rodríguez L., Oller I., Klamerth N., Agüera A., Rodríguez E. M. and Malato S. (2013). Application of solar AOPs and ozonation for elimination of micropollutants in municipal wastewater treatment plant effluents. *Water Research*, **47**, 1521–1528. https://doi.org/10.1016/j.watres.2012.11.002

Qiu S., Ma F., Huang X. and Xu S. (2015). Microcalorimetric study on biodegradation kinetic of micropollutant in wastewater by ammonia-oxidizing bacteria. *Journal of Thermal Analysis and Calorimetry*, **120**, 345–349. https://doi.org/10.1007/s10973-014-4261-x

Roh H., Subramanya N., Zhao F., Yu C. P., Sandt J. and Chu K. H. (2009). Biodegradation potential of wastewater micropollutants by ammonia-oxidizing bacteria. *Chemosphere*, **77**, 1084–1089. https://doi.org/10.1016/j.chemosphere.2009.08.049

Rosal R., Rodríguez A., Perdigón-Melón J. A., Petre A., García-Calvo E., Gómez M. J., Agüera A. and Fernández-Alba A. R. (2010). Occurrence of emerging pollutants in urban wastewater and their removal through biological treatment followed by ozonation. *Water Research*, **44**, 578–588. https://doi.org/10.1016/j.watres.2009.07.004

Rozas O., Baeza C., Núñez K., Rossner A., Urrutia R. and Mansilla H. D. (2017). Organic micropollutants (OMPs) oxidation by ozone: effect of activated carbon on toxicity abatement. *Science of the Total Environment*, **590–591**, 430–439. https://doi.org/10.1016/j.scitotenv.2016.12.120

Sabaliunas D., Webb S. F., Hauk A., Jacob M. and Eckhoff W. S. (2003). Environmental fate of Triclosan in the River Aire Basin, UK. *Water Research*, **37**, 3145–3154. https://doi.org/10.1016/S0043-1354(03)00164-7

Schweizer H. P. (2001). Triclosan: a widely used biocide and its link to antibiotics. *FEMS Microbiology Letters*, **202**, 1–7. https://doi.org/10.1016/S0378-1097(01)00273-7

Shanmuganathan S., Nguyen T. V., Jeong S., Kandasamy J. and Vigneswaran S. (2015). Submerged membrane – (GAC) adsorption hybrid system in reverse osmosis concentrate treatment. *Separation and Purification Technology*, **146**, 8–14. https://doi.org/10.1016/j.seppur.2015.03.017

Sheng C., Nnanna A. G. A., Liu Y. and Vargo J. D. (2016). Removal of Trace Pharmaceuticals from Water using coagulation and powdered activated carbon as pretreatment to ultrafiltration membrane system. *Science of the Total Environment*, **550**, 1075–1083. https://doi.org/10.1016/j.scitotenv.2016.01.179

Sioufi A., Caudal F. and Grimault D. (1977). GLC determination of free and conjugated triclosan in human plasma and urine. *Journal of Pharmaceutical Sciences*, **66**, 1166–1168. https://doi.org/10.1002/jps.2600660830

Snyder S. A., Wert E. C., Rexing D. J., Zegers R. E. and Drury D. D. (2006). Ozone oxidation of endocrine disruptors and pharmaceuticals in surface water and wastewater. *Ozone: Science and Engineering*, **28**, 445–460. https://doi.org/10.1080/01919510601039726

Son H. S., Ko G. and Zoh K. D. (2009). Kinetics and mechanism of photolysis and TiO_2 photocatalysis of triclosan. *Journal of Hazardous Materials*, **166**, 954–960. https://doi.org/10.1016/j.jhazmat.2008.11.107

Son H. S., Khim J. and Zoh K. D. (2010). Degradation of triclosan in the combined reaction of Fe2+ and UV-C: comparison with the Fenton and photolytic reactions. *Environmental Progress and Sustainable Energy*, **29**, 415–420. https://doi.org/10.1002/ep.10422

Song Z., Wang N., Zhu L., Huang A., Zhao X. and Tang H. (2012). Efficient oxidative degradation of triclosan by using an enhanced Fenton-like process. *Chemical Engineering Journal*, **198–199**, 379–387. https://doi.org/10.1016/j.cej.2012.05.067

Suarez S., Dodd M. C., Omil F. and von Gunten U. (2007). Kinetics of triclosan oxidation by aqueous ozone and consequent loss of antibacterial activity: relevance to municipal wastewater ozonation. *Water Research*, **41**, 2481–2490. https://doi.org/10.1016/j.watres.2007.02.049

Sudhakaran S., Maeng S. K. and Amy G. (2013). Hybridization of natural systems with advanced treatment processes for organic micropollutant removals: new concepts in multi-barrier treatment. *Chemosphere*, **92**, 731–737. https://doi.org/10.1016/j.chemosphere.2013.04.021

Tadkaew N., Hai F. I., McDonald J. A., Khan S. J. and Nghiem L. D. (2011). Removal of trace organics by MBR treatment: the role of molecular properties. *Water Research*, **45**, 2439–2451. https://doi.org/10.1016/j.watres.2011.01.023

The Council of the European Communities (1976). Council Directive: on pollution caused by certain dangerous substances discharged into the aquatic environment of the Community. Official Journal of the European Community, 23–29.

Theydan S. K. and Ahmed M. J. (2012). Optimization of preparation conditions for activated carbons from date stones using response surface methodology. *Powder Technology*, **224**, 101–108. https://doi.org/10.1016/j.powtec.2012.02.037

Tixier C., Singer H. P., Canonica S. and Müller S.R. (2002). Phototransformation of triclosan in surface waters: a relevant elimination process for this widely used biocidelaboratory studies, field measurements, and modeling. *Environmental Science and Technology*, **36**, 3482–3489. https://doi.org/10.1021/es025647t

Tohidi F. and Cai Z. (2016). Adsorption isotherms and kinetics for the removal of triclosan and methyl triclosan from wastewater using inactivated dried sludge. *Process Biochemistry*, **51**, 1069–1077. https://doi.org/10.1016/j.procbio.2016.04.018

Tong Y., Mayer B. K. and McNamara P. J. (2016). Triclosan adsorption using wastewater biosolids-derived biochar. *Environmental Science: Water Research and Technology*, **2**, 761–768. https://doi.org/10.1039/c6ew00127k

Trinh T., van den Akker B., Coleman H. M., Stuetz R. M., Drewes J. E., Le-Clech P. and Khan S. J. (2016). Seasonal variations in fate and removal of trace organic chemical contaminants while operating a full-scale membrane bioreactor. *Science of the Total Environment*, **550**, 176–183. https://doi.org/10.1016/j.scitotenv.2015.12.083

Wang J. and Wang S. (2016). Removal of pharmaceuticals and personal care products (PPCPs) from wastewater: a review. *Journal of Environmental Management*, **182**, 620–640. https://doi.org/10.1016/j.jenvman.2016.07.049

Wang Y., Zhu J., Huang H. and Cho H. H. (2015). Carbon nanotube composite membranes for microfiltration of pharmaceuticals and personal care products: capabilities and potential mechanisms. *Journal of Membrane Science*, **479**, 165–174. https://doi.org/10.1016/j.memsci.2015.01.034

Wang Y., Ma J., Zhu J., Ye N., Zhang X. and Huang H. (2016). Multi-walled carbon nanotubes with selected properties for dynamic filtration of pharmaceuticals and personal care products. *Water Research*, **92**, 104–112. https://doi.org/10.1016/j.watres.2016.01.038

Wang Y., Roddick F. A. and Fan L. (2017a). Direct and indirect photolysis of seven micropollutants in secondary effluent from a wastewater lagoon. *Chemosphere*, **185**, 297–308. https://doi.org/10.1016/j.chemosphere.2017.06.122

Wang S., Yin Y. and Wang J. (2017b). Enhanced biodegradation of triclosan by means of gamma irradiation. *Chemosphere*, **167**, 406–414. https://doi.org/10.1016/j.chemosphere.2016.10.028

Wang Y., Huang H. and Wei X. (2018). Influence of wastewater precoagulation on adsorptive filtration of pharmaceutical and personal care products by carbon nanotube membranes. *Chemical Engineering Journal*, **333**, 66–75. https://doi.org/10.1016/j.cej.2017.09.149

Westerhoff P., Yoon Y., Snyder S. and Wert E. (2005). Fate of endocrine-disruptor, pharmaceutical, and personal care product chemicals during simulated drinking water treatment processes. *Environmental Science and Technology*, **39**, 6649–6663. https://doi.org/10.1021/es0484799

Wijekoon K. C., McDonald J. A., Khan S. J., Hai F. I., Price W. E. and Nghiem L. D. (2015). Development of a predictive framework to assess the removal of trace organic chemicals by anaerobic membrane bioreactor. *Bioresource Technology*, **189**, 391–398. https://doi.org/10.1016/j.biortech.2015.04.034

Yu J. C., Kwong T. Y., Luo Q. and Cai Z. (2006). Photocatalytic oxidation of triclosan. *Chemosphere*, **65**, 390–399. https://doi.org/10.1016/j.chemosphere.2006.02.011

Yuval A., Eran F., Janin W., Oliver O. and Yael D. (2017). Photodegradation of micropollutants using V-UV/UV-C processes; Triclosan as a model compound. *Science of the Total Environment*, **601–602**, 397–404. https://doi.org/10.1016/j.scitotenv.2017.05.172

Zepp R. G. and Cline D. M. (1977). Rates of direct photolysis in aquatic environment. *Environmental Science and Technology*, **11**, 359–366. https://doi.org/10.1021/es60127a013

Chapter 14

Existing treatment – globally in full scale plants

Qaisar Mahmood, Muhammad Saqib Khan and Nadia Riaz

Department of Environmental Sciences, COMSATS University Islamabad, Abbottabad Campus 22060, Pakistan

ABSTRACT

Pharmaceutical wastewater (PWW) treatment is challenging due to its complex nature and the presence of recalcitrant constituents. Over the last decade, the existence of pharmaceutical residues in wastewater has increased. Urban wastewater, sewage leachates, hospital waste, household waste, and livestock rearing are some of the leading sources of pharmaceutical constituents in wastewater. Besides, PWW also contains obsolete and expired antibiotics due to incomplete metabolic activity. PWW's constituents threaten the sustainability of diverse ecosystem functions; reports show significant toxic effects on microbes involved in their biodegradation. PWW's prevalent type of pollutants include analgesics, hormones, antibiotics, therapeutic agents, and steroidal medicines used in the care of humans and other animals. Such contaminants continue to circulate in the environment once released, can eventually enter the human food chain. Many factors, such as human metabolic activity, plant stabilization, and/or human consumption rate, determine the presence of specific pollutants in the environment or their ultimate treatment at wastewater treatment plants (WWTPs). Currently, various techniques, physical, biological, chemical, are in practice for PWW treatment. Biological treatment involves the use of plants, microbes, and fungi to treat pharmaceutical waste. During biological treatment, the waste converted or degraded into some other non-toxic forms. Both aerobic and anaerobic types of

 Pharmaceutical Wastewater Treatment Technologies: Concepts and Implementation Strategies
Editor(s): Nadeem A Khan, Sirajuddin Ahmed, Viola Vambol and Sergij Vambol
doi: 10.2166/9781789061338_0353

microbial species are effective in treating PWW. For the selection of the best treatment technique, the better choice would be the PWW technology developed at its generating source. This chapter summarizes the existing technologies for the treatment of PWW, particularly biological wastewater treatment methods.

Keywords: Activated sludge, Advanced oxidation process, Advance wastewater treatment, Pharmaceutical wastewater, Sewage sludge disinfection, Wastewater screening.

Contents

14.1 Introduction 355
14.1.1 Antibiotic waste 356
14.1.2 Hormonal waste 356
14.1.3 Improper disposal of analgesics 356
14.2 Treatment Options for PWW 356
14.2.1 Biological treatment of pharmaceutical wastewater 357
14.2.2 Anaerobic treatment 358
14.2.2.1 UASB bioreactors 358
14.2.2.2 Fixed film bioreactors 358
14.2.2.3 Anaerobic membrane bioreactor 359
14.2.2.4 Chain of processes in anaerobic membrane bioreactor 359
14.2.2.5 Anaerobic sequencing batch reactor (AnSBR) 360
14.2.2.6 Advantages and disadvantages of a batch system 360
14.2.2.7 Hybrid anaerobic and aerobic treatment 361
14.2.2.8 Hybrid anaerobic + chemical/physical systems 362
14.2.3 Aerobic treatment 362
14.3 Fungal Treatment 363
14.4 Bacterial Treatment 363
14.5 Transformation of Pharmaceuticals in Plants 364
14.6 Tertiary Treatment via Membrane Bioreactor (MBR) 365
14.7 Advanced Wastewater Treatment 365
14.7.1 Membrane technology 365
14.7.2 Activated carbon 366
14.7.3 Membrane distillation 366
14.7.4 Advanced oxidation processes (AOP) 366
14.7.4.1 Ozone/hydrogen peroxide treatment 367
14.7.4.2 Fenton's oxidation treatment 367
14.7.4.3 Photo-catalysis 367
14.7.4.4 Electrochemical oxidation/degradation 367
14.7.4.5 Ultrasound irradiation 368
14.7.4.6 Wet air oxidation 368
14.7.5 Hybrid technologies 368
14.7.5.1 Hybrid chemical synthesis technologies wastewater process 369

14.7.5.2 Hybrid fermentation technologies wastewater process 369
References 369

14.1 INTRODUCTION

As the human population is growing, the incidence of diseases is ever escalating, forcing people to use antibiotics of various origins. Due to the complex nature of pharmaceutical wastewater (PWW), their treatment is very difficult. Additionally, the presence of recalcitrant constituents increases their treatment cost. Considering the very high cost of their treatment by conventional methods, more innovative and affordable methods are being continuously sought by various researchers. Currently, the incidence of COVID-19 has been observed as a pandemic affecting the lives of billions of people. This pandemic has necessitated the use of many antibiotics which may pollute the environment.

The existence of pharmaceutical residues in wastewater has been increasing over the past decade and various researchers are monitoring and identifying various pharmaceutical industry-related contaminants (Kümmerer, 2008; Rana *et al.*, 2017). Many factors, such as human metabolic activity, plant stabilization and/or human consumption rate, determine the presence of specific contaminants in the environment or their ultimate treatment at wastewater treatment plants (WWTPs) Rana *et al.* (2017). The amount of PWW is constantly increasing and its constituents are very difficult to degrade by various biological means. In addition, PWW constituents have significant toxic effects on microbes involved in their biodegradation Rana *et al.* (2017). PWW's constituents also threaten the sustainability of diverse ecosystem functions. It has been pointed out that the key sources of different pharmaceutical constituents are urban wastewaters, sewage leachates, hospital waste, household waste and livestock rearing (Bound & Voulvoulis, 2005; Kümmerer & Al-Ahmad, 2010; Rana *et al.*, 2017; Ternes, 1998); It was noted that the human use of various antibiotics was also a major cause of PWW containing various obsolete and expired antibiotics, in addition to incomplete metabolic pathways of antibiotics used that are discarded along feces (Bound & Voulvoulis, 2005; Ternes, 1998). Such an incomplete antibiotic metabolism adds up in the sum of residues of PWW in wastewaters Rana *et al.* (2017).

It was pointed out that a better choice would be the PWW technology developed at its generating source Spina *et al.* (2012). The main objective of any technology should be its low treatment and maintenance costs and biological wastewater treatments are considered as one such technology (El-Gohary *et al.*, 1995; Ng *et al.*, 1989; Patil *et al.*, 2013; Rosen *et al.*, 1998). Even in developed nations which can afford expensive treatments, PWW (Ileri *et al.*, 2003; Khan *et al.*, 2002) looks for low cost treatment options. PWW's prevalent types of pollutants include analgesics, hormones, antibiotics, therapeutic agents, and steroidal medicines used in the care of humans and other animals (Heberer, 2002). Such contaminants, once released into the environment, continue to circulate in the environment, and can eventually enter the human food chain (Heberer, 2002).

14.1.1 Antibiotic waste

Although many antibiotics can be effectively treated or metabolized by human metabolic pathways, or cattle enzymes, or even can be biodegraded by environmental microbes, there are still several antibiotics which are recalcitrant to biodegradation and persist even after wastewater has been treated in WWTPs. Such antibiotics are normally present in the effluents of WWTPs. The examples of recalcitrant antibiotics include ibuprofen, oxytetracycline, tetracycline, norfloxacin, and ciprofloxacin (Halling-Sørensen *et al.*, 2000). Low cost solutions are highly desirable for the treatment of such antibiotics. Some plant species like *Lupinus albus* have been reported to remediate acetaminophen in aqua cultures (Halling-Sørensen *et al.*, 2000).

14.1.2 Hormonal waste

Human urine and municipal wastewaters from many regions have been reported to contain some hormones, particularly steroids (Cartinella *et al.*, 2006). Such hormones normally come from contraceptive drugs and other therapies. The presence of such hormones in the municipal wastewaters creates many issues. For example, these steroids may disrupt the endocrine glands of many aquatic animals such as fish (Cartinella *et al.*, 2006). The treatment of hormone-rich PWW has been rarely reported by any researcher however, a few plant species like *Arabidopsis thaliana* and *Salix exigua* have the ability to metabolize synthetic estrogen (Huang & Sedlak, 2001).

14.1.3 Improper disposal of analgesics

NSAIDs (nonsteroidal anti-inflammatory drugs) and paracetamol are commonly used as medicines for pain killers or pain relievers. NSAID-like compounds are also prevalent in many urban wastewaters, which even after passing through WWTPs further pollute freshwater bodies such as ponds, lakes, dams etc. Many individuals typically use aspirin or paracetamol to relieve pain, eventually these reach the disposal site with other solid domestic waste and ultimately pollute the soil. The plant species such as *Phragmites australis* have shown the ability to metabolize ibuprofen pain killers (Kotyza *et al.*, 2010).

14.2 TREATMENT OPTIONS FOR PWW

Currently, many treatment options are available for the PWW. Various treatment technologies used for PWW are mentioned in Table 14.1. The following section will describe an overview of various technologies.

Table 14.1 Various options for pharmaceutical wastewater treatment.

Treatment	Classification
Any physical treatment	Primary treatment
Aerobic Anaerobic	Secondary biological treatment
Membrane technology Activated carbon Membrane distillation	Tertiary treatments
Ozone/hydrogen peroxide treatment Fenton oxidation Photo catalysis Electrochemical oxidation/degradation Ultrasound irradiation Wet air oxidation	Advanced oxidation processes
Mixed primary, secondary and tertiary treatments	Hybrid technologies

14.2.1 Biological treatment of pharmaceutical wastewater

Biological treatment involves the use of plants, microbes and fungi to treat pharmaceutical waste. During biological treatment the waste is converted or degraded into some other non-toxic form. Many factors affect the biological degradation of diverse PWW-related contaminants including:

- chemical structure;
- toxic effects caused by contaminant;
- the concentration of pollutant;
- capability of the microbes;
- physico-chemical environment;
- retention time (RT); and
- the amount of other chemical compounds.

Various types of microbes are essential for environmental remediation of complex organic pollutants involving either stabilization or mineralization (Misal *et al.*, 2011). In addition, it is found that different groups of microbes consisting of mixed bacterial and fungal species were the most effective in treating toxic PWW pollutants (Rosen *et al.*, 1998). Both aerobic and anaerobic types of microbial species are effective in treating PWW Deegan *et al.* (2011). Different researchers observed the efficiency of full-scale and high rate biological reactors to treat PWW. A membrane aerated biofilm reactor (MABR) was employed to treat PWW which displayed very high efficiency (90%) for COD reduction (Wei *et al.*, 2012) while its nitrogen removal was also very high at 98%. Some researchers (Sirtori *et al.*, 2009) tried an integrated type of biological and non-biological treatment systems employing photo-Fenton along immobilized biomass to treat PWW and accomplished a 95% decrease in DOC (Sirtori *et al.*, 2009).

Biological treatment methods have historically been used to handle PWW, aerobic and anaerobic processes. Aerobic systems include activated sludge, membrane batch reactors and series batch reactors (Raj & Anjaneyulu, 2005). Anaerobic approaches include anaerobic sludge reactors, anaerobic film reactors and anaerobic filters.

14.2.2 Anaerobic treatment

Anaerobic treatment (AT) not only accomplishes waste stabilization, but also helps to recover bioenergy in the form of biogas which can supplement the national fuel demands. AT was previously employed to wastewater, ranging from high to low strength waste, mainly from food processing, textile, distilleries, paper and pulp, and tanneries. The major problems associated with the anaerobic treatment of wastewaters is the souring of reactors upon high loading (low pH) and slow microbial doubling rate, and long hydraulic retention time (HRT). Different anaerobic reactor designs have been used in various researches which include contact process, fluidized bed, up flow anaerobic sludge blanket reactor (UASB), and fixed film reactor (AFFR) (Rao *et al.*, 2005).

14.2.2.1 UASB bioreactors

UASB bioreactors have been widely employed to treat various types of wastewaters from industries and municipal sources (Tóth *et al.*, 2011). The feature of granulation of sludge in UASB is the most important characteristic and is crucial for the success of UASB reactors (Fang *et al.*, 1994). Granulation is accomplished by compaction and self-immobilization due to which the microbes are held inside UASB and are not prone to be lost as mixed liquor suspended solids (MLSS) (Akunna & Clark, 2000). Other benefits of using UASB include the absence of mixing requirement, reuse of sludge and survival capability due to high mass loading (Rana *et al.*, 2017). Oktem *et al.* (2008) employed a hybrid UASB bioreactor to treat PWW and obtained promising results in reducing its toxicity and complexity.

14.2.2.2 Fixed film bioreactors

In FFB reactors biofilms are produced on the supporting material (Rana *et al.*, 2017). This bioreactor is easy to construct and handle as mechanical mixing is not needed. Moreover, these kinds of bioreactors are stable, starve resistant and absorb shock load (Rajeshwari *et al.*, 2000). The support materials are of various kinds, including charcoal, coconut coir, polyacrylate nitrile acryl amide, glass bead, polyurethane foam, nylon fiber, red drain clay, waste tyre rubber, splintered glass, corrugated plastic (Acharya *et al.*, 2008; Borja *et al.*, 1996; Garcia *et al.*, 1995; Pérez *et al.*, 1997). Wastewater from the bulk drug industry was treated using anaerobic FFB by (Raj & Anjaneyulu, 2005) which contained high suspended solids and concluded that the AFFR could be effectively employed to treat PWW containing high BOD, COD and TSS.

14.2.2.3 Anaerobic membrane bioreactor

Anaerobic membrane bioreactors (AMRs) are also a good choice for treating PWW and biodegradable recalcitrant (Ng *et al.*, 2010). AMR is emerging as a promising alternative technology for treating aerobic wastewater. Compared with aerobic processes, it has low energy input as well as lower sludge production. The development of AMR provides a promising solution, when combined with an anaerobic process and membrane filtration, the complete retention of biomass may produce a solid free effluent. AMR is capable of producing methane from the use of a large fraction of organics in wastewater and can contribute to energy production. Footprint and room can also be reduced while delivering a highly improved quality effluent (Ng *et al.*, 2010). However, adopting this technology on a large scale is still hampered by several issues which include membrane fouling, recovery of dissolved CH_4 and low COD. Membrane fouling is a key bottleneck for their marketing (Ng *et al.*, 1989).

14.2.2.4 Chain of processes in anaerobic membrane bioreactor

AMR technology is promising to treat wastewater and resource recovery such as bioenergy in the form of renewable methane, fertilizing elements contained therein, and reuse water to control environmental pollution and energy sustainability (Charfi *et al.*, 2017).

Anaerobic wastewater treatment involves a sequential process, beginning with hydrolysis. During this process, the breakage of complex organic matter results in small water-soluble compounds. The complex molecules containing proteins, carbohydrates and fats are converted by facultative and/or obligate anaerobic hydrolytic bacteria into monosaccharides, fatty acids and amino acids. The hydrolysis rate depends on different process factors such as the temperature, pH, type and mass of the hydrolyzing concentration (Visvanathan & Abeynayaka, 2012).

Acidogenesis is a second step in the process of anaerobic treatment. During this process soluble sugars, amino acids and fatty acids are converted into short chain volatile fatty acids (VFAs: formic, butyric, propionic, and acetic acid), alcohols, aldehydes, hydrogen and carbon dioxide. The characteristics of acidogenic products vary according to the environmental conditions and bacterial type (Visvanathan & Abeynayaka, 2012).

Acetogenesis VFAs, alcohols, and hydrogen are converted into acetate in the third step of the anaerobic treatment process. Two groups of bacteria are involved: (1) obligate H_2 producing bacteria for the oxidation of butyrate, propionate and long chain fatty acids to acetate; and (2) H_2 and CO_2 intake bacteria or acetogenic bacteria, to produce acetate by homoacetogens. Homoacetogens also utilize sugar and alcohols, generating acetate and H_2. Acetogenic bacteria include Clostridium, Acetobacterium, and Sporomusa (Ketheesan & Stuckey, 2015).

Methanogenesis, the production of methane in anaerobic digestion, is a two-way process. Methane is first produced from acetate by aceticlastic methanogens such as methanosarcina and methanosaeta. Approximately 70% methane is produced via aceticlastic methanogenesis. Methane production through hydrogenotrophic methanogens (group of archaea) is the second way of production, which converts carbon dioxide and hydrogen to methane and carbon dioxide (Xie *et al.*, 2017). The production rate of methane depends on the availability of methane formers and substrate (Hussain & Dubey, 2017).

14.2.2.5 Anaerobic sequencing batch reactor (AnSBR)

The anaerobic sequencing batch reactor (ANSBR) is a conceptually simple wastewater treatment technology: During the process, a batch reactor is filled with organic waste; biomass, methane, and carbon dioxide are produced through biochemical reactions while biomass is allowed to settle; clarified supernatant is drawn off; and the process is repeated. Granulate biomass was observed in ANSBRs, allowing high biomass concentrations (30 g L^{-1}) A potential problem with the ANSBR comes from the dynamic microbial community responsible for degrading the waste constituents, especially when handling high strength wastewater. A typical operational strategy has been to fill the reactor quickly (with a fill-time-to-react ratio [F:R]) as sugars can lead to rapid accumulation of volatile fatty acids (VFAs) since acid organisms produce VFAs faster than acetogenic and methanogenic organisms can consume (Charfi *et al.*, 2017).

The anaerobic batch sequencing reactor (AnSBR) is a type of activated sludge process in which five process stages occur in a single reactor, namely filling, reacting, settling, decanting and idling in the respective time ratio without oxygen in dark conditions. AnSBR has been tested for the effect of various doses of antibiotic sulfamethoxazole (SMX) on the use of volatile fatty acids (VFA), and other parameters (Cetecioglu *et al.*, 2015). Their findings have suggested anaerobic treatment with a concentration of up to 40 mg/L of SMX for PWW was appropriate. Lower levels of SMX exerted toxic effects on the microbial community, resulting in inhibition of substrate/COD usage and biogas production, and reactor activity failure. In addition, the effect of various doses of antibiotic SMX on microbial community composition was investigated using AnSBR (Cetecioglu *et al.*, 2015). It has been concluded that through homoacetogenesis coupled with hydrogenotrophic methanogenesis the microbial community preserves the stability of the system under high levels of antibiotics and long-term operation (Shi *et al.*, 2017).

14.2.2.6 Advantages and disadvantages of a batch system

Naturally, the concentration of contaminants in wastewater varies with time/space, and thus feed wastewater has a potentially unstable state behavior. However, conventional treatment schemes for wastewater are all unrealistically designed to operate as stable state systems. This is because it was always assumed that steady state conditions would be necessary to keep effluent concentrations constant and within the limits allowed. It was thought that the incorporation of an equalization tank would have the potential to dampen the impact of the unstable behavior of the system but could not compensate for variations in mass flow rates. Thus these systems are unregulated unstable state systems instead of being run as a stable state system for which they are intended. Such unregulated unstable state systems fail to meet the requirements of the constant state. In practice, the factors known to be effective in controlling the structure and function of microbial aggregates (e.g., activated sludges, biofilms) are difficult to maintain in continuous flow systems. Conversely, the batch system is an unstable state controlled system. Batch systems were first utilized in activated sludge processes as a means of controlling filamentous organisms. The frequent shift of activated sludge between aerobic, anoxic and anaerobic zones allows for the creation of microbial communities capable of nitrification, denitrification and increased absorption of biological phosphates. Short-term unstable state conditions, if properly selected and controlled, tend to be an important tool for the long-term maintenance of quasi state conditions. It is difficult to maintain factors known to be effective in controlling the structure and function of microbial aggregates in continuous flow systems. Since continuous flow systems are primarily designed for maximum loading, these biological systems often operate under sub-optimal conditions of control, even if they are controlled under unstable conditions. A batch reactor can alleviate certain continuous flow reactor deficiencies. The batch reactor can vary its effectiveness over time by volume. The key feature of SBR technology from a microbiological point of view is the transition in a process between the feast and famine. Interactions between different microorganisms are optimized in such fluctuating conditions, especially for the large, diverse and effective microbial population.

14.2.2.7 Hybrid anaerobic and aerobic treatment

Single anaerobic system performance varies greatly, generally achieving efficient COD reduction for PWW. The disparity in the efficiency of anaerobic treatment could be explained by the different characteristics of wastewater, which include salinity, contamination and refractory compounds which are hardly biodegradable under anaerobic conditions. A combination of anaerobic and aerobic process is thus found to be a viable and promising option for more complete oxidation and stable performance in the treatment. Basically, a sequential anaerobic–aerobic process consists of an anaerobic stage where most organic pollutants are removed, and an aerobic stage as a polishing step to further remove the organic residues in the anaerobic effluent and to achieve ammonia reduction through nitrification. UASB + aerobic continuous stirred tank reactor (CSTR) is a popular combination for treating PWW. At an initial sulfamerazine concentration of 90 mg/L, the treatability of sulfamerazine in the sequential UASB and CSTR achieved COD removal efficiency of 97%. COD from readily degradable organics did not limit the anaerobic biodegradation process, while CODs from slowly degradable organics and inert microbial products significantly reduced the performance of anaerobic ABR reactor systems. A strong linear correlation between acute toxicity, CODs from slowly degradable organics and inert microbial products, while a weak correlation has been detected between acute toxicity and CODs from inert compounds (Sponza & Demirden, 2010).

A further extension of CSTR, together with UASB-AF, is known as the anaerobic two-phase digester (Demirer & Chen, 2005). Using this hybrid UASB system, the group investigated various growth rates and optimum pH needed for acid and methanogenic microorganisms, and different reactor conditions requirements. However, a subsequent aerobic cycle is required to meet the discharge requirements, and it has been found that MBR is effective to reach the discharge limit (Shi *et al.*, 2017).

14.2.2.8 Hybrid anaerobic + chemical/physical systems

In addition to hybrid anaerobic systems using biological processes, other hybrid anaerobic systems would also take into account processes other than biological, such as chemical and physical processes, in order to achieve enhanced performance of treatment. A novel combination method comprising a UASB, a novel micro-aerobic hydrolysis acidification reactor (NHAR), a cyclic activated sludge method (CASS) and a biological contact oxidation tank (BCOT) to treat both organics and ammonia in chemical synthesis-based PWW was investigated for the simultaneous removal of multiple pollutants (Chen *et al.*, 2014). The removal of COD, ammonia, and amoxicillin was all >90%, and the main mechanism for removing amoxicillin was proposed as biodegradation, adsorption, and hydrolysis. Meanwhile, applying a sequential biocatalyst to the UASB (SBA-UASB) has been found to be effective in the removal of nitrogen compared to treatment with a single UASB system. The biocatalyst supplement was able to remove nitrogen of up to 9.4 kg $N/m^3/d$ compared to the non-biocatalyst supplement, which was able to remove up to 0.178 kg $N/m^3/d$ with an initial nitrogen charge of up to 10.4 kg $N/m^3/d$ (Tang *et al.*, 2017).

14.2.3 Aerobic treatment

In aerobic processes, the aerobic batch sequencing reactor (ASBR) and the active sludge cycle were widely used for industrial wastewater treatment. This cycle consists of a main settling tank, an intermediate retention drill, two storage tanks and an aerobic tank. Some researchers have published on PWW aerobic biological treatment (LaPara *et al.*, 2001). The aerobic treatment of PWW in a biological reactor was investigated by Khan and Mostafa (Khan & Mostafa, 2011). For the treatment of PWW, a laboratory-scale batch type of integrated aerobic biological treatment plant was designed and operated, and its performance assessed. A cylindrical open tank was used as a reactor and air moved through the tank bottom. The average HRT in this study was 15 days. Every day, they collected the treated water samples and tested it for the following parameters: COD, TSS, pH, conductivity, and TDS to determine the plant's efficiency. Approximately 75% COD removal was achieved using 15-day HRT. Analysis of the treated wastewater indicate that the parameters pH, BOD, COD, TSS, TDS and color were found within the acceptable limits prescribed, indicating the plant's efficiency. This treatment plant has a high potential for eliminating COD and TSS and can be considered as a possible wastewater treatment system for industrial treatment. To sum up, such a plant is environmentally friendly; minimizes the use of chemicals, is cost-effective and easy to use for industrial effluent treatment. Elmolla and Chaudhuri (2012) studied optimizing SBR operating conditions for treating high-strength mixed PWW and achieved BOD removal of 94% and COD elimination of 83% at 24 h HRT. Adishkumar *et al.* (2012) researched the combined solar photo-Fenton process with an aerobic sequencing batch reactor for PWW treatment and achieved 98% COD removal. Altaf and Ali (2010) performed experiments on the treatment of PWW using a sequential batch reactor (SBR) and developed a biological method for measuring the relative toxicity of 2.0 mg/L dissolved oxygen at various pH levels (6.62, 6.69, 6.79 and 6.9) for 7, 14 and 21 days. They observed a substantial reduction in levels of BOD, COD, oil and grease, TDS, TSS, and ammonia that meet the National Environmental Quality Standards (NEQS) requirements.

14.3 FUNGAL TREATMENT

There are a variety of fungal strains that play an important role in industrial wastewater treatment but have drawbacks due to the presence of the long growth cycle and the production of spores. Spina *et al.* (2012) compared fungal treatment to activated sludge for treatment of industrial PWW. They used a fungal strain for this analysis, Bjerkandera adusta MUT 2295, through which they achieved a COD reduction of 91% compared with activated sludge, which reduced COD by 78%.

14.4 BACTERIAL TREATMENT

Some bacterial strains such as Pseudomonas, Enterobactor, Streptomonas, Aeromonas, Acinetobactor and Klebsiella have a COD reduction of up to 44% (Ghosh *et al.*, 2004). Chaturvedi *et al.* (2006) isolated 15 bacteria of the rhizosphere, showing a color reduction of 76% and a reduction of BOD and COD of 85–86% within 30 days. Usage of *Escherichia coli* and methanogenic consortium cells for the removal of toxic contaminants has been reported. Numerous bacteria are known to degrade phenolic and complex organic compounds from the Arthrobacter, Comamonas, Rhodococcus, and Ralstonia genera in particular. Some species of Clostridium (fermenting bacteria) can degrade resorcinol (Kavitha & Beebi 2003). Soil bacteria, especially Pseudomonas predominantly in rhizospheric soil, possess a special potential for phenolic biodegradation. Duffner *et al.* (2000) proposed the thermophilic Bacillus thermoglucosidasius A7 degradation of phenol/cresol, which degrades phenol at 65°C via the meta-cleavage pathway. Bacteria P. fluorescence have also been documented to be capable of degrading phenol and chlorophenols (Agarry & Solomon, 2008). Das *et al.* (2012) investigated the control of the parameters of pharmaceutical effluents by bioremediation.

14.5 TRANSFORMATION OF PHARMACEUTICALS IN PLANTS

Metformin and iopromide are both widely detected in water bodies and are not completely soluble in usually controlled WWTPs. Both compounds clearly belong to critical PPCP micropollutants that require special attention, as their incomplete degradation products can cause secondary environmental pollution. Typha latifolia has been described as a tolerant species for many pollutants and is thus widely used for phytoremediation in developed wetlands. Water hyacinth, a floating aquatic plant, has been described as one of the world's most productive photosynthetic plants, with ten plants producing 600,000 more in an eight-month growing season and covering 0.4 ha (1 acre) of natural freshwater soil entirely. Water hyacinth studies show that it enhances wastewater effluent from oxidation ponds (Polprasert & Khatiwada, 1998) and integrated treatment systems (Liao & Chang, 2011). Studies conducted by Ayman Allam *et al.* (2015), using Duckweed (Lemna gibba) to treat drainage water containing pharmaceuticals shows that duckweed treatment systems provide a cost-effective, less energetic treatment system. The duckweeds are small floating aquatic macrophytes that grow in fresh waters and on the nutrient-rich surface. Contaminated water phytoremediation using duckweed species is promising due to its ability to grow at a wide range of temperature, pH and nutrient (N&P) levels in areas where land is healthy for use. Previous studies investigated the ability of duckweed to cure organic chemicals, including pharmaceuticals (Richards *et al.*, 2004; Schröder *et al.*, 2007). The uptake of organic chemicals by duckweed is based on a complex combination of processes that are both abiotic and plant driven.

Similarly, tetracycline, oxytetracycline, ibuprofen, ciprofloxacin, and norfloxacin remain in drainage even after treatment. Environmental cleaning technology has to be explored, and some plants have been documented to carry out remediation for it, such as Lupinus albus, which remedies acetaminophen under hydroponic conditions (Kotyza *et al.*, 2010). It has been reported that wastewater from municipal waste and human urinary waste have in them sources of hormones, particularly steroids. Plants named Salix exigua and Arabidopsis thaliana have been reported to extract synthetic estrogen and other derivatives from solutions which are thus indicative of their phytoremediation properties (Franks, 2006). NSAIDs, that is non-steroidal anti-inflammatory drugs, and paracetamol, are basically analgesics or painkillers. Even after treatment procedures these bioactive compounds and their metabolic products are part of wastewater streams and river waters. Unused daily medicinal products such as aspirin, paracetamol, etc. are disposed of just as domestic waste, which causes undesirable effects such as renal toxicity, hepatotoxicity, etc., before entering the water table and the food chains. Many plants remediate these toxic wastes from the soil, for example, *Phragmites australis* is known for eliminating ibuprofen contamination from soil (Kotyza *et al.*, 2010).

14.6 TERTIARY TREATMENT VIA MEMBRANE BIOREACTOR (MBR)

The MBR was used for the treatment of industrial wastewater, household wastewater and municipal wastewater on a large scale (Yang *et al.*, 2006). This is used via porous membrane filtration for the biological degradation of wastewater compounds and the physical separation of biomass and treated water. The MBR has several advantages such as complete removal of suspended solids, compact plant size, high degradation rate, operational flexibility, low sludge production rate, disinfection and odor control, extended retention time of the microorganism, and treatment of recalcitrant and harmful organic and inorganic contaminants. Chang *et al.* (2008) achieved the bioreactor membrane treatment of PWW and observed the elimination of COD and BOD by 95 and 99% respectively. Several investigators have used Zee Weed MBR technology for PWW treatment (Noble, 2006). It is likely to become one of the best technologies used to clean up and protect our environment.

14.7 ADVANCED WASTEWATER TREATMENT

Advanced wastewater treatment of PWW is normally employed as the pre-treatment or primary treatment technology to progress the efficiency of elimination of pollutants by secondary treatment. Such technologies include activated carbon-based adsorption, membrane distillation, and membrane technology.

14.7.1 Membrane technology

Membrane implementation in water treatment is increasing sharply. It is well-known that low-pressure membranes are capable of removing microbial constituents without increasing the number of by-products of disinfection, thereby allowing compliance with the rules adopted in response to the amendments to the 1986 Surface Water Treatment Law (Snyder, 2007). If desalination or reuse of water is the goal, low-pressure membrane technology is normally employed as pre-treatment procedures prior to reverse osmosis (RO). An investigation using

RO membranes showed 95% removal of diclofenac (Kimura *et al.*, 2005). Microfiltration/ultrafiltration systems are highly endorsed under the condition of limited space accompanied by variable quality of feedwater. Nanofiltration and ultrafiltration techniques have been used to remove micropollutants and natural organic matter (NOM) in wastewater reclamation and drinking water.

14.7.2 Activated carbon

Owing to its very large surface area, greater than 1000 m^2 per gram, along other features such as well-established pore assembly and surface chemical features, activated carbon (AC) has been extensively investigated to adsorb many organic compounds. It was confirmed that AC was the best choice to eliminate ibuprofen from waste. The AC cycle therefore has an advantage in terms of fast input of raw materials for carbon production. Powdered activated carbon (PAC) or granular activated carbon (GAC) is used in the AC process. Cyr *et al.* (2002) treated PWW by using GAC columns and accomplished 99% elimination of total Hg (organic + inorganic) and about 90% of copper removal from a PWW based on chemical synthesis. Recently, many studies have been published which used AC as pre-treatment to treat many kinds of wastewaters (Stoquart *et al.*, 2012).

14.7.3 Membrane distillation

Membrane distillation is a very powerful separation technology with interesting properties. Currently, membrane distillation is effectively operated to accomplish separation of excessive salts from drinking water (Gryta, 2012). The technology is operated under ordinary environmental conditions without input of extra energy. Membrane distillation offers good quality water; however, one of its chief shortcomings is the fouling of membranes used in the process.

14.7.4 Advanced oxidation processes (AOP)

As the nature of compounds present in PWW is very stable and recalcitrant, individual treatment technologies used for PWW are not effective enough to deal with such compounds. Moreover, half-treated or incompletely treated organic compounds may result in the contamination of freshwater bodies upon discharge and furthermore would cause many toxicological effects for aquatic life (Kümmerer *et al.*, 2000).

AOPs are methods that have been considered robust and used for aqueous medium oxidation. These contain highly reactive chemical compounds like hydroxyl radicals to degrade target contaminants. The important AOPs may be heterogeneous or homogeneous photocatalyst and ultraviolet (UV) or solar irradiation: electrooxidation, wet air oxidation, and Fenton and photo-Fenton processes, ultrasound irradiation and microwave treatment, typically functioning at about 2450 MHz in either a monomodic or multimode vessel type. AOPs may be used either alone or in conjunction with other physico-chemical and biological processes, depending on the nature of PWW and the treatment objectives or transformation.

14.7.4.1 Ozone/hydrogen peroxide treatment

Ozone is a strong oxidizing agent that either degrades under aqueous medium to produce hydroxyl radicals which are more powerful oxidizing agents than ozone itself, thus inducing the indirect oxidation or selectively reacting with many functional groups of pollutants through an electrophilic mechanism (Klavarioti *et al.*, 2009). PWW comprises various types of recalcitrant organics, such as toluene, nitroaniline, nitrophenols, phenols, trichloromethyl propanol (TCMP), and some other compounds which are difficult to be degraded. For these compounds, treatment with ozone/hydrogen peroxide may seem a proven technology for treating such pollutants from PWW. Ozonation was widely used in antibiotic removal (Helmig *et al.*, 2005).

14.7.4.2 Fenton's oxidation treatment

Fenton's reagent involves the reaction of ferrous or ferric ions to hydrogen peroxide through a free radical chain reaction that produces hydroxyl radicals. It is a catalytic heterogeneous reaction where iron acts as a catalyst (Tekin *et al.*, 2006). Because iron is an abundant element, it is the most viable process for treating wastewater. The removal of COD in a pharmaceutical effluent containing chloramphenicol, paracetamol, and COD of some 112 000 mg/L was observed at over 95%. Fenton processes, however, suffer a major disadvantage from pH dependence and iron sludge generation. As a pre-reatment technology, the Fenton process can be better applied to convert the non-biodegradable pharmaceutical effluent into biodegradable and thus make effluent treatment more effective through biological processes.

14.7.4.3 Photo-catalysis

Photo-catalysis is the amplification of photochemical transformation through the action of catalysts such as TiO_2 or Fenton's reagent. Rutile TiO_2 is the catalyst most widely used in all pharmaceutical photocatalytic studies. Photo-catalysis is the most suitable process for high-COD effluents and the complete transformation of highly refractory organic contaminants.

14.7.4.4 Electrochemical oxidation/degradation

The electrochemical process is based on the in situ production of hydroxyl radical (•OH) as the key oxidant, the second strongest oxidizing agent known after fluorine, with such a high standard reduction potential ($E°(•OH/H_2O) = 2.8$ V vs SHE) that it can react with most organic contaminants by hydroxylation or dehydrogenation without reaction.

Simulated, electrochemically treated waste with pharmaceutical residues such as diclofenac, carbamazepine, propranolol, ibuprofen, and ethinyl-estradiol has shown complete degradation. With the appropriate combination of current density and flow rate, TOC removal was observed at nearly 100%. Electro-coagulation coupled with photo-catalysis demonstrated an efficiency of 86% COD removal in wastewater based on chemical synthesis.

14.7.4.5 Ultrasound irradiation

Ultrasound irradiation is a relatively recent method used in treating wastewater. Sonochemical reactions are caused at frequencies causing cavitation (25 kHz), due to high-intensity acoustic irradiation of liquids. Cavitation thus serves as a means of focusing the diffused energy of ultrasound into micro-reactors at the same time as radicals are released. Most estrogenic compounds were eliminated from contaminated waters by ultrasonic irradiation, with a drop of 80–90% COD within 40–60 minutes of treatment.

14.7.4.6 Wet air oxidation

Wet air oxidation is a thermochemical process in which hydroxyl radicals and other active oxygen species develop at high temperatures (200–320°C) and pressures (2–20 MPa). Recent research has demonstrated the applicability of this mechanism to effectively eliminate COD. Catalytic wet air oxidation of a chemical synthesis wastewater with a COD of 7–12 g/L showed removal of total organic matter and improved process with heterogeneous copper catalyst loading and high temperature. A study conducted using the heterogeneous Fe_2O_3/SBA15 Nano-catalyst exhibited high ability to remove TOC and degrade COD. This technique can also be used as a pre-treatment process which makes the wastewater suitable for biological treatment.

14.7.5 Hybrid technologies

Hybrid solutions are the variations of one or more traditional/advanced treatment technologies for pharmaceutical pollutants to be completely eradicated. The need for hybrid technologies stems from the fact that all substances cannot be eliminated by any of the single-treatment technologies. There are a range of hybrid technologies that were used to treat refractory contaminants as well as to reduce treatment process costs. Using the conventional filtration stage, the technology basically eliminates any solid matrix and the sludge is removed for incineration. Instead, the clear wastewater is treated by different process combinations.

14.7.5.1 Hybrid chemical synthesis technologies wastewater process

Wastewater from the chemical synthesis process usually produces high concentrations of organic contaminants ranging from reagents to intermediates and end products. To cope with such a matrix, many researchers have used the combination of advanced treatment methods along with biological treatment methods. Chen *et al.* (2014) used a two-step anaerobic digestion (TPAD) system and a corresponding MBR, TPAD system consisting of a blanket-anaerobic up-flow anaerobic sludge filter (UASBAF) and a continuous stirred tank reactor (CSTR) acting as the acid and methanogenic phases. The integrated pilot plant removed COD by 99%, and the MBR increased the neutral range pH. TPAD-MBR combination can be successfully applied to wastewater that is based on chemical synthesis.

14.7.5.2 Hybrid fermentation technologies wastewater process

Fermentation process wastewater consists mainly of fermentation broth, mycelia and nutrients added to the cell culture. There are also some organic solvents that are added for interest in recovering the API. Helmig *et al.* (2005) treated API formulating waste consisting of estrogens with a hybrid treatment technology consisting of ozonation pre-treatment and aerobic treatment, that is bioreactor membrane technology. It obtained more than 90% COD and TSS removal while total removal was achieved through MBR. Cokgor *et al.* (2005) investigated the treatment of penicillin from PWW. They employed ozonation as pre-treatment along synthetic biomass to accomplish 30% reduction in COD for biologically activated sludge. Ozonation accomplished 34% removal of COD and 24% TOC, and finally COD was treated by activated sludge. Tylosin and avilamycin containing wastewater were treated by Chelliapan and Sallis (2011) using a hybrid upflow anaerobic stage reactor (UASR). UASR has been widely used as a substitute of pre-treatment for COD reduction (up to 70–75%) for avilamycin macrolide and tylosin antibiotics containing PWW. The coagulation stage of Fenton oxidation (pre-treatment) followed by aerobic biological degradation in the batch SBR caused COD removal of 45–50% and the biological treatment reduced COD to 98%.

REFERENCES

Acharya B. K., Mohana S. and Madamwar D. (2008). Anaerobic treatment of distillery spent wash – A study on upflow anaerobic fixed film bioreactor. *Bioresource Technology*, **99**(11), 4621–4626.

Adishkumar S., Sivajothi S. and Rajesh Banu J. (2012). Coupled solar photo-Fenton process with aerobic sequential batch reactor for treatment of pharmaceutical wastewater. *Desalination and Water Treatment*, **48**(1–3), 89–95.

Agarry S. and Solomon B. (2008). Kinetics of batch microbial degradation of phenols by indigenous Pseudomonas fluorescence. *International Journal of Environmental Science and Technology*, **5**(2), 223–232.

Akunna J. C. and Clark M. (2000). Performance of a granular-bed anaerobic baffled reactor (GRABBR) treating whisky distillery wastewater. *Bioresource Technology*, **74**(3), 257–261.

Allam A., Fleifle A., Tawfik A., Yoshimura C. and El-Saadi A. (2015). A simulation-based suitability index of the quality and quantity of agricultural drainage water for reuse in irrigation. *Science of the Total Environment*, **536**, 79–90.

Altaf M. S. and Ali T. A. (2010). Waste water treatment using sequential batch reactor and development of microbiological method for the analysis of relative toxicity. *Pakistan Journal of Nutrition*, **9**(6), 574–576.

Borja R., Sánchez E., Martín A. and Jiménez A. (1996). Kinetic behaviour of waste tyre rubber as microorganism support in an anaerobic digester treating cane molasses distillery slops. *Bioprocess Engineering*, **16**(1), 17–23.

Bound J. P. and Voulvoulis N. (2005). Household disposal of pharmaceuticals as a pathway for aquatic contamination in the United Kingdom. *Environmental Health Perspectives*, **113**(12), 1705–1711.

Cartinella J. L., Cath T. Y., Flynn M. T., Miller G. C., Hunter K. W. and Childress A. E. (2006). Removal of natural steroid hormones from wastewater using membrane contactor processes. *Environmental Science and Technology*, **40**(23), 7381–7386.

Cetecioglu Z., Ince B., Gros M., Rodriguez-Mozaz S., Barceló D., Ince O. and Orhon D. (2015). Biodegradation and reversible inhibitory impact of sulfamethoxazole on the utilization of volatile fatty acids during anaerobic treatment of pharmaceutical industry wastewater. *Science of the Total Environment*, **536**, 667–674.

Chang C.-Y., Chang J.-S., Vigneswaran S. and Kandasamy J. (2008). Pharmaceutical wastewater treatment by membrane bioreactor process – A case study in southern Taiwan. *Desalination*, **234**(1–3), 393–401.

Charfi A., Thongmak N., Benyahia B., Aslam M., Harmand J., Amar N. B., Lesage G., Sridang P., Kim J. and Heran M. (2017). A modelling approach to study the fouling of an anaerobic membrane bioreactor for industrial wastewater treatment. *Bioresource Technology*, **245**, 207–215.

Chaturvedi S., Chandra R. and Rai V. (2006). Isolation and characterization of Phragmites australis (L.) rhizosphere bacteria from contaminated site for bioremediation of colored distillery effluent. *Ecological Engineering*, **27**(3), 202–207.

Chelliapan S. and Sallis P. J. (2011). Application of anaerobic biotechnology for pharmaceutical wastewater treatment. *IIOAB Journal*, **2**, 13–21.

Chen J. L., Ortiz R., Steele T. W. and Stuckey D. C. (2014). Toxicants inhibiting anaerobic digestion: a review. *Biotechnology Advances*, **32**(8), 1523–1534.

Cokgor E., Karahan O., Arslan-Alaton I., Saruhan H. and Orhon D. (2005). Biological treatability of raw and ozonated synthetic penicillin formulation effluent. *Water Science and Technology*, **52**(10–11), 89–96.

Cyr P. J., Suri R. P. S. and Helmig E. D. (2002). A pilot scale evaluation of removal of mercury from pharmaceutical wastewater using granular activated carbon. *Water Research*, **36**, 4725–4734.

Das M. P., Bashwant M., Kumar K. and Das J. (2012). Control of pharmaceutical effluent parameters through bioremediation. *Journal of Chemical and Pharmaceutical Research*, **4**(2), 1061–1065.

Deegan A., Shaik B., Nolan K., Urell K., Oelgemöller M., Tobin J. and Morrissey A. (2011). Treatment options for wastewater effluents from pharmaceutical companies. *International Journal of Environmental Science and Technology*, **8**(3), 649–666.

Demirer G. and Chen S. (2005). Two-phase anaerobic digestion of unscreened dairy manure. *Process Biochemistry*, **40**(11), 3542–3549.

Duffner F. M., Kirchner U., Bauer M. P. and Müller R. (2000). Phenol/cresol degradation by the thermophilic Bacillus thermoglucosidasius A7: cloning and sequence analysis of five genes involved in the pathway. *Gene*, **256**(1–2), 215–221.

El-Gohary F., Abou-Elela S. and Aly H. (1995). Evaluation of biological technologies for wastewater treatment in the pharmaceutical industry. *Water Science and Technology*, **32**(11), 13–20.

Elmolla E. S. and Chaudhuri M. (2012). The feasibility of using combined Fenton-SBR for antibiotic wastewater treatment. *Desalination*, **285**, 14–21.

Fang H., Chui H. and Li Y. (1994). Microbial structure and activity of UASB granules treating different wastewaters. *Water Science and Technology*, **30**(12), 87–96.

Franks C. G. (2006). Phytoremediation of Pharmaceuticals with Salix Exigua. University of Lethbridge, Faculty of Arts and Science, Lethbridge, Alta.

Garcia A., Rivas H. M., Figueroa J. L. and Monroe A. L. (1995). Case history: pharmaceutical wastewater treatment plant upgrade, SmithKline Beecham Pharmaceuticals Company. *Desalination*, **102**(1–3), 255–263.

Ghosh M., Verma S., Mengoni A. and Tripathi A. (2004). Enrichment and identification of bacteria capable of reducing chemical oxygen demand of anaerobically treated molasses spent wash. *Journal of Applied Microbiology*, **96**(6), 1278–1286.

Gryta M. (2012). Effectiveness of water desalination by membrane distillation process. *Membranes*, **2**(3), 415–429.

Halling-Sørensen B., Lützhøft H.-C. H., Andersen H. R. and Ingerslev F. (2000). Environmental risk assessment of antibiotics: A comparison of mecillinam, trimethoprim and ciprofloxacin. *Journal of Antimicrobial Chemotherapy*, **46**(suppl. 1), 53–58.

Heberer T. (2002). Tracking persistent pharmaceutical residues from municipal sewage to drinking water. *Journal of Hydrology*, **266**(3–4), 175–189.

Helmig E. G., Fettig J. D., Cordone L., Schoenberg T. H., DeMarco M. J. and Suri R. (2005). API removal from pharmaceutical manufacturing wastewater – results of process development, pilot-testing, and scale-up. *Proceedings of the Water Environment Federation*, **16**, 207–226.

Huang C. H. and Sedlak D. L. (2001). Analysis of estrogenic hormones in municipal wastewater effluent and surface water using enzyme-linked immunosorbent assay and gas chromatography/tandem mass spectrometry. *Environmental Toxicology and Chemistry: An International Journal*, **20**(1), 133–139.

Hussain A. and Dubey S. K. (2017). Specific methanogenic activity test for anaerobic degradation of influents. *Applied Water Science*, **7**(2), 535–542.

Ileri R., Sengil I., Kulac S. and Damar Y. (2003). Treatment of mixed pharmaceutical industry and domestic wastewater by sequencing batch reactor. *Journal of Environmental Science and Health, Part A*, **38**(10), 2101–2111.

Kavitha G. and Beebi S. K. (2003). Biodegradation of phenol in a packed bed reactor using peat media. *Asian Journal of Microbiology, Biotechnology and Environmental Science*, **5**(2), 157–159.

Ketheesan B. and Stuckey D. C. (2015). Effects of hydraulic/organic shock/transient loads in anaerobic wastewater treatment: a review. *Critical Reviews in Environmental Science and Technology*, **45**(24), 2693–2727.

Khan M. Z. H. and Mostafa M. (2011). Aerobic treatment of pharmaceutical wastewater in a biological reactor. *International Journal of Environmental Sciences*, **1**(7), 1797–1805.

Khan S., Khan A. M. and Khan M. N. (2002). Investigation of pollutants load in waste water of Hayatabad Industrial Estate, Peshawar, Pakistan. *Pakistan Journal of Applied Science*, **2**(4), 457–461.

Kimura K., Hara H. and Watanabe Y. (2005). Removal of pharmaceutical compounds by submerged membrane bioreactors (MBRs). *Desalination*, **178**(1–3), 135–140.

Klavarioti M., Mantzavinos D. and Kassinos D. (2009). Removal of residual pharmaceuticals from aqueous systems by advanced oxidation processes. *Environment International*, **35**(2), 402–417.

Kotyza J., Soudek P., Kafka Z. and Vaněk T. (2010). Phytoremediation of pharmaceuticals – preliminary study. *International Journal of Phytoremediation*, **12**(3), 306–316.

Kümmerer K. (2008). Pharmaceuticals in the Environment: Sources, Fate, Effects and Risks. Springer Science and Business Media, Berlin, Heidelberg.

Kümmerer K. and Al-Ahmad A. (2010). Estimation of the cancer risk to humans resulting from the presence of cyclophosphamide and ifosfamide in surface water. *Environmental Science and Pollution Research*, **17**(2), 486–496.

Kümmerer K., Al-Ahmad A. and Mersch-Sundermann V. (2000). Biodegradability of some antibiotics, elimination of the genotoxicity and affection of wastewater bacteria in a simple test. *Chemosphere*, **40**(7), 701–710.

LaPara T. M., Nakatsu C. H., Pantea L. M. and Alleman J. E. (2001). Aerobic biological treatment of a pharmaceutical wastewater: effect of temperature on COD removal and bacterial community development. *Water Research*, **35**(18), 4417–4425.

Liao H.-C. and Chang H.-H. (2011). The optimal approach for parameter settings based on adjustable contracting capacity for the hospital supply chain logistics system. *Expert Systems with Applications*, **38**(5), 4790–4797.

Misal S. A., Lingojwar D. P., Shinde R. M. and Gawai K. R. (2011). Purification and characterization of azoreductase from alkaliphilic strain Bacillus badius. *Process Biochemistry*, **46**(6), 1264–1269.

Ng K.-K., Lin C.-F., Lateef S. K., Panchangam S. C., Hong P.-K. A. and Yang P.-Y. (2010). The effect of soluble microbial products on membrane fouling in a fixed carrier biological system. *Separation and Purification Technology*, **72**(1), 98–104.

Ng W., Yap M. G. and Sivadas M. (1989). Biological treatment of a pharmaceutical wastewater. *Biological Wastes*, **29**(4), 299–311.

Noble J. (2006). GE ZeeWeed MBR technology for pharmaceutical wastewater treatment. *Membrane Technology*, **9**, 7–9.

Oktem Y. A., Ince O., Sallis P., Donnelly T. and Ince B. K. (2008). Anaerobic treatment of a chemical synthesis-based pharmaceutical wastewater in a hybrid upflow anaerobic sludge blanket reactor. *Bioresource Technology*, **99**(5), 1089–1096.

Patil P., Kulkarni G., Kore S. S. and Kore V. (2013). Aerobic sequencing batch reactor for wastewater treatment: a review. *International Journal of Engineering Research and Technology*, **2**(10), 534–550.

Pérez M., Romero L. I., Nebot E. and Sales D. (1997). Colonisation of a porous sintered-glass support in anaerobic thermophilic bioreactors. *Bioresource Technology*, **59**(2–3), 177–183.

Polprasert C. and Khatiwada N. R. (1998). An integrated kinetic model for water hyacinth ponds used for wastewater treatment. *Water Research*, **32**(1), 179–185.

Raj D. S. S. and Anjaneyulu Y. (2005). Evaluation of biokinetic parameters for pharmaceutical wastewaters using aerobic oxidation integrated with chemical treatment. *Process Biochemistry*, **40**(1), 165–175.

Rajeshwari K., Balakrishnan M., Kansal A., Lata K. and Kishore V. (2000). State-of-the-art of anaerobic digestion technology for industrial wastewater treatment. *Renewable and Sustainable Energy Reviews*, **4**(2), 135–156.

Rana R. S., Singh P., Kandari V., Singh R., Dobhal R. and Gupta S. (2017). A review on characterization and bioremediation of pharmaceutical industries' wastewater: an Indian perspective. *Applied Water Science*, **7**(1), 1–12.

Rao A. G., Naidu G. V., Prasad K. K., Rao N. C., Mohan S. V., Jetty A. and Sarma P. (2005). Anaerobic treatment of wastewater with high suspended solids from a bulk drug industry using fixed film reactor (AFFR). *Bioresource Technology*, **96**(1), 87–93.

Richards S. M., Wilson C. J., Johnson D. J., Castle D. M., Lam M., Mabury S. A., Sibley P. K. and Solomon K. R. (2004). Effects of pharmaceutical mixtures in aquatic microcosms. *Environmental Toxicology and Chemistry: An International Journal*, **23**(4), 1035–1042.

Rosen M., Welander T., Löfqvist A. and Holmgren J. (1998). Development of a new process for treatment of a pharmaceutical wastewater. *Water Science and Technology*, **37**(9), 251–258.

Schröder P., Navarro-Aviñó J., Azaizeh H., Goldhirsh A. G., DiGregorio S., Komives T., Langergraber G., Lenz A., Maestri E. and Memon A. R. (2007). Using phytoremediation technologies to upgrade waste water treatment in Europe. *Environmental Science and Pollution Research International*, **14**(7), 490–497.

Shi X., Leong K. Y. and Ng H. Y. (2017). Anaerobic treatment of pharmaceutical wastewater: a critical review. *Bioresource Technology*, **245**, 1238–1244.

Sirtori C., Zapata A., Oller I., Gernjak W., Agüera A. and Malato S. (2009). Decontamination industrial pharmaceutical wastewater by combining solar photo-Fenton and biological treatment. *Water Research*, **43**(3), 661–668.

Snyder L. (2007). Complementary and Alternative Medicine: Ethics, the Patient, and the Physician. Springer Science & Business Media, Humana Press, Switzerland.

Spina F., Anastasi A. E., Prigione V. P., Tigini V. and Varese G. (2012). Biological treatment of industrial wastewaters: a fungal approach. *Chemical Engineering Transactions*, **27**, 175–180.

Sponza D. T. and Demirden P. (2010). Relationships between chemical oxygen demand (COD) components and toxicity in a sequential anaerobic baffled reactor/aerobic completely stirred reactor system treating Kemicetine. *Journal of Hazardous Materials*, **176**(1–3), 64–75.

Stoquart C., Servais P., Bérubé P. R. and Barbeau B. (2012). Hybrid membrane processes using activated carbon treatment for drinking water: A review. *Journal of Membrane Science*, **411**, 1–12.

Tang C.-J., Duan C.-S., Yu C., Song Y.-X., Chai L.-Y., Xiao R., Wei Z. and Min X.-B. (2017). Removal of nitrogen from wastewaters by anaerobic ammonium oxidation (ANAMMOX) using granules in upflow reactors. *Environmental Chemistry Letters*, **15**(2), 311–328.

Tekin H., Bilkay O., Ataberk S. S., Balta T. H., Ceribasi I. H., Sanin F. D., Dilek F. B. and Yetis U. (2006). Use of Fenton oxidation to improve the biodegradability of a pharmaceutical wastewater. *Journal of Hazardous Materials*, **136**(2), 258–265.

Ternes T. A. (1998). Occurrence of drugs in German sewage treatment plants and rivers. *Water Research*, **32**(11), 3245–3260.

Tóth A. J., Gergely F. and Mizsey P. (2011). Physicochemical treatment of pharmaceutical process wastewater: distillation and membrane processes. *Periodica Polytechnica Chemical Engineering*, **55**(2), 59–67.

Visvanathan C. and Abeynayaka A. (2012). Developments and future potentials of anaerobic membrane bioreactors (AnMBRs). *Membrane Water Treatment*, **3**(1), 1–23.

Wei X., Li B., Zhao S., Wang L., Zhang H., Li C. and Wang S. (2012). Mixed pharmaceutical wastewater treatment by integrated membrane-aerated biofilm reactor (MABR) system – A pilot-scale study. *Bioresource Technology*, **122**, 189–195.

Xie S., Wickham R. and Nghiem L. D. (2017). Synergistic effect from anaerobic co-digestion of sewage sludge and organic wastes. *International Biodeterioration and Biodegradation*, **116**, 191–197.

Yang S.-Y., Hsiung S.-K., Hung Y.-C., Chang C.-M., Liao T.-L. and Lee G.-B. (2006). A cell counting and sorting system incorporated with a microfabricated flow cytometer chip. *Measurement Science and Technology*, **17**, 2001–2009.

Chapter 15

Future scope – expected treatments for Phcs in future

Viola Vambol[1], Nadeem A Khan[2], Sergij Vambol[3] and Sirajuddin Ahmed[2]

[1]Educational and Scientific Department of Occupational Safety and Health, National Scientific and Research Institute of Industrial Safety and Occupational Safety and Health, Ukraine
[2]Department of Civil Engineering, Jamia Millia Islamia, New Delhi
[3]Life Safety and Law Department, Kharkiv Petro Vasylenko National Technical University of Agriculture, Ukraine

ABSTRACT

This chapter is the last chapter of the book where the most up-to-date research results have been presented. The main problems that scientists face when conducting research are highlighted. Attention is focused on the key results obtained by researchers in the chapters of the book. The future prospect of effective wastewater management, including those containing pharmaceutical ingredients, is seen in close collaboration between natural resource management bodies, research institutes and criminal regulatory authorities.

Keywords: Environmental risk water protection, Legislative and criminal law mechanisms, Pharmaceutical ingredients, Polluted water, Treatment techniques.

Contents

15.1 Summarized Results 376
15.2 Analysis and Conclusions 376
15.3 Future Prospects 379
References 381

Editor(s): Nadeem A Khan, Sirajuddin Ahmed, Viola Vambol and Sergij Vambol
doi: 10.2166/9781789061338_0375

15.1 SUMMARIZED RESULTS

This final chapter sets out the summarized results of the current research and studies presented in the chapters of this book.

The most up-to-date research results were systematized and published due to fruitful international cooperation of the scientific community, practitioners of regulatory bodies and other persons whose scientific interests lie in the modern achievements field of pharmaceutical wastewater management; of identifying sources of danger of penetration of active pharmaceutical ingredients into the environment; of ensuring the protection of water bodies from the penetration of pollutants, including pharmaceuticals; of legislating the pharmaceutical effluents' disposal and the development of rules and laws regarding the waste disposal containing active pharmaceutical ingredients.

There are many problems that researchers face in the field of water protection and wastewater management. Among these problems are: a survey of the population regarding behaviour with unused drugs (researchers often encounter a lack of understanding of the population of priority importance and responsibility for methods of removing these highly toxic substances); sampling in rivers and wastewater prior to its treatment (as a rule, there are many barriers for researchers to take qualitative sampling, both of a technological nature and of a permissive nature, that is, there is availability of admission or permit) and many others.

This book contains the results of various international, national and regional research projects, as well as personal research and research by research teams.

15.2 ANALYSIS AND CONCLUSIONS

It is now known that the physico-chemical and biological properties of pharmaceutical microcontaminants distinguish their biological strength when they disperse in the environment (Amy *et al.*, 1996; Anctil, 2008; Barbăneagră, 2003). The distinctive chemical structures of drugs, carbohydrates, proteins, organic molecules, and other chemically complex components are large polymers that show their extensive characteristics depending on their chemical composition (Amy *et al.*, 1996). Water bodies, including surface water, wastewater from wastewater treatment plants, and groundwater, were widely exposed to pharmaceuticals that ultimately end up in drinking water supplies (Belecciu, 2017; Benotti *et al.*, 2009; der Beak *et al.*, 2016; Dhingra *et al.*, 2020; Mazhar *et al.*, 2020).

Homeowners are important sources of unused pharmaceuticals that, after the expiration date, are turned into waste and must be disposed of. This is because residents get rid of unused pharmaceuticals by disposing of them in household waste (landfills) or in the sewer. It was incorrectly assumed that chemical accumulations ended up at wastewater treatment plants or landfills, since these accumulations penetrate drinking water sources through leaching (Directive,

2000/60/EC, 2000; Falås *et al.*, 2012). This is evidenced by the results of the detection in water bodies in some countries of high concentrations of more than 100 types of various pharmaceutical preparations (Belecciu, 2017). This is an important, but not the only, reason for the active pharmaceutical ingredients to enter the environment. A significant contribution to this problem is made by wastewater from specialized institutions of the medical industry (clinics, stationary and mobile hospitals, pharmaceutical enterprises, etc.) (Fatta-Kassinos *et al.*, 2011; Hamoda *et al.*, 2004); of the agricultural enterprises, fisheries and livestock farms; of the treatment plants that annually produce millions of tons of sewage sludge.

The excessive use of antimicrobials is causing concern and concern worldwide; as it contributes to higher concentrations of enter pathogenic viruses in hospital wastewater. That is, sewage (as well as its sludge at sewage treatment plants) delivers not only active pharmaceutical ingredients to the environment, but also viruses and bacteria that are resistant to antibiotics. Antimicrobial substances in wastewater make them more resistant. In addition, hospital wastewater is a source of silver and mercury (due to the use of X-ray machines), high oil content, fat (due to the preparation of meals served to patients), detergents from laundries and pharmaceutical residues through the urine of patients. All this obliges to assess the environmental risk for the aquatic ecosystem in order to outline priority environmental protection measures.

Among the large range of drugs, it is important to prioritize to prevent their entry into the environment and to develop methods to neutralize them. The selection of criteria should be objective, reasonable and sufficient. If the number of criteria is too large, then almost all known preparations will be assigned to priority pharmaceutical compositions, which will create significant difficulties in the development of purification methods. At the same time, with a very limited number of criteria, some hazardous drugs will be ignored and will contribute to environmental degradation. In both cases, the goal – to prevent environmental degradation under the influence of active pharmaceutical ingredients – will not be achieved. That is, a lot of the work that is carried out to achieve the goal will not be effective. It was an extensive review of the literature on various priority criteria, and then a discussion with dozens of experts, including doctors, pharmacists, pharmacologists, and academics, experts from non-governmental organizations and regulatory bodies (the Pharmacopoeia Commission, the Health Department, and the Pollution Control Committee), that identified 12 criteria as appropriate. These criteria are justified by expert recommendations, logical and scientific conclusions.

Using 12 criteria, nine high-priority drugs for therapy were established, namely diclofenac, ibuprofen, erythromycin, ofloxacin, diuretic fructemide, diazepam, carbamazepine, sorafenib, and simvastatin. These drugs are most often consumed by patients and have a high potential hazard to environmental components. In addition, the European Commission on the

Water Framework Directive has already included diclofenac in the list of priority substances, the discharge of which into the environment is controlled (Kettab, 2001; Khan *et al.*, 2020a). Thus, in-depth knowledge about the characteristics of hospital runoffs allows you to focus on developing the most effective treatment for such runoffs in order to preserve the environment and public health.

In addition, the widespread reuse of water can help compensate for the water scarcity that exists in many countries of the world (Khan *et al.*, 2020b). For example, the water shortage in Algeria is increasing significantly, and the resource is becoming less and less regenerated. Currently, water is used on a competitive basis between human, agricultural and industrial needs, seeking limited accessibility. The use of non-traditional water in agriculture can contribute to the creation of new irrigated land or become a reserve resource for existing perimeters (Khan *et al.*, 2021). This is practiced in developed and developing countries (Khan *et al.*, 2019a) because in wastewater nutrients are contained in large quantities, albeit in different proportions. Among these substances are nitrogen, phosphorus, sometimes potassium, zinc, boron and sulphur, which are an important quality parameter for the development of these waters in agriculture and landscape management (Khan *et al.*, 2019b). There are two reasons for this. The first reason is that due to the presence of pharmaceutical antimicrobial substances, the aquatic environment becomes resistant to viruses, bacteria and other contaminants of biological origin. The second reason is those viruses, bacteria, hemilitons, heavy metals and others hazardous pollutants reliably linger on agricultural products that contaminates the food chain. In addition to this, modern human needs have ensured that new pollutants enter the urban wastewater: hormones and pharmaceutical compounds (as mentioned above), any cosmetics, additives for plastics, new pesticides and their metabolites, etc. Some of them have been identified as endocrine disruptors. Therefore, the reuse of wastewater involves its mandatory treatment and decontamination.

It is environmental engineering that occupies the first position in the protection of environmental components (Lin *et al.*, 2008; Mahmoud, 2020; Merwade, 2007; Mozaffari *et al.*, 2020; Nidheesh & Gandhimathi, 2012). However, this is not the only tool for managing the quality of the components of nature. The problem of proper environmental protection needs integrated waste management (pollutants). One of the directions should be the provision of legal mechanisms to counteract illegal encroachment on water resources. Criminal legal aspects have a special place in relation to causing harm to water bodies. In this regard, authorities should establish specific discharge standards for high-priority pharmaceuticals and chemicals and establish adequate and effective liability measures for violating these standards. There are very few countries in the world that do not have regulatory documents governing the discharge of pharmaceutical compounds into water bodies. Moreover, offenses against water are observed around the world.

The study carried out on the example of Ukraine indicates that the main numbers of offenses are violations of water protection rules. It should be emphasized that there is a mismatch between the number of recorded criminal offenses in the field of water protection and convictions. There are significantly fewer convictions. In addition, there is a tendency to increase the number of criminal offenses in the field of protection of water bodies. This means that legal protection in general, and criminal law in particular, is a kind of protection against potential offenders committing criminal offenses against the water resources of our state. A similar situation is observed in many other countries.

15.3 FUTURE PROSPECTS

Effective wastewater management, including those containing pharmaceutical ingredients, can be achieved through close collaboration between natural resource management authorities, research institutions, and penal regulatory authorities (Figure 15.1).

The situation prevailing in many countries of the world speaks of the small influence of the law on the water bodies' protection (rivers, lakes, seas). As the crimes number is steadily growing, many criminal cases are being closed on rehabilitative and non-rehabilitative grounds in relation to the perpetrators of these crimes, and fewer people are held accountable for violations against water resources, which indicates a lack of work in searching and identifying the perpetrators of crimes of this category. This is often due to the inability of legal authorities to either establish simple and easily applicable laws and regulations, or to comply with these rules. Also, the reason for exemption from criminal

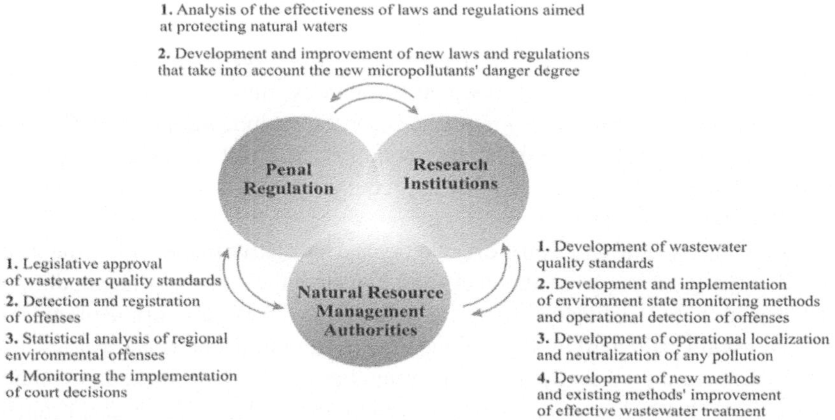

Figure 15.1 Scheme of interaction between government bodies and research institutions for the effective wastewater management, including those containing pharmaceutical ingredients.

liability in many cases is the fact that it is difficult to demonstrate the subjective aspect of a crime against the environment, but this is one of the key elements in the formation of the crime. For example, the guilt of a crime specified in article 229 of the Criminal Code of the Republic of Moldova (Qu *et al.*, 2013) is expressed in both forms provided for by criminal law, both intentionally and recklessly, but this is not explicitly stated in the content of the article (Şener *et al.*, 2017). In addition, the lack of continuous interaction between government bodies of natural resources (in this case, water resources) and law enforcement agencies contributes to offenses against water bodies and their concealment through liability under articles with less significant penalties (as a rule, civil liability occurs with the simultaneous payment of fines, the amount of which is insignificant).

Integrated regional control, taking into account engineering, technical and legal aspects, can positively affect the quantitative reduction of facts of concealment of crimes. First, this is possible because two independent state structures record an offense (but not one that could ignore or hide this fact). Second, modern engineering and technical solutions make it possible to quickly obtain information on the ecological state of water bodies through systematic environmental monitoring (Sirés & Brillas, 2012; Sokolov *et al.*, 2018; Vambol *et al.*, 2017).

With such an integrated approach, among the tasks of the regulatory bodies for natural resources is the development and state registration of clear and stringent quality standards for wastewater before it is discharged into natural water or reused. At the same time, norms regarding the content of pharmaceutical compounds in wastewater are mandatory, since at present this type of pollutant has remained without attention by most countries of the world. Also, natural resource management authorities must ensure the collection of statistical information, which is filled with data on the revealed violations and data on the enforcement of court decisions.

The engineering and technical departments of the regulatory authorities for natural resources, together with research institutions, must constantly search for new methods and improve existing methods for the prompt and justified effective localization and neutralization of any pollution, as well as methods for the effective wastewater treatment so that the requirements of state quality standards have been met.

The judicial system, based on the regulatory and legislative framework of the state, should facilitate the implementation of work to restore the purity of water bodies, as well as work to prevent their subsequent pollution at the expense of the offender's material resources.Regarding the pollutant neutralization methods and wastewater treatment, there are currently a sufficient number of effective methods (Vambol *et al.*, 2018, 2019; Xie *et al.*, 2015; Ziarati *et al.*, 2020; Zwiener & Frimmel, 2003), but all these methods must be adapted to specific conditions. Among these conditions, the main ones are the quantitative and qualitative composition of wastewater, as well as the pollution intensity.

REFERENCES

Amy G., Debroux J. F., Arnold R. and Wilson L. G. (1996). Preozonation for enhancing the biodegradability of wastewater effluent in a potable-recovery soil aquifer treatment (SAT) system. *Revue des Sciences de l'Eau*, **9**(3), 365–380.

Anctil F. (2008). L'eau et ses enjeux (Water encounters). University of Laval, Quebec, p. 228.

Barbăneagră A. (2003). Codul penal al Republicii Moldova. Comentariu (Criminal Code of the Republic of Moldova. Comment). Chişinău, Arc, p. 859.

Belecciu Ş. (2017). The subjective aspect of water pollution offence art. 229 of the Criminal Code of the Republic of Moldova. *Acta Universitatis George Bacovia. Juridica*, **6**(2). Available from: www.ugb.ro/Juridica/Issue12ROEN/8._Latura_subiectiva_a_infractiunii_de_poluare_a_apei.Stefan_Belecciu.EN.pdf

Benotti M. J., Trenholm R. A., Vanderford B. J., Holady J. C., Stanford B. D. and Snyder S. A. (2009). Pharmaceuticals and endocrine disrupting compounds in U.S. drinking water. *Environmental Science and Technology*, **43**(3), 597–603.

der Beak T. A., Weber F. A., Bergmann A., Hickmann S., Ebert I., Hein A. and Kuster A. (2016). Pharmaceuticals in the environment – Global occurrences and perspectives. *Environmental Toxicology and Chemistry*, **35**(4), 823–835.

Dhingra A., Khan N. A., Ahmed S., Gautam S., Vambol S., Vambol V. and Kovalenko S. (2020). Investigation of medical institutions in India as a source of surface water pollution. World Review of Science, Technology and Sustainable Development (in press).

Directive 2000/60/EC of the European parliament and of the council of 23 October 2000. *Official Journal of the European Communities*. Available from: https://eur-lex.europa.eu/resource.html?uri=cellar:5c835afb-2ec6-4577-bdf8-756d3d694eeb.0004.02/DOC_1&format=PDF

Falås P., Andersen H. R., Ledin A. and la Cour Jansen J. (2012). Occurrence and reduction of pharmaceuticals in the water phase at Swedish wastewater treatment plants. *Water Science and Technology*, **66**, 783–791.

Fatta-Kassinos D., Meric S. and Nikolaou A. (2011). Pharmaceutical residues in environmental waters and wastewater: current state of knowledge and future research. *Analytical and Bioanalytical Chemistry*, **399**(1), 251–275.

Hamoda M. F., Al-Ghusain I. and Al-Mutairi N. Z. (2004). Sand filtration of wastewater for tertiary treatment and water reuse. *Desalination*, **164**(3), 203–211.

Kettab A. (2001). Water resources in Algeria: strategies, investments and vision. *Desalination*, **136**(1–3), 25–33.

Khan N. A., Ahmed S., Vambol S., Vambol V. and Farooqi I. (2019a). Field hospital wastewater treatment scenario. *Ecological Questions*, **30**(3), 57–69.

Khan N. A., Khan S. U., Ahmed S., Farooqi I. H., Hussain A., Vambol S. and Vambol V. (2019b). Smart ways of hospital wastewater management, regulatory standards and conventional treatment techniques: A short review. *Smart and Sustainable Built Environment*. https://doi.org/10.1108/SASBE-06-2019-0079

Khan A. H., Khan N. A., Khan S. A., Dhingra A., Khan S. U., Mohammadi A. A., Changani F., Yousefi M., Alam S., Vambol S., Vambol V., Khursheed A. and Ali I. (2020a). Application of advanced oxidation processes followed by different treatment technologies for hospital wastewater: A case study of multispecialty hospitals. *Journal of Cleaner Production*, **269**, 122411–122424.

Khan N. A., Ahmed S., Farooqi I. H., Ali I., Vambol V., Changani F., Yousefi M., Vambol S., Khan S. U. and Khan A. H. (2020b). Occurrence, sources and conventional treatment techniques for various antibiotics present in hospital wastewaters: A critical review. *Trends in Analytical Chemistry*, **129**, 115921–115935.

Khan N. A., Ahmed S., Farooqi I. H., Vambol S., Vambol V. and Koloskov V. (2021). Operational parameters optimisation for simultaneous removal of two drugs in hospital wastewater – RSM approach. *International Journal of Environment and Waste Management* (in press).

Lin A. Y. C., Yu T. H. and Lin C. F. (2008). Pharmaceutical contamination in residential, industrial, and 673 agricultural waste streams: Risk to aqueous environments in Taiwan. *Chemosphere*, **74**(674), 131–141. DOI: 10.1016/j.chemosphere.2008.08.027.

Mahmoud A. E. D. (2020). Nanomaterials: Green synthesis for water applications. In: Handbook of Nanomaterials and Nanocomposites for Energy and Environmental Applications, O. Kharissova, L. Martínez and B. Kharisov (eds). Springer, Cham.

Mazhar M. A., Khan N. A., Ahmed S., Khan A. H., Hussain A., Changani F. and Vambol V. (2020). Chlorination disinfection by-products in Municipal drinking water – A review. *Journal of Cleaner Production*, **273**, 123159–123171.

Merwade V. M. (2007). An automated GIS procedure for delineating river and lake boundaries. *Transactions in GIS*, **11**(2), 213–231.

Mozaffari N., Mozaffari N., Mohammad Elahi S., Vambol S., Vambol V., Khan N. A. and Khan N. (2020). Kinetics study of CO molecules adsorption on Al_2O_3/Zeolite composite films prepared by roll-coating method. *Surface Engineering*. Available from: www.tandfonline.com/doi/full/10.1080/02670844.2020.1768628

Nidheesh P. V. and Gandhimathi R. (2012). Trends in electro-Fenton process for water and wastewater treatment: An overview. *Desalination*, **299**, 1–15.

Qu X., Brame J., Li Q. and Alvarez P. J. J. (2013). Nanotechnology for a safe and sustainable water supply: enabling integrated water treatment and reuse. *Accounts of Chemical Research*, **46**(3), 834–843.

Şener Ş., Şener E. and Davraz A. (2017). Evaluation of water quality using water quality index (WQI) method and GIS in Aksu River (SW-Turkey). *Science of the Total Environment*, **584**, 131–144.

Sirés I. and Brillas E. (2012). Remediation of water pollution caused by pharmaceutical residues based on electrochemical separation and degradation technologies: A review. *Environment International*, **40**, 212–229.

Sokolov D., Sobyna V., Vambol S. and Vambol V. (2018) Substantiation of the choice of the cutter material and method of its hardening, working under the action of friction and cyclic loading. *Archives of Materials Science and Engineering*, **94**(2), 49–54.

Vambol S., Bogdanov I., Vambol V., Suchikova Y., Lopatina H. and Tsybuliak N. (2017). Research into effect of electrochemical etching conditions on the morphology of porous gallium arsenide. *Eastern-European Journal of Enterprise Technologies*, **6/5**(90), 22–31.

Vambol S., Vambol V., Sobyna V., Koloskov V. and Poberezhna L. (2018). Investigation of the energy efficiency of waste utilization technology, with considering the use of low-temperature separation of the resulting gas mixtures. *Energetika*, **64**(4), 186–195.

Vambol S., Vambol V., Sundararajan M. and Ansari I. (2019). The nature and detection of unauthorized waste dump sites using remote sensing. *Ecological Questions*, **30**(3), 43–55. DOI: 10.12775/EQ.2019.018.

Xie H., Yao G. and Liu G. (2015). Spatial evaluation of the ecological importance based on GIS for environmental management: A case study in Xingguo county of China. *Ecological Indicators*, **51**, 3–12.

Ziarati P., Vambol V. and Vambol S. (2020). Use of inductively coupled plasma optical emission spectrometry detection in determination of arsenic bioaccumulation in *Trifolium pratense* L. from contaminated soil. *Ecological Questions*, **31**(1), 15–22. DOI: 10.12775/EQ.2020.003.

Zwiener C. and Frimmel F. (2003). Short-term tests with a pilot sewage plant and biofilm reactors for the biological degradation of the pharmaceutical compounds clofibric acid, ibuprofen, and diclofenac. *Science of the Total Environment*, **309**, 201–211.

Index

A

Adsorption 20, 26, 56, 58, 62–63, 185–193, 207, 221, 233–234, 261, 264, 286, 289–290, 293–294, 327–329, 331–333, 341–344, 362, 365

Advanced oxidation processes 193, 232, 281, 290, 327–329, 357, 366

Advanced treatment technology 328

Antibiotics 7, 9, 21, 26–27, 29–33, 53, 64, 73–77, 80–82, 85–87, 97–99, 101, 135, 147–148, 152, 158–159, 162–163, 169–170, 183, 185, 188, 190, 205, 207, 223–224, 226, 228, 230–232, 273–274, 276, 282, 284, 286–287, 289–290, 353, 355–356, 360, 369, 377

Aquatic 1–2, 5–11, 17–18, 20–21, 28–29, 34–35, 47–51, 53, 55–58, 60, 63–65, 69, 71–73, 76–78, 81–84, 88, 98–101, 107, 115–116, 129, 131, 146, 148, 150–151, 155, 158–164, 166, 168–170, 222, 225–230, 232, 236, 239, 242–243, 245–246, 251, 264, 271, 273–274, 307, 327, 336, 356, 364, 366, 377–378

Aquatic environment 1, 6, 8, 65, 98

Aquatic, environment 239

Arabidopsis thaliana and Salix exiguous 356

B

Biotic systems 17–18

C

Contaminants 18, 31, 48–50, 69, 74, 76–77, 81–82, 87–88, 96–97, 103, 108, 116–117, 119–120, 145–147, 149, 153, 155, 158, 162, 169, 183,187, 194, 221, 223, 227, 232, 236, 273, 282, 287–288, 294, 305, 337, 339, 353, 355–357, 360, 363, 365–369, 376, 378

Conventional 8, 10–11, 33, 63, 76, 97, 147, 149, 162, 181, 190, 205–206, 221, 233, 236, 242, 258, 260, 274, 277–278, 282–283, 287, 291–294, 305–306, 314–315, 319, 321, 328, 338, 341, 343, 355, 360, 368

Conventional legislation 116, 118, 121, 130, 134, 136–137, 160, 231, 305

D

Dynamics of crimes against water resources 114, 138

E

Eco-toxicity 47, 56, 58–60, 64, 108, 116, 145–146, 169, 232

Environment 1–8, 10–11, 17–22, 26–36, 47–51, 53, 55–56, 58, 60–61, 63–65, 69, 71–73, 75–76, 81, 88, 95–101, 103, 107–108, 113–116, 118–129, 131–138, 145, 147–151, 153–154, 158–163, 170, 181, 184–188, 190–192, 205, 207–208, 221–232, 234–236, 239, 242–243, 245–246, 248–252, 254–255, 265, 268, 271, 273–274, 282–283, 294, 305–307, 309, 311, 314, 316, 318, 321–322, 327–328, 336–338, 353, 355–357, 359, 363–366, 375–378, 380

Environmental consequences 2, 4, 11

Environmental impact 129, 221, 231, 234, 274

Environmental risk 159, 192, 223, 305, 316, 321, 338, 375, 377

Environmental technologies 239

Eradication 171, 239, 242, 244, 321

Existing Treatment 150, 158, 329, 353

F

Freshwater 31–32, 34, 69–71, 73, 76–77, 80, 115, 131–133, 308, 312, 314, 356, 364, 366

G

Genotoxicity 50, 146, 169

Ground water reservoir 69–70

H

Health and environmental risks 305

Hospitals 7–8, 18, 30–31, 48–51, 53–54, 58–60, 64–65, 69, 72, 74–75, 77, 87, 95–101, 103–104, 108, 117, 135, 156, 158–163, 169–170, 205–207, 225, 229–230, 242, 279, 283, 377

Hospital wastewater 6, 51, 82–83, 97–98, 100, 103, 105–106, 129, 135, 146, 156, 158–162, 169–170, 277, 279–280, 285, 287, 289–290, 377

Hospitals contaminants 31, 159

I

Ideal destination 305

L

Legislation 116, 118, 121, 130, 134, 136–137, 160, 231, 305

Legislative and criminal law mechanisms 114, 375

M

Membrane 20, 85, 87, 150, 183–184, 191, 205, 207, 224, 233–234, 257, 264, 275, 277, 283–284, 286–288, 291–292, 294, 327–329, 338–343, 357–359, 365–366, 369

Membrane bioreactor 85, 87, 283–284, 288, 291–292, 327–329, 339, 359, 365

O

On-site treatment 272–274, 287

P

Pharmaceuticals 6–8, 18–22, 26–29, 31–34, 36, 47–48, 50–51, 53–57, 59–65, 69–71, 73–78, 80–81, 84–85, 87–88, 95–98, 101–102, 104, 108, 113, 116, 135, 145–149, 151, 153–157, 159, 162–164, 166, 168–171, 184, 187, 190–191, 196, 221–222, 224, 228–230, 232–233, 236, 251, 271, 273–274, 278–294, 339, 364, 376, 378

Pharmaceutical compounds 1, 8, 18, 20, 27–28, 50, 95, 98, 117–118, 129, 139, 145, 148–151, 153–156, 158, 160, 170, 182–186, 190, 193, 195, 197, 203,

205–207, 223, 226, 251, 271, 273, 280, 290, 333, 378, 380
Pharmaceutical ingredients 18, 22, 26, 33, 36, 71, 74, 76, 114, 129–130, 137–138, 207, 375–377, 379
Pharmaceutical products 8, 11, 151, 221–222, 225, 227–229, 231–233, 236, 251, 328
Pharmaceutical residues 8, 48–51, 54, 57, 60, 64–65, 95–97, 99–100, 102–103, 105–108, 147–148, 151, 158, 161, 163, 169, 171, 229, 231–232, 251, 271, 281, 283, 285, 289, 353, 355, 368, 377
Pharmaceutical Wastewater 1–2, 6–8, 11, 17–18, 35, 69, 72, 75–76, 82–84, 147, 151, 221, 271–272, 274, 278, 287, 353–355, 357, 376
Pharmaceutically active compounds 6, 9, 146–147, 183, 189, 285
Photocatalytic degradation 182–183, 195
Polluted water 3, 18, 114, 182, 193, 203, 244–245, 306, 308, 375
Pollution 3–4, 7, 10, 18, 20, 32, 47, 49, 51–52, 69, 71–73, 80–81, 84, 87, 95, 100, 103, 108, 114–118, 120–123, 128–130, 132–135, 137–138, 160–161, 163, 170, 189, 205, 207–208, 221, 223, 227–228, 230–231, 234–235, 239, 241–242, 244, 246–247, 250, 252–255, 257–258, 264–265, 289, 306, 309–310, 316, 359, 364, 377, 380
Primary producers 69–70, 81, 88

R

Recalcitrant 11, 184, 290–291, 293, 353, 355–356, 359, 365–367
Risk quotient 145–146, 155, 157**S**

S

Sanitization 239, 281
Sources and penetration mechanisms 18
Suppressed immune system 146, 166

T

Terrestrial hazards of chemical wastewater 2
Toxic compounds 2–3, 5, 170
Treatment 3–11, 18, 20, 22, 26–28, 30–33, 36, 49–51, 55, 58, 63–64, 69–71, 73–75, 84–87, 95–100, 104, 108, 117, 119–120, 129–132, 134–139, 149–151, 153–154, 156–160, 162–163, 170–171, 181–189, 191, 193, 195, 203, 205–207, 221–236, 239–243, 247, 250, 252–255, 257–259, 261–268, 271–294, 305, 310–313, 315–317, 322, 327–329, 333, 336, 338–339, 341–343, 353–369, 375–378, 380
Treatment techniques 149–150, 153, 182, 221, 239, 241, 263–265, 273, 328, 375
Triclosan 19, 31, 80, 155–156, 170, 291, 327–328, 336

W

Wastewater 1–11, 17–18, 20, 26–33, 35–36, 49, 51, 55, 58, 60, 63, 65, 69–78, 80–87, 97–100, 103, 105–106, 114, 116–120, 129–132, 134–139, 146–151, 154, 156–163, 169–171, 182–184, 186–191, 195, 205–207, 221, 223–224, 226, 228–229, 232–236, 239–240, 242–258, 260–261, 263–265, 268, 271–294, 305–320, 322, 327–331, 333–335, 337–340, 342–343, 353–369, 375–380
Wastewater treatment 2–3, 5, 8–10, 18, 20, 26–28, 31, 36, 63, 69–70, 73–74, 84–86, 117, 131, 136–137, 139, 149–151, 154, 156–157, 160, 170–171, 183–184, 186–189, 191, 195, 205–207, 229, 232–236, 239, 242, 250, 252, 254–255, 257, 264–265, 271–280, 282, 285–288, 290–291, 294, 305, 322, 328, 333, 338, 353–355, 357, 359–360, 362–363, 365, 376, 380
Wastewater treatment plant 8, 18, 20, 26–28, 31, 36, 73–74, 117, 136, 156–157, 160,

183, 229, 233, 236, 239, 254–255, 272–274, 277, 279–280, 286, 290, 328, 353, 355, 376
Wastewater treatment plant 8, 18, 20, 26–28, 31, 36, 73–74, 117, 136, 156–157, 160,
183, 229, 233, 236, 239, 254–255, 272–274, 277, 279–280, 286, 290, 328, 353, 355, 376
Water protection 113–115, 121–122, 127, 137–138, 375–376, 379

IWA Publishing's authorised EU representative for General Product Safety Regulations is Diane D'Arras, 15 rue Duret, 75116 Paris, France, e-mail: safety@iwap.co.uk.

Printed and bound by CPI Group (UK) Ltd, Croydon, CR0 4YY
16/06/2026
02140380-0001